Bio-inspired Computation in Unmanned Aerial Vehicles

Haibin Duan • Pei Li

Bio-inspired Computation in Unmanned Aerial Vehicles

Springer

Haibin Duan
Beihang University (formerly Beijing University of Aeronautics and Astronautics, BUAA)
Beijing
China, People's Republic

Pei Li
Beihang University (BUAA)
Beijing
China, People's Republic

ISBN 978-3-642-41195-3 (hardcover)
ISBN 978-3-642-41196-0 (eBook)
ISBN 978-3-662-51344-6 (softcover)
DOI 10.1007/978-3-642-41196-0
Springer Heidelberg New York Dordrecht London

Printed on acid-free paper

Springer is part of Springer Science+Business Media (www.springer.com)

Preface

Bio-inspired computation, short for biologically inspired computation, is the use of computers to model the living phenomena, and simultaneously the study of life to improve the usage of computers, which has attracted a lot of researchers' attention. A variety of bio-inspired computation models have been proposed and applied to solve many real-world problems successfully, such as ant colony optimization (ACO), particle swarm optimization (PSO), artificial bee colony (ABC) and differential evolution (DE). Although rigorous theoretical analysis for most of the bio-inspired computation methods has not been conducted systematically, and the current study in this field is still in the experimental and preliminary application stage, the bio-inspired computation methods have already found their applications in many typical fields. Some of the phenomena are also known as swarm intelligence, inspired by the social behaviour of gregarious insects and other animals. The emergent behaviour of multiple unsophisticated agents interacting among themselves and with their environment leads to a functional strategy that is useful to achieve complicated goals in an efficient manner. There exist a number of desirable properties in this kind of model, which include feedback, self-organization, adaptation to changing environments, and multiple decentralized interactions among agents to work collaboratively as a group in completing complex tasks.

Unmanned aerial vehicle (UAV), colloquially known as a drone, is an aircraft without a human pilot on board. Its flight is controlled either autonomously by computers in the vehicle or under the remote control of a pilot on the ground or in another vehicle. UAV offers advantages for many applications compared with their manned counterparts. They preserve human pilots of flying in dangerous conditions that can be encountered not only in military applications but also in other scenarios involving operation in bad weather conditions, or near to buildings, trees, civil infrastructures and other obstacles. While recent technological advances have enabled the development of unmanned vehicular systems and recent implementations have proven the UAV's benefits in both military and civilian applications, the full benefit of unmanned systems will be utilized when they can operate autonomously. Typical application domains of UAVs include reconnaissance and surveillance missions in an urban environment, target tracking and evasive manoeuvres, search and rescue operations, homeland security, etc. In recent years a significant shift of focus has occurred in the field of autonomous UAVs as researchers have begun to investigate problems involving multiple rather than single UAV. Systems consisting of multiple UAVs performing complex missions present new challenges to the control community. UAVs must possess attributes of autonomy in order to function effectively in a 'system of systems' configuration. Coordinated and collaborative control of UAV swarms demands new and novel technologies that integrate modelling, control, communications and computing concerns into a single architecture.

From the computational point of view, bio-inspired computation models are largely stochastic search algorithms. They are useful for undertaking distributed and multimodal optimization problems. The search process is robust and efficient in maintaining diversity. A mechanism to impose a form of forgetting is also adopted in some swarm intelligence algorithms such that the solution space can be explored in a comprehensive manner. Thus, the algorithms are able to avoid convergence to a locally optimal solution and, at the same time, arrive at a global optimized solution with a high probability. We can learn more than the optimization algorithms from the bio-inspired algorithms. The interaction among the agents and feedback mechanism are the basic elements that result in the emergence of dynamic patterns at the colony level. The most interesting properties of these self-organized patterns are robustness (the ability for a system to perform without failure under a wide range of conditions) and flexibility (the ability for a system to readily adapt to new, different or changing requirements). Adaptation must happen fast enough for UAVs to provide benefits in case of environmental change, and the autonomy should be constructed so that these lessons can be shared with other autonomous systems that have not yet encountered that situation. Yet, even in a hostile, dynamic, unstructured and uncertain environment, this learning must not adversely affect safety, reliability or the ability to collaborate with the operator or other autonomous systems. Although such capabilities are not currently available, the emergence mechanism of robustness and flexibility in biological colony may provide us many entirely new threads.

This monograph, divided into eight chapters, mainly includes our recent work relevant to UAV control issues in which we have taken advantage of bio-inspired computation, such as path planning and replanning for single UAV and multiple UAVs, formation flight control and formation configuration, heterogeneous cooperative control for multiple UAVs/unmanned ground vehicles (UGVs) and vision-based surveillance and navigation problems. Chapter 1 discusses the development of UAV and emphasizes the role bio-inspired intelligence plays in achieving higher autonomous capability. Chapter 2 introduces four representative bio-inspired algorithms, which are ACO, PSO, ABC and DE. We explain the biological inspiration, principle and implementation procedures of the algorithms in detail. Then in Chapter 3 we deal with the modelling problem of UAVs and give a brief introduction of the controller design method. Chapter 4 deals with the path planning problem using bio-inspired algorithms, both for single UAV and multiple UAVs, both in the two-dimensional scenario and three-dimensional scenarios. This chapter mainly contains three sections. First, a chaotic ABC approach is proposed for two-dimensional path planning. Then path planning is extended to a three-dimensional scenario through an improved ACO added a path smoothing strategy. Section 4.5 deals with coordinated path re-planning for multiple UAVs using the method of Max-Min adaptive ACO. In Chap. 5, we mainly deal with three significant aspects of formation control: formation control, close formation and formation configuration. Chapter 6 discusses multiple UAVs/UGVs heterogeneous cooperation and cooperative search of multiple UAVs based on differential evolution. Chapter 7 describes bio-inspired computation algorithms involving vision-based surveillance and navigation. Chapter 8 discusses the opportunities for the development of UAVs and points out the potential challenges for achieving higher autonomous capability. By incorporating the bio-inspired intelligence into UAVs, it is possible to enhance their ability to understand and adapt to the environment and the ability to cooperate with other autonomous systems.

Special thanks are due to several members of our research team-ANT Research Group: Guanjun MA, Senqi LIU, Hao LI, Yaxiang YU, Xiangyin ZHANG, Chunfang XU, Changhao SUN, Yunpeng ZHANG, Fang LIU, Yimin DENG, Qinan LUO, Yingcai BI, Shuangtian LI, Qifu ZHANG, Olukunle Kolawole SOYINKA, Jiaqian YU, Xiaohua WANG, Junnan LI, Weiren ZHU, Lu GAN, Zenghu ZHANG, Yan XU, Huaxin QIU, Cong ZHANG, Fei YE, Cong LI and Ziwei ZHOU for their diligent work and contributions in the related fields. The authors would also like to extend their thanks and appreciations to Ms. Li SHEN, the physical sciences and engineering editor of Springer, and the editorial assistant, Ms. Jessie GUO, for their kind help and assistance and to Springer's copy editors for their reading of this entire manuscript and their instructive comments.

We have had the benefit of the collaboration of coworkers and discussions with international partners, from whom we have learned a great deal. Among them are Prof. Zongji CHEN, Prof. Bo Hu LI, and Prof. Shiyin QIN of Beihang University(BUAA); Prof. Ben M. CHEN and Prof. Kay Chen TAN of National University of Singapore; Prof. Qinping ZHAO of State Key Laboratory of Virtual Reality Technology and Systems of China; Prof. Xingui HE of Peking University; Prof. Ming LI and Prof. Yanming FAN of Shenyang Aircraft Design Research Institute; Prof. Derong LIU of Institute of Automation of CAS; Prof. Marco DORIGO of Universite' Libre de Bruxelles; Prof. Yuhui SHI of Xi'an Jiaotong-Liverpool University; Prof. Yaochu JIN and Prof. Yang GAO of University of Surrey; Prof. Ling WANG of Tsinghua University; Prof. Licheng JIAO and Prof. Maoguo GONG of Xidian University; Prof. Bin XIAN of Tianjin University; Prof. Sung-Kwun OH of The University of Suwon; Prof. Daobo WANG and Prof. Huajun GONG of Nanjing University of Aeronautics and Astronautics; Prof. Wen-Hua CHEN of Loughborough University; Prof. Youmin ZHANG of Concordia University; Prof. Wei REN of University of California, Riverside; and Prof. Delin LUO of Xiamen University. We are indebted to them for their kind help and valuable comments.

The related work reported in this monograph was partially supported by the Natural Science Foundation of China (NSFC) under grants #61273054, #60975072 and #60604009; National Natural Science Foundation of China (NSFC) State Key Program under grant #61333004; National Key Basic Research Program of China (973 Program) under grants #2014CB046401 and #2013CB035503; National High Technology Research and Development Program of China (863 Program) under grant #2011AA040902; National Magnetic Confinement Fusion Research Program of China under grant #2012GB102006, Program for New Century Excellent Talents in University of China under grant #NCET-10-0021; Top-Notch Young Talents Program of China, Beijing NOVA Program, under grant #2007A017; Fundamental Research Funds for the Central Universities of China, Aeronautical Foundation of China, under grants #20115151019, 2008ZC01006 and #2006ZC51039; Open Fund of the State Key Laboratory of Virtual Reality Technology and Systems under grants #VR-2013-ZZ-02 and #VR-2011-ZZ-01; Open Fund of the Provincial Key Laboratory for Information Processing Technology of Suzhou University under grants #KJS1020 and #KJS0821.

The main objectives pursued have been on addressing the question of how to achieve higher autonomous capability by taking advantages of bio-inspired computation. This monograph is intended for researchers, college students and industrial practitioners who may wish to get an insight into the complex nature of and practical solutions to bio-inspired computation in UAV issues. We hope that it helps to promote further research and practice in this promising field. Finally, we also hope that readers enjoy reading this monograph, and, most importantly, that they learn something new by looking at things from a new perspective.

Beihang University (BUAA) Haibin Duan
Beijing, People's Republic of China Pei Li

Contents

Abbreviations

Acronyms

1stICEO	First International Contest on Evolutionary Optimization
ABC	Artificial Bee Colony
ACL	Autonomous Control Levels
ACO	Ant Colony Optimization
ACS	Ant Colony System
AGL	Above Ground Level
AI	Artificial Intelligence
AIS	Artificial Immune System
API	Application Programming Interface
AS	Ant System
ASrank	Rank-based Ant System
AVS	Aerial Video Surveillance
BA	Bat-inspired Algorithm
BFO	Bacteria Foraging Optimization
BHW	Bio-inspired Hardware
CE	Cultural Evolution
CEC	International Conference on Evolutionary Computation
CMOS	Complementary Metal Oxide Semiconductor
COS	Common Operating System
DARPA	Defense Advanced Research Projects Agency
DE	Differential Evolution
DHM	Digital Hormone Model
DoD	Department of Defense
EA	Evolutionary Algorithm
EAS	Elitist Ant System
EC	Emotion Computing

EF	Employed Foragers
EHW	Evolvable Hardware
EP	Evolutionary Programming
EPF	Edge Potential Function
ESM	Extended Search Map
ESs	Evolution Strategies
ETA	Estimated Time of Arrival
FAA	Federal Aviation Administration
FOVs	Fields Of View
FPGA	Field-Programmable Gate Arrays
GA	Genetic Algorithm
GCS	Ground Control Station
GP	Genetic Programming
GSM	Glowworm Swarm Optimization
HCF	Hyper-Cube Framework
HICA	Hybrid Intelligent Control Agent
ICAO	International Civil Aviation Organization
ISR	Intelligence, Surveillance, and Reconnaissance
JUCAS	Joint Unmanned Combat Air Systems
LI-CQPSO	Chaotic Quantum-behaved PSO based on Lateral Inhibition
MALE	UAS Medium-Altitude, Long-Endurance Unmanned Aircraft System
MAV	Micro Aerial Vehicle
MDF	Multi-sensor Data Fusion
MMAS	MAX-MIN Ant System
NASA	National Aeronautics and Space Administration
NCS	Network Control System
O&M	Operations and Maintenance
OODA	Observe, Orient, Decide and Act
PID	Proportional-Integral-Derivative
PSO	Particle Swarm Optimization
QPSO	Quantum-behaved Particle Swarm Optimization
UAS	Unmanned Aircraft System
UAV	Unmanned Aerial Vehicle
UCAV	Unmanned Combat Aerial Vehicle
UF	Unemployed Foragers
UGV	Unmanned Ground Vehicle
RHC	Receding Horizon Control
RPV	Remote Piloted Vehicle
SAR	Synthetic Aperture Radars
SDK	Software Development Kit
SAVS	Small Air Vehicle System
SWAP	Size, Weight, and Power

TPED	Tasking, Production, Exploitation, and Dissemination
TSP	Travelling Salesman Problem
TVD	Target Value Damage
VL	Virtual Leader
VTOL	Vertical Take-Off and Landing

Chapter 1
Introduction

Haibin Duan and Pei Li

Abstract Unmanned Aerial Vehicles (UAVs) offer great advantages for both military and civilian applications due to their potential to execute missions in hazardous scenarios involving operation in bad weather conditions or near to buildings, trees, civil infrastructures, and other obstacles. Inspired from the biological insights about the incredible abilities of social insects to solve the complicated problems, many bio-inspired computation algorithms have been developed by simulating the collective behavior of decentralized, self-organized system in nature. The main objectives pursued have been on addressing the question of how to achieve higher autonomous capability by taking advantage of bio-inspired computation. This chapter mainly focuses on the fundamental concepts of UAVs and bio-inspired computation algorithms as well as bio-inspired intelligence in UAVs. After a brief introduction to the history of UAVs, autonomous capability of UAVs is discussed from the perspective of functional description, metrics of autonomy. Then the general features of bio-inspired computation and four representative bio-inspired algorithms are presented, which are, respectively, ant colony optimization (ACO), particle swarm optimization (PSO), artificial bee colony (ABC), and differential evolution (DE). Finally, a comprehensive review of bio-inspired intelligence in UAVs and the structure of this monograph are provided.

1.1 Unmanned Aerial Vehicle (UAV)

An unmanned aerial vehicle (UAV), commonly known as a drone, is an aircraft without a human pilot on board, whose flight is controlled either autonomously by computers in the vehicle, or under the remote control of a pilot on the ground or in another vehicle.

A UAV is defined as a "powered, aerial vehicle that does not carry a human operator, uses aerodynamic forces to provide vehicle lift, can fly autonomously

The original version of this chapter was revised. A correction to this chapter is available at https://doi.org/10.1007/978-3-642-41196-0_9

H. Duan and P. Li, *Bio-inspired Computation in Unmanned Aerial Vehicles*,
DOI 10.1007/978-3-642-41196-0_1, © Springer-Verlag Berlin Heidelberg 2014

or be piloted remotely, can be expendable or recoverable, and can carry a lethal or nonlethal payload" to distinguish it from missile. There are a wide variety of vehicle shapes, sizes, configurations, and characteristics. Initially, UAVs were simple remotely piloted aircraft, but autonomous control is increasingly being employed. They are deployed predominantly for military applications but also used in a small but growing number of civil applications, such as remote sensing; commercial aerial surveillance; oil, gas, and mineral exploration and production; transport; search and rescue; and forest fire detection, to name just a few (Duan and Liu 2010b; Xu et al. 2010; Zhang and Duan 2012). UAVs are often preferred for missions that are too "dull, dirty, or dangerous" for manned aircraft. Due to the ability to perform dangerous and repetitive tasks in remote and hazardous environments, UAV is very promising for the technological leadership of one nation and essential for improving the security of the society. While recent technological advances have enabled the development of unmanned vehicular systems and recent implementations have proven the UAV's great advantage in both military and civilian applications, the full benefit of unmanned systems will be utilized when they are fully autonomous.

1.1.1 History of UAVs

The earliest attempt at a powered UAV was A. M. Low's "Aerial Target" of 1916. Nikola Tesla described a fleet of unmanned aerial combat vehicles in 1915. A number of remote-controlled airplane advances followed, including the Hewitt-Sperry Automatic Airplane, during and after World War I, including the first scale RPV (Remote Piloted Vehicle), developed by the film star and model airplane enthusiast Reginald Denny in 1935 (Newcome 2004). More were made in the technology rush during World War II; these were used both to train antiaircraft gunners and to fly attack missions. Nazi Germany also produced and used various UAV aircraft during the course of World War II. Jet engines were applied after World War II, in such types as the Teledyne Ryan Firebee I of 1951, while companies like Beechcraft also got in the game with their Model 1001 for the United States Navy in 1955. Nevertheless, they were little more than remote-controlled airplanes until the Vietnam Era.

During the 1973 Yom Kippur War, Soviet-supplied surface to air missile batteries in Egypt and Syria caused heavy damage to Israeli fighter jets. As a result, Israel developed the first modern UAV. Israel pioneered the use of UAVs for real-time surveillance, electronic warfare, and decoys. Images and radar decoying provided by these UAVs helped Israel to completely neutralize the Syrian air defenses at the start of the 1982 Lebanon War, resulting in no pilots downed. The first time drones were used as proof-of-concept of super-agility post-stall controlled flight in combat flight simulations was with tailless, stealth-technology-based three-dimensional thrust vectoring flight control jet steering in Israel in 1987.

The birth of US UAVs (called RPVs at that time) began in 1959 when United States Air Force officers, concerned about losing pilots over hostile territory, began

Fig. 1.1 Aerial demonstrators at the 2005 Naval Unmanned Aerial Vehicle Air Demo

planning for the use of unmanned flights. With the maturing and miniaturization of applicable technologies as seen in the 1980s and 1990s, interest in UAVs grew within the higher echelons of the US military (Fig. 1.1). In the 1990s, the US Department of Defense (DoD) gave a contract to US corporation AAI Corporation of Maryland along with Israeli company Mazlat. The US Navy bought the AAI Pioneer UAV that was jointly developed by American AAI Corporation and Israeli Mazlat; this type of drone is still in use. Many of these Pioneer and newly developed US UAVs were used in the 1991 Gulf War (Wikipedia website, 2013). UAVs were seen to offer the possibility of cheaper, more capable fighting machines that could be used without risk to aircrews. Initial generations were primarily surveillance aircraft, but some were armed, known as an unmanned combat aerial vehicle (UCAV).

The first UAV created by United States was the Pioneer (see *top left* in Fig. 1.2), initially place aboard Iowa-class battleships to provide gunnery spotting. Since its performance was so exemplary, its mission evolved into reconnaissance and surveillance, primarily for amphibious forces, with new models constantly being introduced. As of 2008, the United States Air Force employed 5,331 drones, which is twice the number of manned planes. Out of these, the Predators are the most commendable (see *top right* in Fig. 1.2), which is described as a "Tiger II" MALE UAS (medium-altitude, long-endurance unmanned aircraft system) by the USAF. Unlike other UAVs, the Predator was armed with Hellfire missiles so that it can terminate the target that it locates, serving as a primary UAV used for offensive operations. This was done after Predators sighted Osama Bin Laden multiple times but could not do anything about it other than send back images. In addition, the Predator is capable of orchestrating attacks by pointing lasers at the targets, which is important as it puts a robot in a position to set off an attack. The overall success

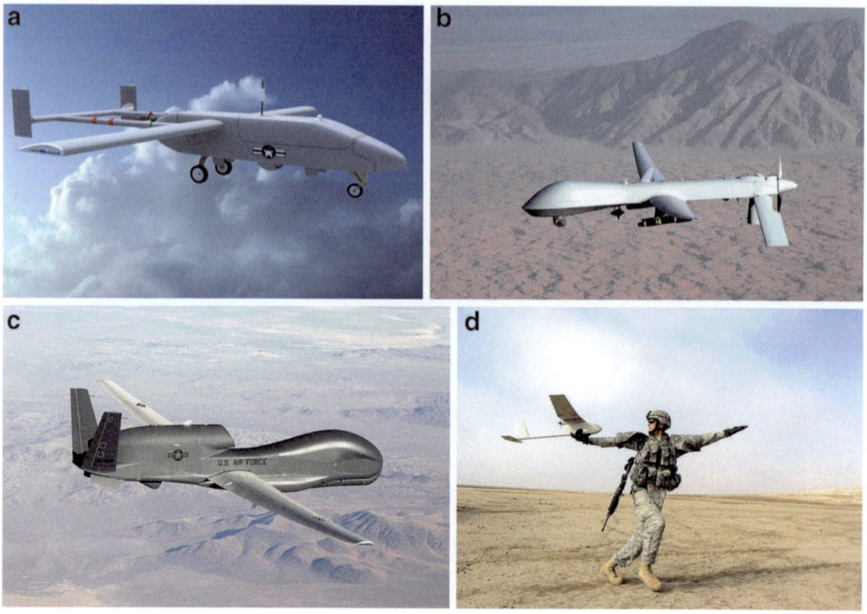

Fig. 1.2 UAVs developed by the United States

of Predator is witnessed by their extensive practical experience. From June 2005 to June 2006 alone, Predators carried out 2,073 missions and participated in 242 separate raids.

In contrast to the Predator, which is remotely piloted via satellites by pilots located 7,500 miles away, the Global Hawk operates virtually autonomously (see *bottom left* in Fig. 1.2). It is used as a high-altitude platform for surveillance and security, which could provide a broad overview and systematic surveillance using high-resolution synthetic aperture radar and long-range electro-optical/infrared sensors with long loiter times over target areas. Surprisingly, Global Hawks have the capability to fly from San Francisco and map out the entire state of Maine, before having to return. In addition, some UAVs have become so small that they can be launched from one's hand and maneuvered through the street. These UAVs, known as Ravens, are especially useful in urban areas such as Iraq, in order to discover insurgents and potential ambushes the next block up. The RQ-11B Raven UAV (see *bottom right* in Fig. 1.2) is launched by hand, thrown into the air like a free flight model airplane. The Raven lands itself by autopiloting to a predefined landing point and then performing a 45 degree slope controlled "Autoland" descent, which can provide day or night aerial intelligence, surveillance, target acquisition, and reconnaissance.

Although intelligence, surveillance, and reconnaissance remain UAVs' predominant mission, their roles have expanded to electronic attack, strike missions, suppression and/or destruction of enemy air defense, network node or communication relay, combat search and rescue, and other areas. As the capabilities of UAVs

Table 1.1 UAV classification on the basis of weight, operational altitude, and mission endurance

Class	Gross weight (lb)	Altitude AGL (ft)	Endurance (h)	Examples
Large	21,500	>10,000	>6	Global Hawk, Predator, K-Max
Mini	55–1,500	5,000–20,000	>5	Shadow, Pioneer, Fire Scout
Small	<55	<1,000	<1.5	T-Hawk, Raven, Desert Hawk
Micro	<2	<500	<1	WASP, BatCam
Nano	<0.4	<250	<0.5	Samarai, Black Hornet, Nano Hummingbird

grow, they are able to perform a multitude of missions. They range in cost from a few thousand dollars to tens of millions of dollars and range in weight from less than one pound to over 40,000 lb. In February 2013, it was reported that the UAVs were used by at least 50 countries, several of which made their own, for example, Iran, Israel, and China.

On the basis of gross weight, operational altitude above ground level (AGL), and mission endurance, UAVs may be classified broadly into large, mini, small, micro-, and nano-classes, as shown in Table 1.1. UAVs typically fall into one of six functional categories despite that multi-role airframe platforms are becoming more prevalent:

1. ***Target and decoy***: Providing ground and aerial gunnery a target that simulates an enemy aircraft or missile
2. ***Reconnaissance***: Providing battlefield intelligence
3. ***Combat***: Providing attack capability for high-risk missions
4. ***Logistics***: UAVs specifically designed for cargo and logistics operation
5. ***Research and development***: Used to further develop UAV technologies to be integrated into field deployed UAV aircraft
6. ***Civil and commercial UAVs***: UAVs specifically designed for civil and commercial applications

1.1.2 Unmanned Aircraft System

The term unmanned aircraft system (UAS) emphasizes the integration of aircraft with other necessary elements to form a system that is able to perform a specific mission, which is far beyond an aircraft itself. This term was first officially used by the Federal Aviation Administration (FAA) of the United States in early 2005 and subsequently adopted by DoD of the United States that same year in their unmanned aircraft system roadmap 2005–2030. It is also used by the International Civil Aviation Organization (ICAO) and other government aviation regulatory organizations.

What makes up a UAS? A typical civilian UAS consists of an unmanned or remotely piloted aircraft, the human element, payload, control elements, and data link communication architecture. A military UAS may also include other elements such as a weapons system platform and the supported soldiers. Obviously, the most fundamental component in UAS is unmanned aircraft, which is vehicle of fixed-wing or rotor-wing vehicles that fly without a human on board. Another element in UAS is the ground control station (GCS), which is a land- or sea-based control center that provides the facilities for human control of unmanned vehicles in the air or in space. GCS varies in physical size and can be as small as a handheld transmitter or as large as a self-contained facility with multiple workstations. Large military UAS usually requires a GCS with multiple personnel to operate separate aircraft systems. Ultimate goal for future UAS operation will be the capability for one crew to operate multiple aircraft from a GCS. How the UAS command and control information is sent and received both to and from the GCS and autopilot, namely, data link, is also an important issue, which follows into two categories in UAS: radio frequency line-of-sight and beyond line-of-sight. In most cases, UASs executing a specific mission require an onboard payload related to surveillance, weapons delivery, communications, aerial sensing, or cargo. UASs are often designed around the intended payload they will employ, and the size and weight of payloads is one of the largest considerations when designing a UAS. Take the missions of surveillance and aerial sensing, for example; sensor payloads come in many different forms for different missions. Conventional sensors for such mission include electro-optical cameras, infrared cameras, synthetic aperture radars (SAR), or laser range finder/designators. The launch and recovery element of the UAS is often considered as one of the most labor-intensive aspects of the UAS operation. Some UASs have very elaborate launch and recovery procedures, whereas others have virtually none. Larger systems have complicated procedures and dedicated personnel that prepare, launch, and recover the UAV. Other support equipment such as ground tugs, fuel trucks, and ground power units are necessary for many kinds of large UASs. Small vertical takeoff and landing (VTOL) UASs tend to have the least complex procedures and equipment when it comes to launch and recovery, most of which consists of only a suitable takeoff and landing area. The last but not the least element of the UAS is the human factor. Human element is essential for the contemporary UAS, which often consists of a pilot, a sensor, and supporting ground crew. With the development of technology, this kind of factor can be replaced with complex systems with higher autonomy, which means that the human element will likely get smaller as technological capability increases in the future.

1.1.3 Autonomy: A Key Enabler

The concept of autonomy is the ability for an unmanned system to execute its mission following a set of preprogrammed instructions without operator intervention. A fully autonomous UAV is able to fly without operator intervention from takeoff to landing. The degree of autonomy in a UAV varies widely from none to full autonomy. On one end of the spectrum, the vehicle is operated completely by remote control with constant operator involvement. Although the vehicle's flight characteristics are stabilized by its own autopilot system, the vehicle would eventually crash without constant involvement of the pilot. On the other end of the spectrum, the vehicle's onboard autopilot controls everything from takeoff to landing, requiring no pilot intervention. The operator can intervene in case of emergencies, overriding the autopilot if necessary to change the flight path or to avoid a hazard. Some of these positions of the operator can be combined into one depending on the complexity of the UAV system. The human factor will likely get smaller as technological capability increases, and eventually automation will require less human interaction.

Autonomy is actually a collection of functions that can be performed without direct intervention by the pilot/crew, which has been present in some form in the early UAVs even before the definition of the information age. The autonomy is included in the system to reduce the burden of pilot/crew. The Sperry autopilot, introduced in 1914, is a well-known example of an early technology advance that added significant autonomous functionality. Today's UAVs often combine remote control and computerized automation. For example, autonomous capability may be introduced into the control and guidance systems to perform low-level human pilot duties, such as speed and flight path stabilization, and simple scripted navigation functions, such as waypoint following. With the advance of time and technology, more complex and capable manifestations of autonomy have continued to emerge over a near-century-long evolution.

In the context of intelligent vehicles, autonomy is usually defined as the ability to make decisions without human intervention. To this end, the goal of autonomy is to make vehicles as smart as possible and capable of acting like humans. Someone holds the opinion that the development in the field of artificial intelligence in the 1980s and 1990s such as expert systems, neural networks, machine learning, natural language processing, and vision can bring opportunities and benefits for enhancing the autonomous capabilities of UAVs. However, since the mode of technological development in the field of autonomy has mostly followed a bottom-up approach, such as hierarchical control systems, and recent advances have been largely driven by the practitioners in the field of control science, not computer science. So there is reason to believe that autonomy has been and probably will continue to be considered an extension of the controls field.

The field of UAV autonomy is a recently emerging field, benefiting from the more and more broad prospect both in military application and civil application, especially the military application. Compared to the manufacturing of UAV flight hardware, the market for autonomy technology is far from mature and developed. So autonomy has been and may continue to be the bottleneck for future UAV developments. The future development of UAV market could be largely driven by advances to be made in the field of autonomy. Autonomy technology that is important to UAV development falls under the following categories:

- ***Sensor fusion:*** Combining information from different sensors for use on board the vehicle
- ***Communications:*** Handling communication and coordination between multiple agents in the presence of incomplete and imperfect information
- ***Path planning:*** Determining an optimal path for vehicle to go while meeting certain objectives and mission constraints, such as obstacles or fuel requirements
- ***Trajectory generation (sometimes called motion planning):*** Determining an optimal control maneuver to take to follow a given path or to go from one location to another
- ***Trajectory regulation:*** The specific control strategies required to constrain a vehicle within some tolerance to a trajectory
- ***Task allocation and scheduling:*** Determining the optimal distribution of tasks among a group of agents, with time and equipment constraints
- ***Cooperative tactics:*** Formulating an optimal sequence and spatial distribution of activities between agents in order to maximize chance of success in any given mission scenario

1.1.3.1 Functional Description

To distinguish the automatic system from autonomous system in terms of systems-design paradigms, it is appropriate to clarify these two terms from the perspective of the definition. Automatic systems are fully preprogrammed and act repeatedly and independently of external influence or control. An automatic system can be described as self-steering or self-regulating and is able to follow an externally given path while compensating for small deviations caused by external disturbances. However, the automatic system is not able to define the path according to some given goal or to choose the goal dictating its path. By contrast, autonomous systems are self-directed toward a goal in that they do not require external control, but rather are governed by laws and strategies that direct their behavior. In the early stage, the control algorithms are created and tested by teams of human operators and software developers. Then autonomous systems can develop modified strategies for themselves by which they select their behavior if machine learning or other intelligent method is utilized. In other words, an autonomous system is self-directed by choosing the behavior it follows to reach a human-directed goal. In addition, autonomous systems may even be able to optimize the behavior in a goal-directed manner in uncertain and complicated situations. The special feature of an autonomous system is its ability to be goal directed in unpredictable situations. This ability is a significant improvement in capability compared to the capabilities of automatic systems. An autonomous system is able to make a decision based on a set of given rules and/or limitations. It is able to determine what information is important in making a decision. It is capable of a higher level of performance compared to the performance of a system operating in a predetermined manner. To sum up, autonomy implies self-governance and self-direction, while autonomic implies self-management.

For instance, the goals of a system may be to find a particular phenomenon using its onboard sensors. The system may have autonomy to decide between certain parameters. However, ultimately the task may fail if the system cannot cope with uncertain dynamic changes in the environment. So, autonomous system should ensure that it is fault tolerant and be able to operate under fault conditions. From this perspective, the autonomic and self-management initiatives may be considered specialized forms of autonomy, that is, the autonomy is specifically to heal, protect, configure, and optimize the system.

1.1.3.2 Metrics of Autonomy

According to (Francis 2012), autonomy in aerospace arena can be categorized in a few of ways. It is a common perspective to segregate different functions on the basis of the nature of the missions, such as those that are unique to the vehicle as compared to those that are applicable to the mission-level activities and are therefore vehicle agnostic. Functions such as vehicle stabilization, flight control, maneuvering flight, and basic autoland are related to the features of a vehicle, so they belong to the vehicle-level autonomy. Functions such as auto navigation, route planning, mission objective determination, and flight plan contingencies are not dependent on the characteristic of the vehicles, whereas they are only dependent on the mission type and thus belong to mission-level autonomy. According to this view, dynamic trajectory generation and collision avoidance also fall into this category. Another dimension of autonomous functionality relates to the number of participating vehicle entities. In that regard, one can consider single vehicle operation, multivehicle operations, and, in an extreme, many-vehicle operations (so-called swarms).

It is also a metric to categorize autonomy on the basis of decision logic employed by a system. At one end of this scale are highly deterministic and very objective decisions such as those consistent with rule-based systems. Note that physics and other hard science-driven decisions such as digital computer-based stabilization and control, auto takeoff, and autoland are, for the most part, rule based. This kind of functions can be very reliable and precise, often exceeding the capabilities of even the best human pilots. Autonomous control systems must perform well under significant uncertainties with uncertain and dynamic environment, and they must be able to compensate for system failures without external intervention. So it is necessary to introduce some stochastic features and other forms of predictable uncertainty to handle the uncertainty. The other end of the scale is defined by highly subjective conditions or far less deterministic circumstances, such as those which may be encountered in an ill-defined, adversarial environment. Besides, the nature of the decision logic is also important in that it drives the strategy and approach taken in all the other elements of the decision cycle.

Another way to characterize autonomy is one that couples its functionality and degree of situational difficulty. Two major variables which facilitate this decomposition are complexity and uncertainty. Autonomous functions which derive

Table 1.2 Autonomous levels

Level	Name	Description
1	Human operated	A human operator makes all decisions. The system has no autonomous control of its environment although it may have information-only responses to sensed data
2	Human delegated	The vehicle can perform many functions independently of human control when delegated to do so. This level encompasses automatic controls, engine controls, and other low-level automation that must be activated or deactivated by human input and must act in mutual exclusion of human operation
3	Human supervised	The system can perform a wide variety of activities when given top-level permissions or direction by a human. Both the human and the system can initiate behaviors based on sensed data, but the system can do so only if within the scope of its currently directed tasks
4	Fully autonomous	The system receives goals from humans and translates them into tasks to be performed without human interaction. A human could still enter the loop in an emergency or change the goals, although in practice there may be significant time delays before human intervention occurs

from objective criteria, such as that found in physical laws, may be inherently complex, but are not often subject to high levels of uncertainty. Alternatively, autonomous decision-making which involves more subjective circumstances or is associated with highly uncertain conditions is far more difficult to model or quantify. These situations often exceed today's state of the art for artificial intelligence methods. They are best dealt with by knowledgeable and trained human crews. Contingency management problems, especially those that entail elements of the mission-level environment, often fall into this latter category.

Table 1.2 and Fig. 1.3 graphically depict these various decompositions, which was developed by the US Air Force Research Laboratory and adopted by the DoD (Weatherington and Deputy 2005). Actually, as a framework for defining Autonomous Control Levels (ACL) for air vehicles, it is far from adequate to describe the many dimensions of autonomous functionality.

Autonomous capabilities have been enabled by advances in computer science, artificial intelligence, cognitive and behavioral sciences, machine training and learning, and communication technologies. Rule-based autonomy has made great progress, such as basic waypoint navigation, auto takeoff, and autoland. Even autonomous damage-tolerant flight control, such as loss of portions of wings, stabilizers, or other vehicle control elements, has been demonstrated in a real-world flight environment. The limiting factor in these rule-based cases is usually not the machine-based logic employed in decision-making, but rather the inadequacy of sensors or other sources of information needed to establish situation awareness and/or the context for the needed decision. Far less progress has been made in those domains involving more subjective decision criteria, extreme complexity, or higher levels of uncertainty in the key decision variables.

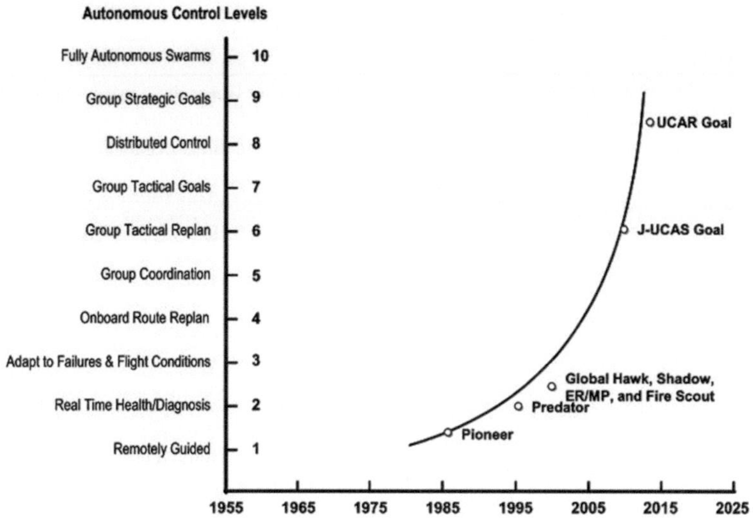

Fig. 1.3 UAV Roadmap 2005

1.2 Bio-inspired Computation

1.2.1 Definition of Swarm

Swarming is not about numbers, but about the interactions between individuals in a group, which is a particular emergent behavior of simple autonomous individuals (Clough 2002). In a Rand report written about swarming, swarming is looked at in two ways: the group interaction among individuals and swarming tactics that could be done by military groups. The second statement describes a process like this: originally dispersed, the groups come together to accomplish an objective, then disperse again, somewhat akin to guerilla war tactics. The report then goes on to describe a system that uses decentralized control to command a swarm of low-cost precision munitions. Webster's Dictionary gives the formal definition for "swarms," which is illustrated in Table 1.3. Bruce Clough provides a more formal definition from technical perspective: a collection of autonomous individuals relying on local sensing and reactive behaviors interacting such that a global behavior emerges from the interactions (Clough 2002).

The term swarm is often used for an aggregation of animals such as fish schools, birds flocks, and insect colonies such as ant, termite, and bee colonies performing collective behavior, which shows great robustness and cost-effectiveness. In this process, only local rules are utilized through interactions between self-organized agents. It is possible for us to take advantage of these ideas to develop autonomous UAV control systems. To this end, we should make it clear what the core advantage of this collective behavior is and when it is appropriate to use swarming. So let's go

Table 1.3 Formal definition of "swarm" in Webster's Dictionary

No	Definition
1	a: (1) a great number of honeybees emigrating together from a hive in company with a queen to start a new colony elsewhere. (2) a colony of honeybees settled in a hive
	b: an aggregation of free-floating or free-swimming unicellular organisms, usually used of zoospores
2	a: a large number of animate or inanimate things massed together and usually in motion
	b: a number of similar geological features or phenomena close together in space or time
3	to form and depart from a hive in a swarm
4	a: to move or assemble in a crowd
	b: to hover about in the manner of a bee in a swarm

Table 1.4 Comparison of swarms and teams

Attribute	Swarm	Team
Temporal	Reactive	Predictive
Composition	Homogenous	Heterogeneous
Interrelationships	Simple	Complex
Predictability	Probabilistic	Deterministic
Individual	Expendable	Critical
Efficiency	Low	High

back to examine the attributes of it by comparison of swarms and teams found in nature, which is shown as follows (Table 1.4):

- ***Temporal:*** Swarms are composed of individuals that can only react, not predict, while teams own the capability of predicting, which requires onboard models and plans.
- ***Composition:*** Swarms are composed of homogeneous or very limited heterogeneous individual types or roles, while teams are very heterogeneous with distinct roles assigned to distinct individuals.
- ***Interrelationships:*** Swarms interact with other agents using simple rules, with little knowledge of the global pattern, while teams exhibit much more complicated social behaviors. Besides, messages are not targeted to any particular individual, but are broadcasted. Teams communicate with higher-level semantic messages, usually directed at a particular individual.
- ***Predictability:*** Swarms are probabilistic; their performance can be described by distributions. They are not deterministic. Teams, on the other hand, are deterministic, with a known objective at a known time achieved via a deterministic plan.
- ***Individual worth:*** Swarms get their robustness from the similarity of individuals. This ensures that, if one of the interactions fails or if one of the insects misses its task, their failure is quickly compensated by the other insects. So individual worth is low. Teams on the other hand are composed of specialized individuals; loss of one can cripple critical team functions, and so individual worth is high.

- ***Efficiency:*** Swarms in nature are inefficient compared to ways humans might decide to do the same things. One of the benefits of teaming is that it maximizes efficiency by synergistic use of heterogeneous resources.

We should note that the inefficiency of swarms is not necessarily a drawback for the agents, since what they lack in efficiency they make up for in robustness. Their redundancy can survive attrition, where even a loss of a single team member could spell failure. Inefficiency leaves room for improvement in times of need, allows attrition when things go wrong, but it requires more resources. Agents interact with each other, and they have no big picture of the whole colony. However, trial and error over millions of years has developed the simple behaviors such that they interact to the good of the collective.

To obtain swarming ability, all one needs is an appropriate set of individual reactive behaviors supported by local sensing that will interact to develop the group behavior of interest and a simple architecture to manage those behaviors in the individual, which can be summarized as the following four aspects:

- Emergent behavior of the interacting individuals in the swarm
- Simple reactive behaviors in the individuals
- Behavior architecture in the individuals allowing switches between reactive behaviors
- Local sensing with individual variability

1.2.2 General Features of Bio-inspired Computation

Swarm intelligence, as a scientific discipline including research fields such as swarm optimization or distributed control in collective robotics, was born from biological insights about the incredible abilities of social insects to solve the complicated problems. Their colonies range in size from a few animals to millions of individuals, which shows fascinating behaviors that combine efficiency with both flexibility and robustness. Examples of complex and sophisticated behaviors are numerous and diverse among social insects, such as traffic management in the foraging process, dynamic task allocation between individuals, and the building of nest, to name just a few.

As we have explained, the colony often exhibits amazing and complex collective behavior. However, the individual always executes simple actions and follows simple rules. Of course, as a masterpiece of God, insects are highly elaborated machines to some extent, with the ability to modulate their behavior on the basis of the processing of many sensory inputs. Nevertheless, as pointed out by Seeley (2002), the complexity of an individual insect in terms of cognitive or communicational abilities may be high in an absolute sense, while remaining not sufficient to effectively supervise a large system and to explain the complexity of all the behaviors at the colony scale. It is often the case that a single insect can't find an efficient solution to a complex problem, while the whole colony is able to find an optimal solution very easily by taking advantage of wisdom of the population.

We have known that individual insects don't need any representation, any map, or explicit knowledge of the global structure they produce. A single insect can't assess a global situation, to centralize information about the state of its entire colony and then to control the tasks to be done by the other workers, which means there is no supervisor in the colony. Actually, a social insect colony is rather like a decentralized system consisting of autonomous agents that are distributed in the environment and that follow simple probabilistic stimulus-response rules. The rules that govern interactions among insects are executed only according to the local information. These interactions ensure the propagation of information through the colony, and they also organize the activity of each individual. Due to these sophisticated interaction networks, social insects can solve many kinds of problems and respond to external challenges in a very flexible and robust way. Behind this "organization without an organizer" are several hidden mechanisms which enable insect societies, whose members only deal with partial and noisy information about their environment, to cope with uncertain situations and to find solutions to complex problems, which is called self-organization by the researchers. We should note that other biological systems share similar collective properties such as colonies of bacteria or amoeba, fish schools, bird flocks, sheep herds, or even crowds of human beings.

The collective decisions in ants rely on self-organization that appears to be a major component of a wide range of collective behaviors in social insects, from the thermoregulation of bee swarms to the construction of nests in ants and termites. Self-organization is a set of dynamic mechanisms whereby structures appear at the global level of a system from interactions among its lower-level components, without being explicitly coded at the individual level. It relies on four basic ingredients (reprinted from Garnier et al. (2007), with permissions from Springer Nature):

Positive feedback (amplification): It often constitutes the basis of morphogenesis in social insects, resulting the execution of simple behavioral "rules of thumb" that promote the creation of structures. Examples of positive feedback include recruitment and reinforcement. For instance, trail recruitment to a food source is a kind of positive feedback which creates the conditions for the emergence of a trail network at the global level.

Negative feedback (for counterbalance and stabilization): It counterbalances positive feedback and helps to stabilize the collective pattern. Take the ant foraging, for example; it may exhibit some forms of negative feedback. It may result from the limited number of available foragers, the food source exhaustion, and the evaporation of pheromone or a competition between paths to attract foragers.

Amplification of fluctuations (randomness, errors, random walks): Collective behaviors emerge although social insects are known to perform stochastic actions.

What's more, randomness is often crucial because it enables discovery of new solutions, and fluctuations can act as seeds from which structures nucleate and grow. For instance, lost foragers have the chance to find new, unexploited food sources and then recruit nest mates to these food sources. In this way, better food sources can be found by utilizing the potential of randomness.

Multiple interactions: Self-organization often requires minimal density of mutually tolerant individuals. Moreover, individuals should be able to makes use of results of their own activities as well as others' activities. In other words, it requires multiple direct or stigmergic interactions among individuals to produce apparently deterministic outcomes and the appearance of large and enduring structures.

In addition to the previously detailed ingredients, self-organization is also characterized by a few key properties:

Dynamic: As stated before, the production of structures as well as their persistence requires permanent interactions between the members of the colony and with their environment. These interactions promote the positive feedbacks that create the collective structures and act for their subsistence against negative feedbacks that tend to eliminate them.

Emergent properties: They display properties that are more complex than the simple contribution of each agent. These properties arise from the nonlinear combination of the interactions between the members of the colony.

Bifurcations: A bifurcation is the appearance of new stable solutions when some of the system's parameters change. This corresponds to a qualitative change in the collective behavior.

Multi-stable: Multi-stability means that, for a given set of parameters, the system can reach different stable states depending on the initial conditions and on the random fluctuations.

As categorized by Garnier et al. (2007), the organization of collective behaviors in social insects can be understood as the combination of the four coordination, cooperation, deliberation, and collaboration functions. Each of these functions emerges at the collective level from the unceasing interactions between the insects. They support the information processing abilities of the colony in two ways: (1) coordination and collaboration shape the spatial, temporal, and social structures that result from the colony's work. The coordination function regulates the spatiotemporal density of individuals, while the collaboration function regulates the allocation of their activities. (2) Cooperation and deliberation provide tools for the colony to face the environmental challenges. The deliberation function represents the mechanisms that support the decisions of the colony, while the cooperation function represents the mechanisms that overstep the limitations of the individuals. Together, these four functions contribute together to the accomplishment of the various collective tasks of the colony.

1.2.3 *Bio-inspired Computation Algorithms*

Bio-inspired computation, short for biologically inspired computation, is the use of computers to model the living phenomena and simultaneously the study of life to improve the usage of computers, which is a major subset of natural computation. By simulating the collective behavior of decentralized, self-organized system from nature, many optimization algorithms have been developed, which could be called as either swarm intelligence or bio-inspired computation from different perspectives. Colonies of social insects have fascinated researchers for many years, and the mechanisms that govern their behavior remained unknown for a long time. Even though the single members of these colonies are non-sophisticated individuals, they are able to achieve complex tasks in cooperation. Coordinated colony behavior emerges from relatively simple actions or interactions between the colonies' individual members. Many aspects of the collective activities of social insects are self-organized and work without a central control. For example, leafcutter ants cut pieces from leaves, bring them back to their nest, and grow fungi used as food for their larvae. Weaver ant workers build chains with their bodies in order to cross gaps between two leaves. The edges of the two leaves are then pulled together and successively connected by silk that is emitted by a mature larva held by a worker. Another example is about the recruitment of other colony members for prey retrieval. Other examples include the capabilities of termites and wasps to build sophisticated nests or the ability of bees and ants to orient themselves in complicated and changing environment.

Optimization techniques inspired by collective behavior have become increasingly popular during the last decade. They are characterized by a decentralized way of working that mimics the behavior of swarms of social insects, flocks of birds, or schools of fish. The advantage of these approaches over traditional techniques is their robustness and flexibility. These properties make swarm intelligence a successful design paradigm for algorithms that deal with increasingly complex problems. In this monograph we focus on four of the most successful examples of optimization techniques inspired by swarm intelligence: ant colony optimization (ACO), particle swarm optimization (PSO), artificial bee colony (ABC), and differential evolution (DE).

1.2.3.1 Ant Colony Optimization

In many ant species, individual ants may deposit a pheromone (a chemical that ants can smell) on the ground while walking. By depositing pheromone, ants create a trail that is used, for example, to mark the path from the nest to food sources and back. Foragers can sense the pheromone trails and follow the path to food discovered by other ants (Duan 2005). Several ant species are capable of exploiting pheromone trails to determine the shortest among the available paths leading to the food (Fig. 1.4).

Deneubourg et al. (1990) and colleagues used a double bridge connecting a nest of ants and a food source to study pheromone trail laying and following behavior

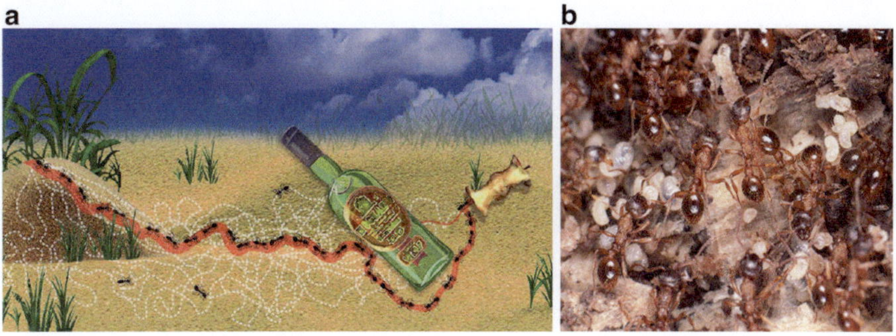

Fig. 1.4 Ant colony optimization

in controlled experimental conditions. They ran a number of experiments in which they varied the ratio between the length of the two branches of the bridge. The most interesting, for our purposes, of these experiments is the one in which one branch was longer than the other. In this experiment, at the start the ants were left free to move between the nest and the food source, and the percentage of ants that chose one or the other of the two branches was observed over time. The outcome was that, although in the initial phase random oscillations could occur, in most experiments all the ants ended up using the shorter branch.

This result can be explained as follows. When a trial starts there is no pheromone on the two branches. Hence, the ants do not have a preference, and they select with the same probability either of the two branches. It can be expected that, on average, half of the ants choose the short branch and the other half the long branch, although stochastic oscillations may occasionally favor one branch over the other. However, because one branch is shorter than the other, the ants choosing the short branch are the first to reach the food and to start their travel back to the nest. But then, when they must make a decision between the short and the long branch, the higher level of pheromone on the short branch biases their decision in its favor. Therefore, pheromone starts to accumulate faster on the short branch, which will eventually be used by the great majority of the ants.

Marco Dorigo and his colleagues introduced the first ACO algorithm in the early 1990s (Dorigo 1992; Dorigo et al. 1996). It should be clear by now how real ants have inspired ant system (AS) and later algorithms: the double bridge was substituted by a graph and pheromone trails by artificial pheromone trails. Also, because we wanted artificial ants to solve problems more complicated than those solved by real ants, we gave artificial ants some extra capacities, like a memory (used to implement constraints and to allow the ants to retrace their solutions without errors) and the capacity for depositing a quantity of pheromone proportional to the quality of the solution produced (a similar behavior is observed also in some real ant species in which the quantity of pheromone deposited while returning to the nest from a food source is proportional to the quality of the food source).

Fig. 1.5 Particle swarm optimization

1.2.3.2 Particle Swarm Optimization

The initial ideas on particle swarms of Kennedy (a social psychologist) and Eberhart (an electrical engineer) were essentially aimed at producing computational intelligence by exploiting simple analogues of social interaction (Kennedy and Eberhart 1995), rather than purely individual cognitive abilities. The first simulations were influenced by Heppner and Grenander's work and involved analogues of bird flocks searching for corn. These soon developed into a powerful optimization method—PSO.

The first particle swarm program was written by modifying a bird-flocking simulation. Two disparate groups of researchers in the 1980s had derived very similar models of the dynamics of bird flocks. Reynolds (1987), working from a computer-graphics perspective (e.g., he wanted to be able to portray realistic bird flock animations on a computer screen), concluded that bird flocking could be simulated using three rules:

Rule 1: **Separation**—steer to avoid colliding with local flockmates
Rule 2: **Alignment**—try to move in the same average direction as local flockmates
Rule 3: **Cohesion**—steer to move toward the perceived center of the local flock (Fig. 1.5)

Heppner, a biologist, made three-dimensional movies of bird flocks and carefully studied the dynamics of their choreography (Heppner and Grenander 1990). His model, though it was developed independently of Reynolds', contained essentially the same three rules, plus a fourth: attraction to a roost. His flocks eventually settled down.

The author had been experimenting with these flocking models and added one more feature, with surprising results. Inspired by Heppner's "attraction to a roost," "cornfield vector" was added, which in the first program was simply a two-dimensional point on the plane of the computer monitor. This point was considered to simulate some food on the ground; birds flying past might see the food or some sign of the presence of food, and most importantly, birds flying past could see that other birds seemed to be zeroing in on some target. Thus, members of the flock were attracted toward positions that other members of the flock had found to be relatively near to food.

The first experiments were shocking. The flock immediately converged on the point on the screen, as if sucked in by a vacuum cleaner. That first algorithm worked as follows:

1. Each bird evaluated its distance from the cornfield.
2. Each bird identified some "neighbors" who were nearby on the display plane.
3. Each bird identified which of its neighbors had come closest to the target point and where that had happened (note that the location of the point was not known, but only the distance from it).
4. If its position was to the right of (above) its own previous best point, then it moved some random amount to the left (down); else it moved a random amount to the right (upward).
5. If its position was to the right of (above) the best neighbor's best point, it moved a random amount to the left (down), otherwise right (upward).

The success of the food-searching program prompted the second set of experiments. The evaluation of distance from an arbitrary point on the screen was replaced by an XOR feedforward neural network. A network was defined with two input nodes, three hidden nodes, and one output, requiring nine connection weights and four biases. Thus optimizing the weights (including biases) meant searching through a thirteen-dimensional space. Weights were initialized with random values, and the program ran iteratively. For testing purposes, two of the weights were graphed on the screen, meaning that the entire flock could be watched as a display of swarming particles. Again, the flock had no difficulty finding an optimal matrix of weights. The plotted points zoomed immediately, it seemed, with no hesitation, toward a configuration that resulted in squared error in the network very near zero.

The PSO developed by Kennedy and Eberhart (1995) comprises a very simple concept, and paradigms can be implemented in a few lines of computer code (Duan et al. 2013b). It requires only primitive mathematical operators and is computationally inexpensive in terms of both memory requirements and speed. Early testing has found the implementation to be effective with several kinds of problems.

1.2.3.3 Artificial Bee Colony

A very interesting swarm in nature is honeybee swarm that allocates the tasks dynamically and adapts itself in response to changes in the environment in a collective intelligent manner. The honeybees have photographic memories; space-age sensory and navigation systems, possibly even insight skills; and group decision-making process during selection of their new nest sites, and they perform tasks such as queen and brood tending, storing, retrieving and distributing honey and pollen, communicating, and foraging. These characteristics are incentive for researchers to model the intelligent behaviors of bees.

Foraging is the most important task in the hive. Many studies have investigated the foraging behavior of each individual bee and what types of external information (such as odor, location information in the waggle dance, the presence of other bees at the source, or between the hive and the source) and internal information (such as remembered source location or source odor) affect this foraging behavior. Foraging process starts with leaving the hive of a forager in order to search food source to gather nectar. After finding a flower for herself, the bee stores the nectar in her honey stomach. Based on the conditions such as richness of the flower and the distance of the flower to the hive, the bee fills her stomach in about 30–120 min, and honey-making process begins with the secretion of an enzyme on the nectar in her stomach. After coming back to the hive, the bee unloads the nectar to empty honeycomb cells, and some extra substances are added in order to avoid the fermentation and the bacterial attacks. Filled cells with the honey and enzymes are covered by wax.

After unloading the nectar, the forager bee which has found a rich source performs special movements called "dance" on the area of the comb in order to share her information about the food source such as how plentiful it is and its direction and distance and recruits the other bees for exploiting that rich source. While dancing, other bees touch her with their antenna and learn the scent and the taste of the source she is exploiting. She dances on different areas of the comb in order to recruit more bees and goes on to collect nectar from her source. There are different dances performed by bees depending on the distance information of the source: round dance, waggle dance, and tremble dance. If the distance of the source to the hive is less than 100 m, round dance is performed, while if the source is far away, waggle dance is performed. Round dance does not give direction information. In case of waggle dance, direction of the source according to the sun is transferred to other bees. Longer distances cause quicker dances. The tremble dance is performed when the foraging bee perceives a long delay in unloading its nectar (Karaboga and Akay 2009).

A honeybee colony needs to divide its workforce so that individuals of appropriate number are allocated for each of the many tasks (Beekman et al. 2007). Bees are specialized in order to carry out every task in the hive. However, there is a controversy about which factors have roles on the specialization of bees, such as their age, hormones (internal factors), and individual predisposition coming from their genetic determination (Dornhaus et al. 1998), and also the allocation of tasks can dynamically change. For example, when food is drought, younger nurse bees will also join to foraging process.

Some approaches have been proposed to model the specific intelligent behaviors of honeybee swarms, and they have been applied for solving various problems. Virtual bee algorithm was developed by Yang (2005) to solve the numerical function optimizations. In the model, a swarm of virtual bees are generated, and they are allowed to move randomly in the phase space, and these bees interact when they find some target nectar. Nectar sources correspond to the encoded values of the function. The solution for the optimization problem can be obtained from the intensity of bee

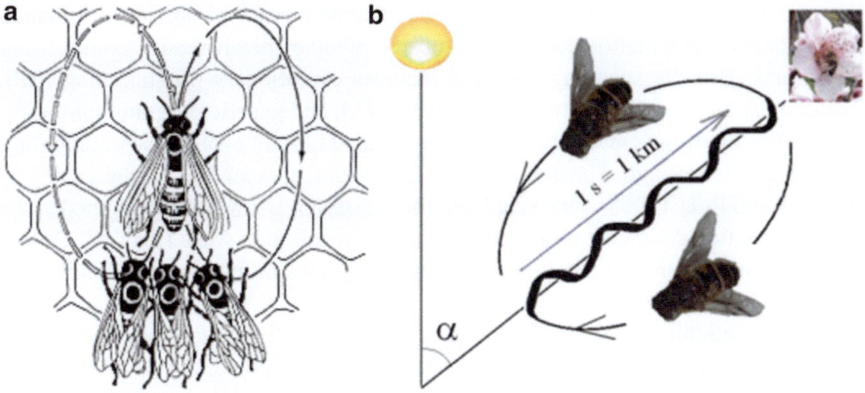

Fig. 1.6 Artificial bee colony optimization

interactions. The bees algorithm was described by Pham et al. (2006) and mimics the foraging behavior of honeybees. In its basic version, the algorithm performs a kind of neighborhood search combined with random search and can be used for both combinatorial optimization and functional optimization. BeeHive algorithm, which has been inspired by the communication in the hive of honeybees, was proposed by Wedde et al. (2004) and applied to the routing in networks. In BeeHive algorithm, bee agents travel through network regions called foraging zones. On their way their information on the network state is delivered for updating the local routing tables. Bee colony optimization was described by Teodorović and Dell'Orco (2005) for the ride-matching problem and for the routing and wavelength assignment in all-optical networks. ABC algorithm simulating foraging behavior of honeybees was invented by Karaboga (2005) (Fig. 1.6). Among the algorithms mentioned above, ABC is the one which has been most widely studied and applied to solve the real-world problems, so far. In this monograph, we will take ABC as the representative algorithm simulating bee swarm intelligence.

1.2.3.4 Differential Evolution

The most common of which is the phenomenon that every species had to adapt their physical structures to fit to the environments they were in and to strengthen their survival ability. By simulating the underlying relation between optimization and biological evolution, an important branch of computational intelligence, the evolutionary computing techniques have been developed to handle the complex optimization problems. Evolutionary computation uses iterative progress, such as growth or development in a population. This population is then selected in a guided random search using parallel processing to achieve the desired end. The paradigm of evolutionary computing techniques dates back to early 1950s, when the idea to use Darwinian principles for automated problem solving originated. Evolutionary computation techniques mostly involve metaheuristic optimization algorithms. Broadly speaking, the field includes evolutionary programming (EP), evolution strategies (ESs), genetic algorithm (GA), and genetic programming (GP).

The DE algorithm emerged as a very competitive form of evolutionary computing more than a decade ago. The first written article on DE appeared as a technical report by Storn and Price (1995). One year later, the success of DE was demonstrated at the First International Contest on Evolutionary Optimization in May 1996, which was held in conjunction with the 1996 IEEE International Conference on Evolutionary Computation (CEC).

DE finished third at the First International Contest on Evolutionary Optimization (1st ICEO), which was held in Nagoya, Japan. DE turned out to be the best evolutionary algorithm for solving the real-valued test function suite of the 1st ICEO (the first two places are not universally applicable but solved the test problems faster than DE). Price presented DE at the Second International Contest on Evolutionary Optimization in 1997, and it turned out as one of the best among the competing algorithms. Two journal articles describing the algorithm in sufficient details followed immediately in quick succession. In 2005 CEC competition on real parameter optimization, on 10-D problems classical DE secured 2nd rank and a self-adaptive DE variant called SaDE secured third rank although they performed poorly over 30-D problems. It is also interesting to note that the variants of DE continued to secure front ranks in the subsequent CEC competitions like CEC 2006 competition on constrained real parameter optimization (first rank), CEC 2007 competition on multi-objective optimization (second rank), CEC 2008 competition on large-scale global optimization (third rank), CEC 2009 competition on multi-objective optimization (first rank was taken by a DE-based algorithm MOEA/D for unconstrained problems), and CEC 2009 competition on evolutionary computation in dynamic and uncertain environments (first rank). We can also observe that no other single search paradigm such as PSO was able to secure competitive rankings in all CEC competitions.

Generally speaking, DE is considered as a kind of evolutionary computation technique. Like other popular EAs, it is a population-based optimization tool. However, different from other EAs, DE generates offspring by perturbing the solutions with a scaled difference of two randomly selected population vectors, instead of recombining the solutions under conditions imposed by a probabilistic scheme. What's more, DE employs a one-to-one spawning logic which allows replacement of an individual only if the offspring outperforms its corresponding parent. DE has been seen as an attractive optimization tool for continuous optimization for the advantages of simple implementation procedure, good performance for continuous optimization problems, less control parameters compared with other method, and the low space complexity.

1.3 Bio-inspired Intelligence in UAVs

1.3.1 Achieve Higher Autonomous Capability

One of the inevitable trends for the future information technology is intelligence, and the effective way to achieve intelligence is to simulate the various intelligent behaviors in nature (Duan et al. 2011b). Many of the adaptive optimization phenomena in nature inspire us that many highly complex optimization problems can be perfectly solved with the self-evolution in organisms and ecological systems. In recent years, some bio-inspired intelligent methods have emerged, which are definitely different from the classical mathematical programming principle. The typical bio-inspired methods include genetic algorithms (GA), ACO, PSO, artificial immune system (AIS), ABC, cultural evolution (CE), emotion computing (EC), and DNA computing. All the bio-inspired intelligent methods are trying to simulate the natural ecosystem mechanisms, which have greatly enriched the modern optimization techniques and provided practical solutions for those complicated combinatorial optimization problems. With the stealthy rise of bio-inspired intelligence era, which is characterized by the simulation of natural and biological mechanisms, a number of bio-inspired intelligence technologies have demonstrated strong vitality and potential for further development in solving the classic NP-C problems and practical applications.

Bio-inspired intelligence has the advantages of strong robustness, good distributed computing mechanism, and easy to combine with other methods. Although the rigorous theoretical analysis for most of the bio-inspired intelligent methods has not been conducted, and the current study in this field is still in the experimental and preliminary application stage, the bio-inspired intelligent methods have already found their applications in many typical fields. These methods can not only solve the one-dimensional static optimization problems but also solve multidimensional dynamic optimization problems. There are also many breakthroughs in the hardware implementation of bio-inspired intelligent methods. All these make the bio-inspired intelligence show strong vitality and broad prospects for further development.

While recent technological advances have enabled the development of unmanned vehicle systems and recent implementations have proven the UAV's benefits in both military and civilian applications, the full benefit of unmanned systems will be utilized when they can operate autonomously. Today, UAVs can not only be used in communications, meteorology, disaster monitoring, agriculture, geology, transportation, and many other civilian fields but also have a wide application in intelligent monitoring and surveillance, artificial interference, bait, military communications, air defense suppression, fighter or cruise missile defense, air-to-air combat, and border patrols. Each component in UAVs is a high-technologically complicated subsystem, and there is a strong dependence and coordination between the various components. Therefore, two prominent features of a typical UAV are intensive high-technology and "system of systems." UAV not only plays an active role in the civilian fields but also can adapt to a long, large-depth combat and even can act as the air combat weapon platform to support five-dimensional integration (land, sea, air, space, and electric) in future high-technological warfare. All these are due to that UAV has the light weight, low cost, zero casualty, high mobility, and good adaptability.

It has been explained that the ultimate goal in the development of autonomy technology for UAVs is to replace the human pilot, which means build it into a fully autonomous system. The special feature of an autonomous system is its ability to be goal directed in unpredictable situations. This ability is a significant improvement in capability compared to the capabilities of automatic systems. An autonomous system is able to make a decision based on a set of rules and/or limitations. It is able to determine what information is important in making a decision. It is capable of a higher level of performance compared to the performance of a system operating in a predetermined manner. Autonomous capabilities have been enabled by advances in computer science (digital and analog), artificial intelligence, cognitive and behavioral sciences, machine training and learning, and communication technologies. In order to achieve these autonomous capabilities in the highly dynamic unmanned system environment, improvement is essential in advanced algorithms that provide robust decision-making capabilities (such as machine reasoning and intelligence), automated integration of highly disparate information, and the computational construct to handle data sets with imprecision, incompleteness, contradiction, and uncertainty. The dynamic, self-organization, coordination, strong robustness, and other characteristics emerged in the process of bio-inspired intelligence can meet the requirements of UAVs under complicated battlefield environments. So, bio-inspired intelligence may provide a promising way to handle the complicated situation for UAVs, thus enhancing the autonomous capability of UAVs.

1.3.2 Enhance Robustness and Flexibility

We have seen that complex colony-level structures and many aspects of the so-called swarm intelligence of social insects can be understood in terms of interaction networks and feedback loops among individuals. These are the basic elements that allow the emergence of dynamic patterns at the colony level. These patterns can be material or social and lead the colony to structure its environment and solve problems. The most interesting properties of these self-organized patterns are robustness (the ability for a system to perform without failure under a wide range of conditions) and flexibility (the ability for a system to readily adapt to new, different, or changing requirements). Robustness results from the multiplicity of interactions between individuals that belong to the colony. This ensures that, if one of the interactions fails or if one of the insects misses its task, their failure is quickly compensated by the other insects. This also promotes stability of produced patterns, whereas individual behaviors are mostly probabilistic.

Flexibility of self-organized systems is well illustrated by the ability of social insects to adapt their collective behaviors to changing environments and to various colony sizes. These adaptations can occur without any change of the behavioral rules at the individual level. For instance, in the case of the selection of the shortest path in ants, a geometrical constraint applied on one of the two alternative paths increases the time needed by the ants to come back to their nest through this path and thus biases the choice toward the other path without any modification of the insects' behaviors.

But flexibility can also rely on the modulation of the individual behavioral rules by some factors produced by the environment or by the activity of the colony. For instance, the presence of an air flow modifies the probability for an ant to pick up and drop corpses of dead ants. This subtle modification of behavioral probabilities deeply modifies the shape of the piles resulting from the ants' aggregating activity. Last, the presence of environmental heterogeneities can modulate the behavior of an insect and thus bias the behavior of the colony toward a particular solution.

All these behavioral modulations provide the opportunity for a colony to develop a wide range of collective behaviors and can also be a powerful lever for evolution to shape and optimize these behaviors in a highly adaptive way. Thanks to the sensitivity of individuals to variations (either environmental or triggered by the colony itself), the colony as a whole is able to perceive these changes and can thus react appropriately in almost every situation.

To operate in complex and uncertain environments, the UAV must be able to sense and understand the environment. This capability implies that the UAV must be able to create a model of its surrounding world by conducting multisensor data fusion (MDF) and converting these data into meaningful information that supports a variety of decision-making processes. The perception system must be able to perceive and infer the state of the environment from limited information and be able to assess the intent of other agents in the environment (Sun and Duan 2012). This understanding is needed to provide future UAVs with the flexibility and adaptability for planning and executing missions in a complex, dynamic world.

While robustness in adaptability to environmental change is necessary, the future need is to be able to adapt and learn from the operational environment because every possible contingency cannot be programmed a priori. This adaptation must happen fast enough to provide benefits within the adversary's decision loop, and the autonomy should be constructed so that these lessons can be shared with other autonomous systems that have not yet encountered that situation. Yet even in a hostile, dynamic, unstructured, and uncertain environment, this learning must not adversely affect safety, reliability, or the ability to collaborate with the operator or other autonomous systems. Although such capabilities are not currently available, recent advancements in computational intelligence (especially neuro-fuzzy systems), neuroscience, and cognition science may lead to the implementation of some of the most critical functionalities of heterogeneous, sensor net-based MDF systems. Especially, we can learn something from the flexibility and robustness of self-organized patterns in swarm intelligence.

1.3.3 Cooperative Control of Multiple UAVs

In recent years a significant shift of focus has occurred in the field of autonomous UAVs as researchers have begun to investigate problems involving multiple rather than single UAV. As a result of this focus on multiple UAV system, multiple UAV coordination has received significant attention. Research involving multiple UAV coordination is being undertaken from coordinated path planning for multiple UAVs (Duan et al. 2009, 2013a) to the controller design of formation flight and formation reconfiguration (Duan et al. 2008; Duan and Liu 2010a; Zhang et al. 2010) or to opening new application domains. This field is still in its infancy, and many exciting new approaches are being explored for different applications.

Modern military systems are becoming increasingly sophisticated, with a mixture of manned and unmanned vehicles being used in complex battlefield environments. Traditional solutions involve a centralized resource allocation, followed by decentralized execution. More modern battlespace management systems are considering the use of cooperative operation of large collections of distributed vehicles, with location computation, global communication connections, and decentralized control actions. It is easy to understand that a group of UAVs is more capable than a single UAV, since the workload can be divided among the group. The individual UAVs can be equipped for different functions in the mission whether it be surveillance and reconnaissance, strike, or battle damage assessment. Surveillance missions can be completed quickly by covering more search area when the group is spread out. They also offer redundancy to ensure that the mission is completed. If one UAV is destroyed by the enemy or drops out because of mechanical failure, the rest of the group can fill in and carry out the mission. By having more than one UAV assigned to a mission, the probability of success dramatically increases.

One of the simplest cooperative control problems is that of formation flight: a set of vehicle fly in a formation, specified by the relative locations of nearby vehicle. This area has received considerable attention in the literature, both as a centralized and as a decentralized problem. One way to approach the formation control problem is to formulate it as an optimization, and another approach to solving this problem is to consider the mechanical nature of the systems and to shape the dynamics of the formation using potential fields. An important issue that arises in formation control is that of string stability, in which disturbances grow as they propagate through a system of vehicles.

As a result of this focus on multiple UAV system, multiple UAV coordination has received significant attention. In particular multiple UAV resource allocation has recently risen to prominence. In order to achieve the desired objective while meeting various tactical/technical constraints, it has to be decided which UAV should execute which task, searching, classifying, attacking, or performing battle damage assessment on potential targets. This is necessary even for relatively simple multiple UAV teams, and the importance of resource allocation grows with the complexity, size, and capacities of the team. The objective of resource allocation in multiple UAVs can be divided into three classes, which are, respectively, maximizing target value damage (TVD), minimizing UAV attrition, and minimizing white target (such as church, school) damage, also called collateral damage. The selection of the proper objective depends on the tactical mission, UAV target information, and the combat situation. There are two types of resource allocation problems: static and dynamic. Static resource allocation means that the assignment may be made at a specific time such that all of the UAVs are committed, while in the dynamic case, resource may be reallocated when the future variation in the battlefield is taken account of.

Take an example to explain the superiority of executing mission using multiple cooperative UAVs. One of the applications that motivates the use of multiple cooperative UAVs and poses several challenges to system engineers (Xargay et al. 2012), both from a theoretical and practical standpoint, is automatic road search for improvised explosive device detection. The mission is initiated by a minimally trained user who scribbles a path on a digital map, generating a precise continuous ground track for the airborne sensors to follow. This ground track is then transmitted over the network to a fleet of small tactical UAVs equipped with complementary visual sensors. An optimization algorithm autonomously generates feasible flight trajectories that maximizes road coverage and accounts for sensor capabilities, field of view, resolution, and gimbal constraints, as well as intervehicle and ground-to-air communication limitations. The fleet of UAVs then starts the cooperative road search. During this phase, the information obtained from the sensors mounted on board the UAVs is shared over the network and retrieved by remote users in near real time. The explosive device detection can thus be done remotely on the ground, based on in situ imagery data delivered over the network.

In this particular mission scenario, a robust cooperative control algorithm for the fleet of UAVs can improve mission performance and provide reliable target discrimination by effectively combining the capabilities of the onboard sensors (Xargay et al. 2012). Flying in a coordinated fashion is what allows for the overlap of the fields of view (FOVs) of multiple sensors to be maximized and for full advantage to be taken of complementary sensor suites. This mission can be completed in three basic steps. Initially, each vehicle is assigned a feasible path with a desired speed profile. Together, the trajectories generated for all vehicles satisfy the mission requirements and the vehicle dynamic constraints, while ensuring collision-free maneuvers. Then, a path-following algorithm ensures that every vehicle follows its own path independently of the temporal assignments of the mission. Finally, the vehicles coordinate their positions along the paths with the remaining vehicles engaged in the mission by exchanging coordination information over the communication network. These three steps are accomplished by judiciously decoupling space and time in the formulation of the trajectory-generation, path-following, and time-coordination problems and by relying on the existing inner-loop controllers for nominal control of the autonomous systems.

As we have explained, swarm strengths are in executing reactive tasks, executing simple tasks, using multiple similar agents, and robustness to attrition. In addition, emergent behavior stability relies on imperfect sensing, which feeds into the strengths of inexpensive sensors. As history shows, simple autonomous systems have successfully used emergent behavior over millions of years to ensure their survival and proliferation. It is easy to draw the conclusion that if one was designing a distributed system of simple autonomous entities, one should look to emergent behavior as the autonomous control method of choice.

Emergent behaviors have very attractive attributes for inclusion into autonomous UAVs:

1. ***Inherently decentralized***: It eliminates the need for any centralized command, an attribute that is desired for some of our distributed UAV control system development.
2. ***Inherently implicit***: It eliminates any requirement to explicitly control the individuals.
3. ***Inherently resilient***: Natural history shows that it has ensured the survival of millions of species over millions of years. It is tolerant to imperfection and variance among individuals.
4. ***Inherently scalable***: Emergent systems are very robust to the addition and subtraction of members, so such systems could work well in high-risk situations.

Of course, emergent behavior has its drawbacks:

1. ***Results are probabilistic.*** One cannot say exactly when an outcome will happen, just that there is a probability distribution that it will occur within some interval which depends on the number of individuals, their micro behaviors, and the environment.
2. ***Execution is nondeterministic.*** One cannot know exactly how each individual will do something or how the particular individual's contribution will contribute to the whole, just that the collective will emerge the behavior "somehow."
3. ***Chaotic behaviors are possible.*** Depending on the micro behaviors, results could be chaotic (small changes in input result in huge differences in output).
4. ***Design is problematic.*** Nonlinear (possibly chaotic) and nondeterministic interacting behaviors relying on environmental aspects cannot be predicted, only observed. Design of such systems is by trial and error (evolution).

1.3.4 Cooperative Control of Heterogeneous Vehicle Groups

UAVs can be used to cover large areas searching for targets. However, sensors on UAVs are typically limited in their accuracy of localization of targets on the ground. On the other hand, unmanned ground vehicles (UGVs) can be deployed to accurately locate ground targets, but they have the disadvantage of not being able to move rapidly or see through such obstacles as buildings or fences. Typical applications include air- and ground-based mapping of predetermined areas for tasks such as surveillance, target detection, tracking, and search and rescue operations. The use of multiple collaborative vehicles is ideally suited for such tasks. And heterogeneous cooperative techniques can widen the application fields of unmanned aerial or ground vehicles and enhance the effectiveness of implementing detection, search, and rescue tasks (Duan and Liu 2010b; Duan et al. 2011a).

Consider the task of reliably detecting and localizing an unknown number of features within a prescribed search area. In this setting, it is highly desired to fuse information from all available sources. It is also beneficial to proactively focus the attention of resources, minimizing the uncertainty in detection and localization. Deploying teams of vehicles working toward this common objective offers several advantages. A key aspect of this task is the synergistic integration of aerial and ground vehicles that exhibit complementary capabilities and characteristics. Fixed-wing aircraft offer broad field of view and rapid coverage of search areas. However, minimum limits on operating airspeed and altitude, combined with attitude uncertainty, place a lower limit on their ability to resolve and localize ground features. Ground vehicles, on the other hand, offer high-resolution sensing over relatively short ranges, with the disadvantage of obscured views and slow coverage. The use of aerial- and ground-based sensor platforms is closely related to other efforts to exploit the truly complementary capabilities of air and ground vehicles.

Automatic target recognition is one of the key tasks for the reconnaissance UAVs or UGVs. There are some key problems to tackle with, such as the number of reconnaissance UAVs or UGVs used in a number of interactions, the network delay for central positioning, and the perceiving ability of attack systems for positioning centers. Dasgupta (2008) mainly focused on the coordination aspects between UAVs to efficiently decide how they are to act by using a swarming mechanism and described algorithms for the different operations performed by the UAVs in the system and for different swarming strategies, which are embedded within software agents located on the UAVs. Grocholsky et al. (2006) described the implementation of a decentralized architecture for autonomous teams of UAVs and UGVs engaged in active perception and also proposed a theoretical framework based on an established approach to the underlying sensor fusion problem. This provided transparent integration of information from heterogeneous sources, and the approach was applied to missions involving searching for and tracking multiple ground targets.

As safety and security has become a critical problem worldwide, aerial video surveillance (AVS) systems based on UAV or UGV have now been playing more and more important roles in not only civil but also military applications. Active vision is a kind of intelligent visual information acquisition mechanism. It is based on actively changing the sensor orientation and location to acquire visual information. Fregene et al. (2005) developed a system- and control-oriented intelligent agent framework called the hybrid intelligent control agent (HICA), as well as its composition into specific kinds of multi-agent systems. HICA is essentially developed around a hybrid control system core so that knowledge-based planning and coordination can be integrated with verified hybrid control primitives to achieve the coordinated control of multiple multimode dynamic systems. The scheme was successfully applied to the control of teams of UAVs and UGVs engaged in a pursuit-evasion war game. Ariyur and Fregene (2008) proposed a generally applicable method for the tracking of UGVs by UAVs. The beauty of this approach is that the only information provided to the UAV guidance system was waypoints required to establish and maintain track of the target. Therefore, no modification of any sort was required on the UAV's guidance and control systems. As long as the vehicle could follow simple waypoint commands in a closed loop, it would work with the proposed method. Another advantage is that it works very well for UAVs that have sensors of fixed attitude and geometry.

1.4 Outline of the Monograph

This monograph is dedicated to the bio-inspired computation in UAVs, which is divided into 8 chapters (Fig. 1.7). Chapter 1 discusses the development of UAVs and emphasizes the role bio-inspired intelligence plays in achieving higher autonomous capability. Chapter 2 introduces four representative bio-inspired algorithms, which are ACO, PSO, ABC, and DE, respectively. The biological inspiration, principle, and implementation procedures of the algorithms are explained in detail. Then in Chap. 3 the parameter identification of flight control system is discussed on the basis of modeling for the UAV. Then a specific kind of controller design problem involving the pendulum-like oscillation in the hover and stare state for a micro aerial vehicle (MAV) is handled.

Chapter 4 deals with path-planning problem using bio-inspired algorithms, for both single UAV and multiple UAVs, both in the 2-dimensional scenario and 3-dimensional scenario. This chapter mainly contains three sections. First, a chaotic ABC approach is proposed for two-dimensional path planning. Then path planning is extended to three-dimensional scenario through an improved ACO added a path smoothing strategy in Sect. 4.4. In Sect. 4.5 coordinated path replanning for multiple UAVs is dealt with by taking advantage of Max–Min adaptive ACO.

Chapter 5 deals with three significant problems in the formation flight problem, which are, respectively, formation control, close formation (tight formation), and formation configuration. First, a chaotic PSO-based nonlinear dual-mode receding horizon control (RHC) method is proposed for solving the constrained nonlinear systems. Then a novel type of control strategy of using hybrid RHC and DE algorithm is proposed based on the nonlinear model of multiple UAVs close formation. Moreover, based on the Markov chain model, the convergence of DE is proved. The formation configuration, which is about diving multiple UAVs to form a new flying formation state, is explained in detail using the RHC-based DE in Sect. 5.4. This global control problem is transformed into several online local optimization problems at a series of receding horizons, while the DE algorithm is adopted to optimize control sequences at each receding horizon.

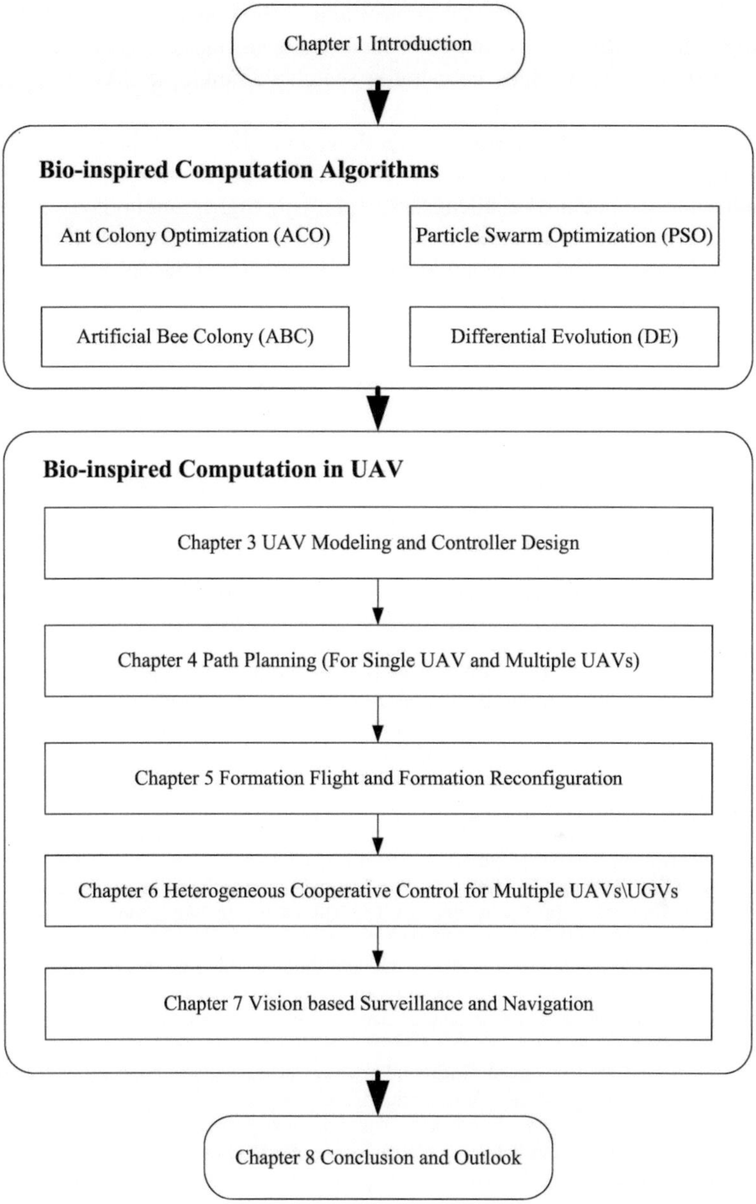

Fig. 1.7 Outline of the monograph

Chapter 6 discusses multiple UAV/UGV heterogeneous cooperation and cooperative search of multiple UAVs based on DE. In multiple UAV/UGV heterogeneous control, two key issues are considered in multiple UGV subgroups, which are Reynolds rule and Virtual Leader (VL). PSO based RHC is proposed for multiple

UGV flocking, and velocity vector control approach is adopted for multiple UAV flocking. Then, multiple UAV and UGV heterogeneous tracking can be achieved by these two approaches. Besides, cooperative search with multiple UAVs is dealt with using receding horizon control based on DE.

Chapter 7 describes bio-inspired algorithms involving vision-based surveillance and navigation. Chapter 8 discusses the opportunities for the development of UAVs and points out the potential challenges for achieving higher autonomous capability. By incorporating the bio-inspired intelligence into the UAS, it is possible to enhance the ability to understand and adapt to the environment and the ability to cooperate with other autonomous systems.

References

Ariyur KB, Fregene KO (2008) Autonomous tracking of a ground vehicle by a UAV. In: Proceedings of American Control Conference, Seattle, WA. IEEE, pp 669–671

Beekman M, Gilchrist AL, Duncan M, Sumpter DJ (2007) What makes a honeybee scout? Behav Ecol Sociobiol 61(7):985–995

Clough BT (2002) UAV swarming? So what are those swarms, what are the implications, and how do we handle them? DTIC Document. Available online: http://www.dtic.mil/dtic/tr/fulltext/u2/a405548.pdf

Dasgupta P (2008) A multiagent swarming system for distributed automatic target recognition using unmanned aerial vehicles. IEEE Trans Syst Man Cybern A Syst Hum 38(3):549–563

Deneubourg J-L, Aron S, Goss S, Pasteels JM (1990) The self-organizing exploratory pattern of the argentine ant. J Insect Physiol 3(2):159–168

Dorigo M (1992) Optimization, learning and natural algorithms. PhD Thesis. Politecnico di Milano, Italy

Dorigo M, Maniezzo V, Colorni A (1996) Ant system: optimization by a colony of cooperating agents. IEEE Trans Syst Man Cybern B Cybern 26(1):29–41

Dornhaus A, Klügl F, Puppe F, Tautz J (1998) Task selection in honeybees-experiments using multi-agent simulation. In: Proceedings of The Third German Workshop on Artificial Life, Bochum. Verlag Harry Deutsch, pp 171–183

Duan H (2005) Ant colony algorithms: theory and applications. Science Press, Beijing, China

Duan H, Liu S (2010a) Non-linear dual-mode receding horizon control for multiple unmanned air vehicles formation flight based on chaotic particle swarm optimisation. IET Control Theory Appl 4(11):2565–2578

Duan H, Liu S (2010b) Unmanned air/ground vehicles heterogeneous cooperative techniques: current status and prospects. Sci China Technol Sci 53(5):1349–1355

Duan H, Luo Q, Yu Y (2013a) Trophallaxis network control approach to formation flight of multiple unmanned aerial vehicles. Sci China Technol Sci 56(5):1066–1074

Duan H, Luo Q, Ma G, Shi Y (2013b) Hybrid particle swarm optimization and genetic algorithm for multi-UAVs formation reconfiguration. IEEE Comput Intell Mag 8(3):16–27

Duan H, Ma G, Luo D (2008) Optimal formation reconfiguration control of multiple UCAVs using improved particle swarm optimization. J Bionic Eng 5(4):340–347

Duan H, Shao S, Su B, Zhang L (2010) New development thoughts on the bio-inspired intelligence based control for unmanned combat aerial vehicle. Sci China Technol Sci 53(8):2025–2031

Duan H, Zhang X, Wu J, Ma G (2009) Max-min adaptive ant colony optimization approach to multi-UAVs coordinated trajectory replanning in dynamic and uncertain environments. J Bionic Eng 6(2):161–173

Duan H, Zhang X, Xu C (2011a) Bio-inspired computing. Science Press, Beijing, China

Duan H, Zhang Y, Liu S (2011b) Multiple UAVs/UGVs heterogeneous coordinated technique based on Receding Horizon Control (RHC) and velocity vector control. Sci China Technol Sci 54(4):869–876

Francis MS (2012) Unmanned Air Systems: challenge and opportunity. J Aircraft 49(6):1652–1665

Fregene K, Kennedy DC, Wang DW (2005) Toward a systems and control oriented agent framework. IEEE Trans Syst Man Cybern B Cybern 35(5):999–1012

Garnier S, Gautrais J, Theraulaz G (2007) The biological principles of swarm intelligence. Swarm Intell 1(1):3–31

Grocholsky B, Bayraktar S, Kumar V, Taylor CJ, Pappas G (2006) Synergies in feature localization by air-ground robot teams. In: Proceedings of 9th International Symposium on Experimental Robotics (ISER'04), Singapore. Springer Berlin Heidelberg, pp 353–362

Heppner F, Grenander U (1990) A stochastic nonlinear model for coordinated bird flocks. In: Krasner S (ed) The ubiquity of chaos. AAAS Publications, Washington, DC, pp 233–238

Karaboga D (2005) An idea based on honey bee swarm for numerical optimization. Techn Rep TR06, Erciyes Univ Press, Erciyes

Karaboga D, Akay B (2009) A survey: algorithms simulating bee swarm intelligence. Artif Intell Rev 31(1–4):61–85

Kennedy J, Eberhart R (1995) Particle swarm optimization. In: Proceedings of IEEE International Conference on Neural Networks. Piscataway. IEEE, pp 1942–1948

Newcome LR (2004) *Unmanned aviation: a brief history of unmanned aerial vehicles*. AIAA, Reston

Pham D, Ghanbarzadeh A, Koc E, Otri S, Rahim S, Zaidi M (2006) The bees algorithm–a novel tool for complex optimisation problems. In: Proceedings of the 2nd Virtual International Conference on Intelligent Production Machines and Systems, Cardiff, UK. Elsevier, pp 454–459

Reynolds CW (1987) Flocks, herds and schools: A distributed behavioral model. Comput Graphics 21(4):25–34

Seeley TD (2002) When is self-organization used in biological systems? Biol Bull 202(3):314–318

Storn R, Price K (1995) Differential Evolution-a simple and efficient adaptive scheme for global optimization over continuous spaces. Techn Rep. International Computer Science Institute, Berkeley, CA

Sun C, Duan H (2012) A restricted-direction target search approach based on coupled routing and optical sensor tasking optimization. Optik 123(24):2226–2229

Teodorović D, Dell'Orco M (2005) Bee colony optimization–a cooperative learning approach to complex transportation problems. In: Proceedings of 16th Mini-EURO Conference and 10th Meeting of EWGT, Poznan. Publishing House of the Polish Operational and System Research, pp 51–60

Weatherington D, Deputy U (2005) Unmanned Aircraft Systems Roadmap, 2005–2030. Available online: http://uav.navair.navy.mil/roadmap05/USRoadmapAug%2005.pdf

Wedde HF, Farooq M, Zhang Y (2004) BeeHive: an efficient fault-tolerant routing algorithm inspired by honey bee behavior. In: Ant colony optimization and swarm intelligence: 4th international workshop, ANTS 2004, Brussels, Belgium. Springer, pp 83–94

Wikipedia website (2013). Unmanned aerial vehicle. https://en.wikipedia.org/wiki/Unmanned_aerial_vehicle. Accessed 22 Jun 2002

Xargay E, Dobrokhodov V, Kaminer I, Pascoal AM, Hovakimyan N, Cao C (2012) Time-critical cooperative control of multiple autonomous vehicles. IEEE Control Syst Mag 32(5):49–73

Xu C, Duan H, Liu F (2010) Chaotic artificial bee colony approach to Uninhabited Combat Air Vehicle (UCAV) path planning. Aerosp Sci Technol 14(8):535–541

Yang XS (2005) Engineering optimizations via nature-inspired virtual bee algorithms. In: Proceedings of First International Work-Conference on the Interplay Between Natural and Artificial Computation, Las Palmas, Canary Islands, Spain. Springer, pp 317–323

Zhang X, Duan H (2012) Differential evolution-based receding horizon control design for multi-UAVs formation reconfiguration. Trans Inst Meas Control 34(2–3):165–183

Zhang X, Duan H, Yu Y (2010) Receding horizon control for multi-UAVs close formation control based on differential evolution. Sci China Inf Sci 53(2):223–235

Chapter 2
Bio-inspired Computation Algorithms

Pei Li and Haibin Duan

Abstract Bio-inspired computation is the use of computers to model the living phenomena and simultaneously the study of life to improve the usage of computers. Swarm behaviors in animal groups such as bird flocks, bees, ants, fish schools, and sheep herds, as well as insects like mosquitoes, ants, and bees, often exhibit incredible abilities to solve complex problems that seem far beyond their capabilities. This chapter mainly focuses on the biological inspiration, principle, and implementation procedures of four popular bio-inspired computation algorithms including ant colony optimization (ACO), particle swarm optimization (PSO), artificial bee colony (ABC), and differential evolution (DE). Special emphasis has been laid on how the biological behavior can be transferred into a technical algorithm. Moreover, description of algorithms in more general terms and the most successful variants of these algorithms are provided. Finally, a brief introduction to other bio-inspired computation algorithms such as glowworm swarm optimization (GSM), bacteria foraging optimization (BFO), bat-inspired algorithm (BA) is presented.

2.1 Introduction

Bio-inspired computation, short for biologically inspired computation, is a way of developing computer systems by taking ideas from the biological world. It is a field of study that loosely knits together subfields related to the topics of connectionism, social behavior, and emergence. Briefly put, it is the use of computers to model the living phenomena and simultaneously the study of life to improve the usage of computers. Bio-inspired computation in an interdiscipline composed of many different fields, such as biology, computer science, physics, mathematics, and genetics. As we have known, biological systems have many advantages over computer systems, as they are able to solve much more complex problems beyond

The original version of this chapter was revised. A correction to this chapter is available at https://doi.org/10.1007/978-3-642-41196-0_9

H. Duan and P. Li, *Bio-inspired Computation in Unmanned Aerial Vehicles*,
DOI 10.1007/978-3-642-41196-0_2, © Springer-Verlag Berlin Heidelberg 2014

the capability of the contemporary computer with far less energy and much higher robustness. Many of the ideas taken from natural processes have been applied to machine learning, leading to new developments in artificial intelligence.

Bio-inspired computation is a major subset of natural computation, which is different from traditional artificial intelligence (AI) in that it often takes a more evolutionary approach to learning. In traditional AI, as the creator, programmers give their program some degree of intelligence in the process of programming. However, bio-inspired techniques often involve the method of specifying a set of simple rules, a set of simple organisms which adhere to those rules, and a method of iteratively applying those rules. In other words, bio-inspired computation takes a more bottom-up, decentralized approach. It is a new approach to enable the intelligence, as the constructed simple system is able to involve into a more complex one. Complexity gets built upon complexity until the end result is something markedly complex and quite often completely counterintuitive from what the original rules would be expected to produce.

By simulating the collective behavior of decentralized, self-organized system from nature, many optimization algorithms have been developed, which could be called as either swarm intelligence or bio-inspired computation from different perspectives. The term swarm intelligence refers to a kind of problem-solving ability that emerges in the interactions among the individuals that follow a simple rule. The concept of a swarm means multiplicity, stochasticity, randomness, and messiness, and the concept of intelligence suggests that the problem-solving method is somehow successful. Swarm behavior can be seen in bird flocks, fish schools, as well as in insects like mosquitoes and midges. Many animal groups such as fish schools and bird flocks clearly display structural order, with the behavior of the organisms so integrated that even though they may change shape and direction, they appear to move as a single coherent entity. Each individual in the group attempts to maintain a minimum distance with other members at all times. This rule has the highest priority and corresponds to a frequently observed behavior of animals in nature. If individuals are not performing an avoidance maneuver, they tend to avoid being isolated from others and to align themselves with their neighbors. A swarm can be viewed as a group of agents cooperating to achieve some purposeful behavior and achieve some goal. This collective intelligence seems to emerge from large groups composed of relatively simple agents. The agents use simple local rules to govern their actions and via the interactions of the entire group, then the swarm achieves its objectives eventually.

There is no supervisor in the colony, which means that there is no central control in the swarm and each individual has a stochastic behavior by taking advantage of her perception in the environment and her neighborhood. The agents use simple local rules to govern their actions, and via the interactions of the entire group, the swarm achieves its objectives. Note that the local rules have no relation to the global pattern. Interactions among the members through the network lead to the emergence behavior, which make the colony be able to cope with complicated situations and to find solutions to complex problems, which is called self-organization by researchers. Self-organization is a crucial feature of a swarm system which results to global-level (macroscopic level) response by means of low-level (microscopic level) interactions. Bonabeau et al. (1999) interpreted the self-organization in swarms through four characteristics, which are respectively positive feedback, negative feedback, fluctuations, and multiple interactions. Positive feedback is a simple behavioral "rules of thumb" that promotes the creation of convenient structures. Recruitment and reinforcement such as trail laying and following in some ant species or dances in bees can be shown as examples of positive feedback. Then we have a negative feedback that counterbalances positive feedback and helps to stabilize the collective pattern. In order to avoid the saturation which might occur in terms of available foragers, food source exhaustion, crowding, or competition at the food sources, a negative feedback mechanism is needed. Fluctuations such as random walks, errors, and random task switching among swarm individuals are vital for creativity and innovation. Randomness is often crucial for emergent structures since it enables the discovery of new solutions. Multiple interactions occur since agents in the swarm use the information coming from the other agents so that the information and data spread to all network.

Millonas (1994) also defined five principles to be satisfied by a swarm to have an intelligent behavior:

1. *The proximity principle*: The swarm should be able to do simple space and time computations.
2. *The quality principle*: The swarm should be able to respond to quality factors in the environment such as the quality of foodstuffs or safety of location.
3. *The principle of diverse response*: The swarm should not allocate all of its resources along excessively narrow channels, and it should distribute resources into many nodes.
4. *The principle of stability*: The swarm should not change its mode of behavior upon every fluctuation of the environment.
5. *The principle of adaptability*: The swarm must be able to change behavior mode when the investment in energy is worth the computational price.

Ethologists have modeled the behavior of a swarm with the features described above in both low level and global level (Crina and Ajith 2006). Recently researchers have been inspired by those models, and they have provided novel problem-solving techniques based on swarm intelligence for solving difficult real-world problems such as network routing, clustering, data mining, job scheduling, and bioinformatics, to name just a few. In the last two decades, especially two approaches based on ant colony described by Colorni et al. (1991) and on fish schooling and bird flocking introduced by Kennedy and Eberhart (1995) have attracted the interest of researchers all over the world. Both approaches have been studied by many researchers, and their variants have been introduced and applied for solving several problems in different areas. In this chapter, we focus on four popular bio-inspired optimization algorithms, which are, respectively, ant colony optimization (ACO), particle swarm optimization (PSO), artificial bee colony (ABC), and differential evolution (DE).

2.2 Ant Colony Optimization

ACO is a metaheuristic for solving hard combinatorial optimization problems (Duan 2005, 2010; Duan et al. 2011). The inspiring source of ACO is the pheromone trail laying and following behavior of real ants, which use pheromones as a communication medium (Fig. 2.1). In analogy to the biological example, ACO is based on indirect communication within a colony of simple agents, called (artificial) ants, mediated by (artificial) pheromone trails. The pheromone trails in ACO serve as a distributed, numerical information, which the ants use to probabilistically construct solutions to the problem being solved and which the ants adapt during the algorithm's execution to reflect their search experience.

The first example of such an algorithm is ant system (AS), which was proposed using as example application the well-known traveling salesman problem (TSP). Despite encouraging initial results, AS could not compete with state-of-the-art algorithms for the TSP. Nevertheless, it had the important role of stimulating further research both on algorithmic variants, which obtain much better computational performance, and on applications to a large variety of different problems. In fact, there exist now a considerable number of applications of such algorithms where world-class performance is obtained. Examples are applications of ACO algorithms to problems such as sequential ordering, scheduling, assembly line balancing, probabilistic TSP, 2D-HP protein folding, DNA sequencing, protein–ligand docking, and packet-switched routing in Internet-like networks. The ACO metaheuristic provides a common framework for the existing applications and algorithmic variants. Algorithms which follow the ACO metaheuristic are called ACO algorithms.

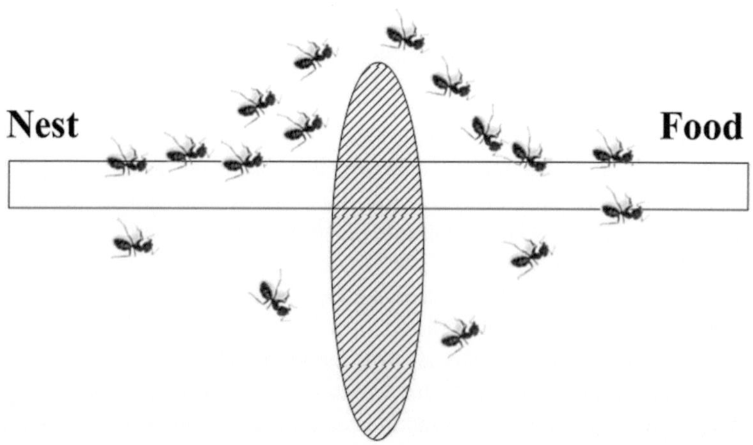

Fig. 2.1 Schematic diagram of ACO shows that ant colony has succeeded in finding the shortest route

The (artificial) ants in ACO implement a randomized construction heuristic which makes probabilistic decisions as a function of artificial pheromone trails and possibly available heuristic information based on the input data of the problem to be solved. As such, ACO can be interpreted as an extension of traditional construction heuristics, which are readily available for many combinatorial optimization problems. Yet, an important difference with construction heuristics is the adaptation of the pheromone trails during algorithm execution to take into account the cumulated search experience.

2.2.1 Biological Inspiration

Marco Dorigo and colleagues introduced the first ACO algorithm in the early 1990s (Dorigo 1992; Dorigo et al. 1996). The development of these algorithms was inspired by the observation of ant colonies. Ants are social insects. They live in colonies, and their behavior is governed by the goal of colony survival rather than being focused on the survival of individuals. The behavior that provided the inspiration for ACO is the ants' foraging behavior and, in particular, how ants can find shortest paths between food sources and their nest. When searching for food, ants initially explore the area surrounding their nest in a random manner. While moving, ants leave a chemical pheromone trail on the ground. Ants can smell pheromone. When choosing their way, they tend to choose, in probability, paths marked by strong pheromone concentrations. As soon as an ant finds a food source, it evaluates the quantity and the quality of the food and carries some of it back to the nest. During the return trip, the quantity of pheromone that an ant leaves on the ground may depend on the quantity and quality of the food. The pheromone trails will guide other ants to the food source. It has been shown by Deneubourg et al. (1990) that the indirect communication between the ants via pheromone trails—known as stigmergy—enables them to find shortest paths between their nest and food sources. This is explained in an idealized setting in Fig. 2.2.

As a first step towards an algorithm for discrete optimization, we present in the following a discretized and simplified model of the phenomenon explained in Fig. 2.2. After presenting the model, we will outline the differences between the model and the behavior of real ants. Our model consists of a graph $G=(V,E)$, where V consists of two nodes, namely, v_s (representing the nest of the ants) and v_d (representing the food source). Furthermore, E consists of two links, namely, e_1 and e_2, between v_s and v_d. To e_1 we assign a length of l_1 and to e_2 a length of l_2 such that $l_2 > l_1$. In other words, e_1 represents the short path between v_s and v_d, and e_2 represents the long path. Real ants deposit pheromone on the paths on which they move. Thus, the chemical pheromone trails are modeled as follows. We introduce an artificial pheromone value τ_i for each of the two links e_i, $i=1,2$. Such a value indicates the strength of the pheromone trail on the corresponding path. Finally, we introduce n_a artificial ants. Each ant behaves as follows: starting from v_s (i.e., the nest), an ant chooses with probability between $p_i = \tau_i/(\tau_1 + \tau_2)$ path e_1 and path e_2

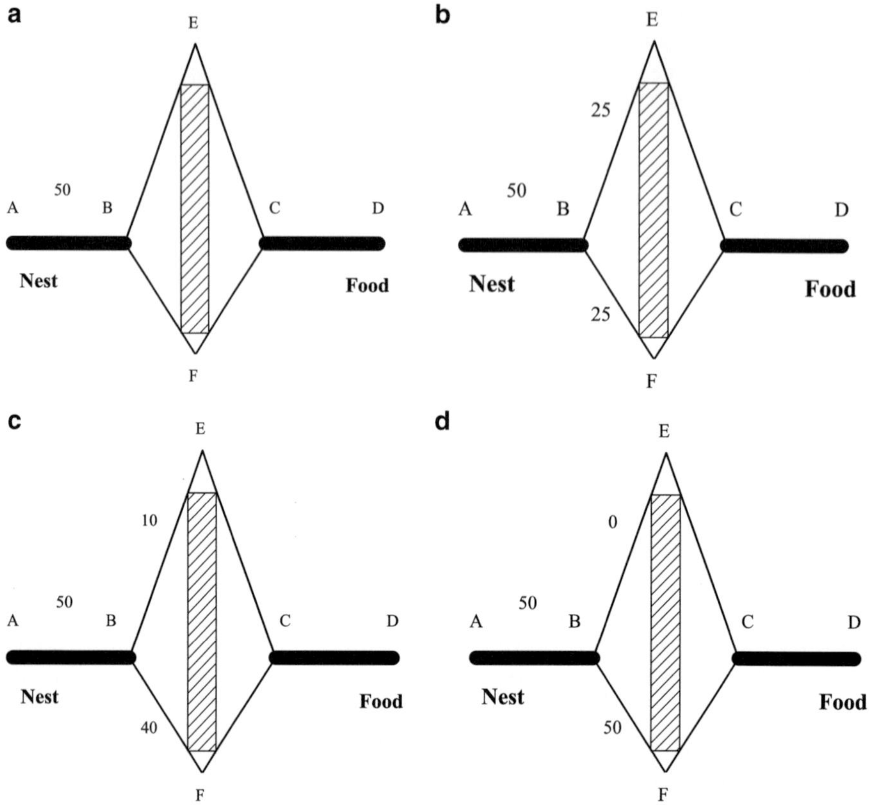

Fig. 2.2 An experimental setting that demonstrates the shortest pathfinding capability of ant colonies

for reaching the food source v_d. Obviously, if $\tau_1 > \tau_2$, the probability of choosing e_1 is higher and vice versa. For returning from v_d to v_s, an ant uses the same path as it chose to reach v_d, and it changes the artificial pheromone value associated to the used edge. More in detail, having chosen edge e_i, an ant changes the artificial pheromone value τ_i as follows:

$$\tau_i \leftarrow \tau_i + \frac{Q}{l_i} \tag{2.1}$$

where the positive constant Q is a parameter of the model. In other words, the amount of artificial pheromone that is added depends on the length of the chosen path: the shorter the path, the higher the amount of added pheromone. The foraging of an ant colony is in this model iteratively simulated as follows: at each step (or iteration), all the ants are initially placed in node v_s. Then, each ant moves from v_s to v_d as outlined above. As mentioned in the caption of Fig. 2.2d, in nature the deposited pheromone is subject to an evaporation over time. We simulate this pheromone evaporation in the artificial model as follows:

$$\tau_i \leftarrow (1-\rho)\,\tau_i,\ i = 1, 2 \tag{2.2}$$

The parameter $\rho \in (0, 1]$ is a parameter that regulates the pheromone evaporation. Finally, all ants conduct their return trip and reinforce their chosen path as outlined above.

In the beginning, all ants are in the nest. There is no pheromone on both paths. Then the foraging starts. In probability, 50 % of ants take the short path, and the other 50 % take the long path to the food source. The ants that have taken the shorter path would arrive earlier at the food source. Therefore, when returning, the probability to take again the short path is higher. The pheromone trail on the short path receives a stronger reinforcement to take this path grows. Finally, due to the evaporation of the pheromone on the long path, the whole colony will, in probability, use the short path in probability.

2.2.2 *Principle of Ant Colony Optimization*

2.2.2.1 The First ACO Algorithm: Ant System

In AS each ant is initially put on a randomly chosen city and has a memory which stores the partial solution it has constructed so far (initially the memory contains only the start city) (Dorigo and Stützle 2003). Starting from its start city, an ant iteratively moves from city to city. When being at a city *i*, an ant *k* chooses to go to a city *j* with a probability given by

$$p_{ij}^{k}(t) = \begin{cases} \dfrac{\left[\tau_{ij}(t)\right]^{\alpha}\left[\eta_{ij}\right]^{\beta}}{\sum\limits_{k \in allowed_k} \left[\tau_{ik}(t)\right]^{\alpha}\left[\eta_{ik}\right]^{\beta}} & if\ \ j \in allowed_k \\ 0 & otherwise \end{cases} \tag{2.3}$$

where $allowed_k = \{N - \text{tabu}_k\}$, α and β are parameters that control the relative importance of trail versus visibility, η_{ij} is the heuristic desirability, and $\eta_{ij} = 1/d_{ij}$ where d_{ij} is the distance between city *i* and city *j* and τ_{ij} is the amount of pheromone trail on edge (*i,j*). If $\alpha = 0$, the selection probabilities are proportional to $[\eta]^{\beta}$, and the closest cities will more likely be selected: in this case AS corresponds to a classical stochastic greedy algorithm (with multiple starting points since ants are initially randomly distributed on the cities). If $\beta = 0$, only pheromone amplification is at work: this will lead to the rapid emergence of a stagnation situation with the corresponding generation of tours which, in general, are strongly suboptimal.

The solution construction ends after each ant has completed a tour. Next, the pheromone trails are updated. In AS this is done by first lowering the pheromone trails by a constant factor (this is pheromone evaporation) and then allowing each ant to deposit pheromone on the arcs that belong to its tour:

$$\tau_{ij}\,(t+n) = (1-\rho)\cdot\tau_{ij}\,(t) + \Delta\tau_{ij} \tag{2.4}$$

$$\Delta\tau_{ij} = \sum_{k=1}^{m} \Delta\tau_{ij}^{k} \tag{2.5}$$

where $1-\rho\{\rho\subset(0,1)\}$ represents the evaporation of trail between time t and $t+n$. The parameter ρ is used to avoid unlimited accumulation of the pheromone trails and enables the algorithm to "forget" previously done bad decisions. Where $\Delta\tau_{ij}{}^{k}$ is the quantity of per unit length of pheromone trail laid on edge (i,j) by the kth ant between time t and $t+n$. In the popular ant-cycle model, it is given by

$$\Delta\tau_{ij}^{k} = \begin{cases} \frac{Q}{L_k}\ if\ k-th \text{ ant uses } (i,j) & \text{in its tour} \\ 0 & \text{otherwise} \end{cases} \tag{2.6}$$

where Q is a constant and L_k is the tour length of the kth ant.

2.2.2.2 Framework of the ACO Metaheuristic

After initialization, the metaheuristic iterates over three phases: construct ant solutions, apply local search, and update pheromones, which are described in detail as follows:

Construct ant solutions: A set of m artificial ants constructs solutions from elements of a finite set of available solution components $C=\{c_{ij}\}$, $i=1,\ldots,n$, and $j=1,\ldots,|D_i|$. A solution construction starts from an empty partial solution $s^p=\varnothing$. At each construction step, the partial solution s^p is extended by adding a feasible solution component from the set $N\,(s^p)\subseteq\mathbb{C}$, which is defined as the set of components that can be added to the current partial solution s^p without violating any of the constraints in Ω. The process of constructing solutions can be regarded as a walk on the construction graph $G_C=(V,E)$.

The choice of a solution component from $N(s^p)$ is guided by a stochastic mechanism, which is biased by the pheromone associated with each of the elements of $N(s^p)$. The rule for the stochastic choice of solution components varies across different ACO algorithms, but, in all of them, it is inspired by the model of the behavior of real ants given in (2.1).

Apply local search: Once solutions have been constructed and before updating the pheromone, it is common to improve the solutions obtained by the ants through a local search. This phase, which is highly problem specific, is optional although it is usually included in state-of-the-art ACO algorithms.

Update pheromones: The aim of the pheromone update is to increase the pheromone values associated with good or promising solutions and to decrease those that are associated with bad ones. Usually, this is achieved by decreasing all the pheromone values through pheromone evaporation and by increasing the pheromone levels associated with a chosen set of good solutions.

2.2.3 Ant System and Its Extensions

Even though the original AS algorithm achieved encouraging results for the TSP problem, it was found to be inferior to state-of-the-art algorithms for the TSP as well as for other combinatorial optimization problems. Therefore, several extensions and improvements of the original AS algorithm were introduced over the years, which show better performance than AS when applied to many optimization problems, such as Elitist AS (EAS) (Dorigo 1992), rank-based Ant System (ASrank) (Bullnheimer et al. 1999), MAX–MIN Ant System (MMAS) (Stutzle and Hoos 1997), and Ant Colony System (ACS) (Dorigo and Gambardella 1997).

A first improvement over AS was obtained by introducing the elitist strategy, which is called EAS. In this variant, the pheromone values are updated using all the solutions that were generated in the respective iteration and the best-so-far solution. It consists in giving the best tour since the start of the algorithm (called T^{gb}, where *gb* stays for global best) a strong additional weight. In practice, each time the pheromone trails are updated, those belonging to the edges of the global-best tour get an additional amount of pheromone. For these edges (2.6) becomes

$$\Delta\tau_{ij}^{gb} = \begin{cases} \frac{e}{L^{gb}(t)} \; if \;\; \text{arc}\,(i,j) \in T^{gb} \\ 0 \qquad\qquad\qquad \text{otherwise} \end{cases} \tag{2.7}$$

The arcs of T^{gb} are therefore reinforced with a quantity of $e \cdot 1/L^{gb}$, where L^{gb} is the length of T^{gb} and *e* is a positive integer.

Another improvement over AS is the ASrank proposed by Bullnheimer et al. (1999), which is an extension of the elitist strategy to some extent. In this approach, the best-so-far solution has the highest influence on the pheromone update at each iteration, while a selection of the best solutions constructed at that current iteration influences the update in accordance with their rankings. It sorts the ants according to the lengths of the tours they generated, and, after each tour construction phase, only the $(\omega - 1)$ best ants and the global-best ant are allowed to deposit pheromone. The *r*th best ant of the colony contributes to the pheromone update with a weight given by $\max\{0, \omega - r\}$, while the global-best tour reinforces the pheromone trails with weight ω. Then (2.4) and (2.5) becomes therefore

$$\tau_{ij}\ (t+n) = (1-\rho)\cdot\tau_{ij}(t) + \sum_{r=1}^{\omega}(\omega-r)\cdot\Delta\tau_{ij}^{r}(t) + \omega\cdot\Delta\tau_{ij}^{gb}(t) \tag{2.8}$$

where $\Delta\tau_{ij}^{r}(t) = 1/L^{r}(t)$ and $\Delta\tau_{ij}^{gb}(t) = 1/L^{gb}$.

As one of the most successful ACO variants, MMAS introduces upper and lower bounds to the values of the pheromone trails, as well as a different initialization of their values. In MMAS, the allowed range of the pheromone trail strength is limited to the interval $[\tau_{\min},\tau_{\max}]$, that is, $\tau_{\min} \le \tau_{ij} \le \tau_{\max}$, $\forall\, \tau_{ij}$, and the pheromone trails are initialized to the upper trail limit, which allows a higher exploration at the start of the algorithm. The value of this bound is updated each time a new best-so-far solution is found by the algorithm. Depending on some convergence measure, at each iteration, either the iteration-best update or the global-best update rule is used for updating the pheromone values. At the start of the algorithm, the iteration-best update rule is used more often, while during the run of the algorithm, the frequency with which the global-best update rule is used increases.

ACS, which was introduced by Dorigo and Gambardella (1997), differs from the original AS algorithm in more aspects than just in the pheromone update. In this approach, the importance of exploitation of information collected by previous ants with respect to exploration of the search space is increased, which is achieved via two mechanisms. ACS improves over AS by increasing the importance of exploitation of information collected by previous ants with respect to exploration of the search space. This is achieved via two mechanisms. First, a strong elitist strategy is used to update pheromone trails, which means only the ant that has produced the best solution is allowed to update pheromone trails, according to a pheromone trail update rule similar to that used in AS:

$$\tau_{ij}\ (t+n) = (1-\rho)\cdot\tau_{ij}(t) + \rho\cdot\tau_{ij}^{best}(t) \tag{2.9}$$

The best ant can be the iteration-best ant, that is, the best in the current iteration, or the global-best ant, that is, the ant that made the best tour from the start of the trial.

Second, ants choose the next city to move to using a so-called pseudorandom proportional rule: with probability q_0, they move to the city j for which the product between pheromone trail and heuristic information is maximum, that is, $j = \arg\max_{j\in N_i^k}\left\{\tau_{ij}(t)\cdot\eta_{ij}^{\beta}\right\}$, while with probability $1-q_0$, they operate a biased exploration in which the probability $p_{ij}{}^{k}(t)$ is the same as in AS. When the parameter q_0 is set to a value close to 1, as it is the case in most ACS applications, exploitation is favored over exploration. It is obvious that, when $q_0 = 0$, the probabilistic decision rule becomes the same as in AS.

Besides, ACS differs from previous ACO algorithms also because ants update the pheromone trails while building solutions as that in ant quantity and in ant density. In practice ACS ants "eat" some of the pheromone trail on the edges they visit. This operation favors exploitation, counterbalancing this way the other two abovementioned modifications that strongly favor exploitation of the collected knowledge about the problem. In this way, the probability that a same path is used by all the ants is decreased. In other words, it helps to avoid to be trapped into local optimum. ACS has been made more performing over other variants also by the addition of local search routines that take the solution generated by ants to their local optimum just before the pheromone update.

Although retaining some of the original biological inspiration, they are less and less biologically inspired and more and more motivated by the need of making ACO algorithms competitive with state-of-the-art algorithms or improve their performance. Nevertheless, many aspects of the original AS remain, such as the need for a colony, the role of autocatalysis, the cooperative behavior mediated by artificial pheromone trails, the probabilistic construction of solutions biased by artificial pheromone trails and local heuristic information, the pheromone updating guided by solution quality, and the evaporation of pheromone trail, which is the same in all ACO algorithms. For more information about the variants of ACO such as hypercube framework (HCF), reader can refer to Dorigo and Blum (2005). Ant algorithms are receiving increasing attention in the scientific community as a promising novel approach to distributed control and optimization.

2.3 Particle Swarm Optimization

The initial ideas on particle swarms of Kennedy (a social psychologist) and Eberhart (an electrical engineer) were essentially aimed at producing computational intelligence by exploiting simple analogues of social interaction (Kennedy and Eberhart 1995) rather than purely individual cognitive abilities. The first simulations were influenced by Heppner and Grenander's work (Heppner and Grenander 1990) and involved analogues of bird flocks searching for corn. These soon developed into a powerful optimization method-PSO.

2.3.1 Biological Inspiration

A number of scientists have created computer simulations of various interpretations of the movement of organisms in a bird flock or fish school. Notably, Reynolds (1987) and Heppner and Grenander (1990) presented simulations of bird flocking. Reynolds was intrigued by the aesthetics of bird-flocking choreography, and Heppner, a zoologist, was interested in discovering the underlying rules that enabled large numbers of birds to flock synchronously, often changing direction suddenly, scattering and regrouping, etc. Both of these scientists had the insight that local processes, such as those modeled by cellular automata, might underlie the unpredictable group dynamics of bird social behavior. Both models relied heavily on

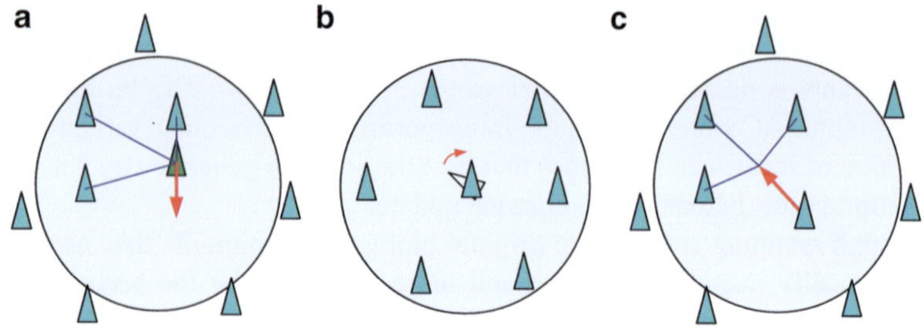

Fig. 2.3 Boid model. (**a**) Separation. Each agent tries to move away from its neighbors if they are too close. (**b**) Alignment. Each agent steers towards the average heading of its neighbors. (**c**) Cohesion. Each agent tries to go towards the average position of its neighbors

manipulation of interindividual distances; that is, the synchrony of flocking behavior was thought to be a function of birds' efforts to maintain an optimum distance between themselves and their neighbors (Fig. 2.3).

It has been believed that social sharing of information among conspeciates offers an evolutionary advantage: this hypothesis was fundamental to the development of PSO. One motive for developing the simulation was to model human social behavior, which is of course not identical to fish schooling or bird flocking. The important difference is its abstractness. Birds and fish adjust their physical movement to avoid predators, seek food and mates, optimize environmental parameters such as temperature, etc. Humans adjust not only physical movement but cognitive or experiential variables as well. We do not usually walk in step and tum in unison (though some fascinating research in human conformity shows that we are capable of it); rather, we tend to adjust our beliefs and attitudes to conform with those of our social peers.

This is a major distinction in terms of contriving a computer simulation, for at least one obvious reason: collision. Two individuals can hold identical attitudes and beliefs without banging together, but two birds cannot occupy the same position in space without colliding. It seems reasonable, in discussing human social behavior, to map the concept of change into the bird/fish analogue of movement. This is consistent with the classic Aristotelian view of qualitative and quantitative change as types of movement. Thus, besides moving through three-dimensional physical space and avoiding collisions, humans change in abstract multidimensional space, collision-free. Physical space of course affects informational inputs, but it is arguably a trivial component of psychological experience. Humans learn to avoid physical collision by an early age, but navigation of n-dimensional psychosocial space requires decades of practice, and many of us never seem to acquire quite all the skills we need.

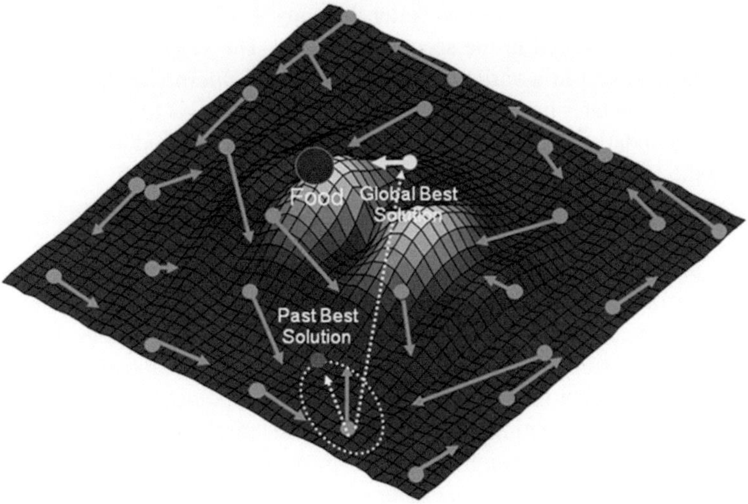

Fig. 2.4 Schematic diagram of PSO © [2002] IEEE (Reprinted, with permission, from Duan and Liu (2010))

2.3.2 *Principle of Particle Swarm Optimization*

2.3.2.1 The Framework of PSO

In PSO, a number of simple entities, the particles, are placed in the search space of some problem or function, and each evaluates the objective function at its current location. Each particle then determines its movement through the search space by combining some aspect of the history of its own current and best (best fitness) locations with those of one or more members of the swarm, with some random perturbations (Duan and Xing 2009; Duan and Liu 2010; Duan et al. 2011). The next iteration takes place after all particles have been moved. Eventually the swarm as a whole, like a flock of birds collectively foraging for food, is likely to move close to an optimum of the fitness function. Each individual in the particle swarm is composed of three D-dimensional vectors, where D is the dimensionality of the search space. These are the current position $\boldsymbol{x}_i$, the previous best position $\boldsymbol{p}_i$, and the velocity $\boldsymbol{v}_i$. Shi and Eberhart (1998) firstly introduced the inertia weights w into the basic PSO model, by adjusting w to improve the performances of the PSO algorithm. Figure 2.4 describes the schematic diagram of PSO.

2.3.2.2 Original Version

The current position $\boldsymbol{x}_i$ can be considered as a set of coordinates describing a point in space. At each iteration of the algorithm, the current position is evaluated as a problem solution. If that position is better than any that has been found so far, then the coordinates are stored in the second vector, $\boldsymbol{p}_i$. The value of the best function result so far is stored in a variable that can be called $pbest_i$ (for "previous best"), for comparison on later iterations. The objective, of course, is to keep finding better positions and updating $\boldsymbol{p}_i$ and $pbest_i$. New points are chosen by adding $\boldsymbol{v}_i$ coordinates to $\boldsymbol{x}_i$, and the algorithm operates by adjusting $\boldsymbol{v}_i$, which can effectively be seen as a step size (Poli et al. 2007).

The particle swarm is more than just a collection of particles. A particle by itself has almost no power to solve any problem; progress occurs only when the particles interact. Problem solving is a population-wide phenomenon, emerging from the individual behaviors of the particles through their interactions. In any case, populations are organized according to some sort of communication structure or topology, often thought of as a social network. The topology typically consists of bidirectional edges connecting pairs of particles, so that if j is in i's neighborhood, i is also in j's. Each particle communicates with some other particles and is affected by the best point found by any member of its topological neighborhood. This is just the vector $\boldsymbol{p}_i$ for that best neighbor, which we will denote with $\boldsymbol{p}_g$. The potential kinds of population "social networks" are hugely varied, but in practice certain types have been used more frequently:

$$\begin{aligned} \boldsymbol{v}_i\,(t+1) &= \boldsymbol{v}_i(t) + c_1 \cdot rand_1\,(\boldsymbol{x}_i(t) - \boldsymbol{p}_i(t)) + c_2 \cdot rand_2\,(\boldsymbol{x}_i(t) - \boldsymbol{p}_g(t)) \\ \boldsymbol{x}_i\,(t+1) &= \boldsymbol{x}_i(t) + \boldsymbol{v}_i\,(t+1) \end{aligned} \tag{2.10}$$

where each individual particle i has the following properties: a position vector in search space $\boldsymbol{x}_i(t)$ at time t, a velocity vector $\boldsymbol{v}_i(t)$ at time t, and a personal best position in search space $\boldsymbol{p}_i(t)$. The personal best position $\boldsymbol{p}_i(t)$ corresponds to the position in search space where particle i had the minimum fitness value $pbest_i$ determined by the objective function (in a minimization problem). The global-best position denoted by $p_g(t)$ represents the position yielding the lowest error among all the $\boldsymbol{p}_i(t)$, which has the best fitness value $gbest$ among all the particles. Two pseudorandom sequences, $rand_1 \sim (0, 1)$ and $rand_2 \sim (0, 1)$ are used to effect the stochastic algorithm nature.

The PSO algorithm consists of repeated application of (2.10). In theory, particles of a swarm may benefit from the prior discoveries and experiences of all the members of a swarm when foraging. The key point of PSO is that particles in the swarm share information with each other, which offers some sort of evolutionary advantage. Therefore, due to the simple concept, easy implementation, and quick convergence, PSO has gained much attention and wide applications in solving continuous nonlinear optimization problems. The process for implementing the original PSO is described as follows:

Step 1 Initialize a population array of particles with random positions and velocities on D dimensions in the search space.

Step 2 For each particle, evaluate the desired optimization fitness function in D variables.

Step 3 Compare particle's fitness evaluation with its $pbest_i$. If current value is better than $pbest_i$, then set $pbest_i$ equal to the current value and $\boldsymbol{p}_i(t)$ equal to the current location $\boldsymbol{x}_i(t)$ in the D-dimensional space.

Step 4 Identify the particle in the neighborhood with the best success so far, and assign its index to the variable g.

Step 5 Change the velocity and position of the particle according to the (2.10).

Step 6 If a criterion is met (usually a sufficiently good fitness or a maximum number of iterations), stop. Otherwise, go to Step 2.

2.3.2.3 Other Variants of PSO

Motivated by the desire to better control the scope of the search, reduce the importance of $V_{\max}$, and perhaps eliminate it altogether, the following modification of the PSO's update equations was proposed (Shi and Eberhart 1998):

$$\boldsymbol{v}_i(t+1) = w \cdot \boldsymbol{v}_i(t) + c_1 \cdot rand_1 (\boldsymbol{x}_i(t) - \boldsymbol{p}_i(t)) + c_2 \cdot rand_2 (\boldsymbol{x}_i(t) - \boldsymbol{p}_g(t))$$

$$\boldsymbol{x}_i(t+1) = \boldsymbol{x}_i(t) + \boldsymbol{v}_i(t+1) \tag{2.11}$$

where w was termed the "inertia weight." If we interpret c_1 and c_2 as the external force, $\boldsymbol{f}_\mathrm{i}$, acting on a particle, then the change in a particle's velocity (i.e., the particle's acceleration) can be written as $_\Delta v_i = \boldsymbol{f}_i + (1-w)\boldsymbol{v}_i$. That is, the constant $1-w$ acts effectively as a friction coefficient, and so w can be interpreted as the fluidity of the medium in which a particle moves. This perhaps explains why researchers have found that the best performance could be obtained by initially setting w to some relatively high value (e.g., 0.9), which corresponds to a system where particles move in a low viscosity medium and perform extensive exploration, and gradually reducing w to a much lower value (e.g., 0.4), where the system would be more dissipative and exploitative and would be better at homing into local optima. It is even possible to start from values of $w > 1$, which would make the swarm unstable, provided that the value is reduced sufficiently to bring the swarm in a stable region.

With (2.11) and an appropriate choice of w and of the acceleration coefficients, c_1 and c_2, the PSO can be made much more stable so much so that one can either do without $V_{\max}$ or set $V_{\max}$ to a much higher value, such as the value of the dynamic range of each variable. In this case, $V_{\max}$ may improve performance, though with use of inertia or constriction techniques, it is no longer necessary for damping the swarm's dynamics.

Though the earliest researchers recognized that some form of damping of the dynamics of a particles (e.g., $V_{\max}$) was necessary, the reason for this was not understood. But when the particle swarm algorithm is run without restraining velocities in some way, these rapidly increase to unacceptable levels within a few iterations. Kennedy (1998) noted that the trajectories of nonstochastic one-dimensional particles contained interesting regularities when $c_1 + c_2$ was between

0.0 and 4.0. Clerc's analysis of the iterative system led him to propose a strategy for the placement of "constriction coefficients" on the terms of the formulas; these coefficients controlled the convergence of the particle and allowed an elegant and well-explained method for preventing explosion, ensuring convergence, and eliminating the arbitrary V_{max} parameter. The analysis also takes the guesswork out of setting the values of c_1 and c_2. Clerc and Kennedy (2002) noted that there can be many ways to implement the constriction coefficient. One of the simplest methods of incorporating it is the following:

$$\boldsymbol{v}_i\,(t+1) = \chi\,(\boldsymbol{v}_i(t) + c_1 \cdot rand_1(\boldsymbol{x}_i(t) - \boldsymbol{p}_i(t)) + c_2 \cdot rand_2(\boldsymbol{x}_i(t) - \boldsymbol{p}_g(t)))$$

$$\boldsymbol{x}_i\,(t+1) = \boldsymbol{x}_i(t) + \boldsymbol{v}_i\,(t+1) \tag{2.12}$$

$$\chi = \frac{2}{\phi - 2 + \sqrt{\phi^2 - 4\phi}} \tag{2.13}$$

When Clerc's constriction method is used, ϕ is commonly set to 4.1, and the constant multiplier χ is approximately 0.7298. The results in the previous velocity being multiplied by 0.7298 and each of the two $(\boldsymbol{p} - \boldsymbol{x})$ terms being multiplied by a random number limited by $0.7298 \times 2.05 \approx 1.49618$. The constricted particles will converge without using any V_{max} at all. However, subsequent experiments and applications concluded that a better approach to use as a prudent rule of thumb is to limit V_{max} to X_{max}, the dynamic range of each variable on each dimension, in conjunction with (2.12) and (2.13). The result is a PSO algorithm with no problem-specific parameters. And this is the canonical particle swarm algorithm of today. Note that a PSO with constriction is algebraically equivalent to a PSO with inertia.

Indeed, (2.11) and (2.12) can be transformed into one another via the mapping $w \leftrightarrow \chi$ and $c_i \leftrightarrow \chi c_i$. So, the optimal settings suggested by Clerc correspond to $w = 0.7298$ and $c_1 = c_2 = 1.49618$ for a PSO with inertia.

2.3.3 *Parameters and Population Topology*

The role of inertia weight w in (2.11) is considered critical for the convergence behavior of PSO. The inertia weight is employed to control the impact of the previous history of velocities on the current one. Accordingly, the parameter w regulates the trade-off between the global (wide-ranging) and local (nearby) exploration abilities of the swarm. A large inertia weight facilitates global exploration (searching new areas), while a small one tends to facilitate local exploration, i.e., fine-tuning the current search area. A suitable value for the inertia weight w usually provides balance between global and local exploration abilities and consequently results in a reduction of the number of iterations required to locate the optimum solution. Initially, the inertia weight is set as a constant. However, some experiment results indicate that it is better to initially set the inertia to a large value, in order to promote global exploration of the search space, and gradually decrease it to get more refined solutions. Thus, an initial value around 1.2 and gradually reducing towards 0 can be considered as a good choice for w. A better method is to use some adaptive approaches (e.g., fuzzy controller), in which the parameters can be adaptively fine-tuned according to the problems under consideration (Grosan and Abraham 2011).

The parameters c_1 and c_2 in (2.11) are not critical for the convergence of PSO. However, proper fine-tuning may result in faster convergence and alleviation of local minima. As default values, usually, $c_1 = c_2 = 2$ are used, but some experiment results indicate that $c_1 = c_2 = 1.49$ might provide even better results. Recent work reports that it might be even better to choose a larger cognitive parameter, c_1, than a social parameter, c_2, but with $c_1 + c_2 \leq 4$.

The first particle swarms evolved out of bird-flocking simulations of a type described by Reynolds (1987) and Heppner and Grenander (1990). In these models, the trajectory of each bird's flight is modified by application of several rules, including some that take into account the birds that are nearby in physical space. So, early PSO topologies were based on proximity in the search space. Kennedy and Mendes studied the various population topologies on the PSO performance. Different concepts for neighborhoods could be envisaged. It can be observed as a spatial neighborhood when it is determined by the Euclidean distance between the positions of two particles or as a sociometric neighborhood (e.g., the index position in the storing array).

The next topology to be introduced, the *gbest* topology (for "global best"), was one where the best neighbor in the entire population influenced the target particle. While this may be conceptualized as a fully connected graph, in practice it only meant that the program needed to keep track of the best function result that had been found and the index of the particle that found it.

The *gbest* is an example of static topology, i.e., one where neighbors and neighborhoods do not change during a run. The *lbest* topology (for "local best") is another static topology, which was introduced in Eberhart and Kennedy (1995). It is a simple ring lattice where each individual was connected to $K = 2$ adjacent members in the population array, with toroidal wrapping (naturally, this can be generalized to $K > 2$). This topology had the advantage of allowing parallel search, as subpopulations could converge in diverse regions of the search space. Where equally good optima were found, it was possible for the population to stabilize with particles in the good regions, but if one region was better than another, it was likely to attract particles to itself. Thus this parallel search resulted in a more thorough search strategy; though it converged more slowly than the *gbest* topology, *lbest* was less vulnerable to the attraction of local optima.

Several classical communications structures from social psychology (Bavelas 1950), including some with small-world modifications, were experimented with in Kennedy (1999). Circles, wheels, stars, and randomly assigned edges were tested on a standard suite of functions. The most important finding was that there were important differences in performance depending on the topology implemented; these differences depended on the function tested, with nothing conclusively suggesting that any one was generally better than any other.

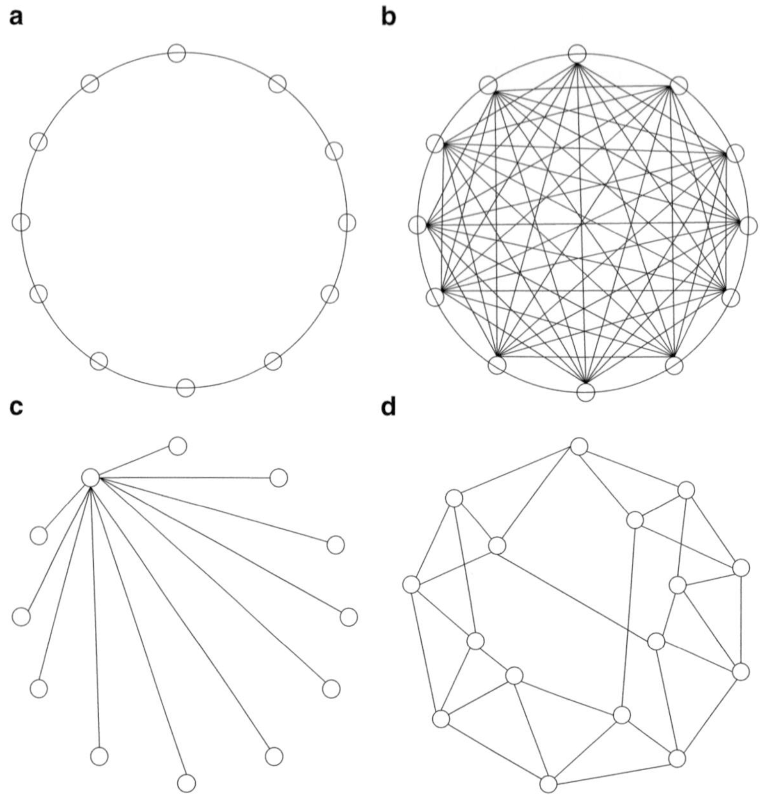

Fig. 2.5 Some neighborhood topologies of PSO. (**a**) Local-best topology. (**b**) Global-best topology. (**c**) Star topology. (**d**) Von Neumann topology

Numerous aspects of the social-network topology were tested in Kennedy and Mendes (2002) (Fig. 2.5). For instance, the effect of including the target particle in its neighborhood (as opposed to allowing only "external" influences) was evaluated, finding, surprisingly, that whether or not the particle belonged to its neighborhood had little impact on behavior. 1343 random graphs were generated and then modified to meet certain criteria, including mean degree, clustering, and the standard deviations of those two measures to introduce homogeneous and heterogeneous structures. Several regular topologies were included, as well; these were the *gbest* and *lbest* versions mentioned above, as well as a von Neumann topology which defined neighborhoods on a grid, a "pyramid" topology, and a handmade graph with clusters of nodes linked sparsely.

One finding that emerged was the relative superiority of the von Neumann structure. This topology possesses some of the parallelism of *lbest*, yet nodes have degree $K = 4$; thus the graph is more densely connected than *lbest* but less densely than *gbest*.

2.4 Artificial Bee Colony

ABC algorithm was originally presented by Karaboga and Basturk (2007), under the inspiration of collective behavior on honeybees, and it has been proved to possess a better performance in function optimization problem, compared with genetic algorithm, DE and PSO. As we know, usual optimization algorithms conduct only one search operation in one iteration, for example, the PSO algorithm carries out global search at the beginning and local search in the later stage. Compared with the usual algorithms, the major advantage of ABC algorithm lies in that it conducts both global search and local search in each iteration, and as a result the probability of finding the optimal parameters is significantly increased, which efficiently avoid local optimum to a large extent.

2.4.1 Biological Inspiration

A very interesting swarm in nature is honeybee swarm that allocates the tasks dynamically and adapts itself in response to changes in the environment in a collective intelligent manner. The honeybees have photographic memories; space-age sensory and navigation systems, possibly even insight skills; and group decision-making process during selection of their new nest sites, and they perform tasks such as queen and brood tending, storing, retrieving and distributing honey and pollen, communicating, and foraging. These characteristics are incentive for researchers to model the intelligent behaviors of bees. Before presenting the algorithms described to use intelligent behaviors and their applications, behavior of the colony is explained.

Bees are social insects living as colonies. There are three kinds of bees in a colony: drones, queen, and workers. Foraging is the most important task in the hive. Many studies (Seeley 1985) have investigated the foraging behavior of each individual bee and what types of external information (such as odor, location information in the waggle dance, the presence of other bees at the source or between the hive and the source) and internal information (such as remembered source location or source odor) affect this foraging behavior. Foraging process starts with leaving the hive of a forager in order to search food source to gather nectar. After finding a flower for herself, the bee stores the nectar in her honey stomach. Based on the conditions such as richness of the flower and the distance of the flower to the hive, the bee fills her stomach in about 30–120 min, and honey-making process begins with the secretion of an enzyme on the nectar in her stomach. After coming back to the hive, the bee unloads the nectar to empty honeycomb cells, and some extra substances are added in order to avoid the fermentation and the bacterial attacks. Filled cells with the honey and enzymes are covered by wax.

After unloading the nectar, the forager bee which has found a rich source performs special movements called "dance" on the area of the comb in order to share her information about the food source such as how plentiful it is and its direction and distance and recruits the other bees for exploiting that rich source. While dancing, other bees touch her with their antenna and learn the scent and the taste of the source she is exploiting. She dances on different areas of the comb in order to recruit more bees and goes on to collect nectar from her source. There are different dances performed by bees depending on the distance information of the source: round dance, waggle dance, and tremble dance. If the distance of the source to the hive is less than 100 m, round dance is performed, while if the source is far away, waggle dance is performed. Round dance does not give direction information. In case of waggle dance, direction of the source according to the sun is transferred to other bees. Longer distances cause quicker dances. The tremble dance is performed when the foraging bee perceives a long delay in unloading its nectar.

Forager bees use a maplike organization of spatial memory for homing and food source search flights. This organization is based on the computations of two experienced vectors or on viewpoints and landmarks. There are two perspectives of which one certainly true is not known. First one is that bees use stimuli obtained during their flights. The second one is that they encode the spatial information in their dances into their maplike spatial memory (Menzel et al. 2006).

A honeybee colony needs to divide its workforce so that the appropriate number of individuals is allocated for each of the many tasks. Bees are specialized in order to carry out every task in the hive. However, there is a controversy about which factors have roles on the specialization of bees, such as their age, hormones, and individual predisposition coming from their genetic determination (Dornhaus et al. 1998), and also the allocation of tasks can dynamically change. For example, when food is drought, younger nurse bees will also join to foraging process. Depending on the swarm intelligent behaviors of a bee swarm noted above, several approaches have been introduced and applied to solve problems.

Karlvon Frisch, a famous Nobel Prize winner, found that in nature, although each bee only performs one single task, yet through a variety of information communication ways between bees such as waggle dance and special odor, the entire colony can always easily find food resources that produce relative high amount of nectar, hence realize its self-organizing behavior.

2.4.2 *Principle of Artificial Bee Colony*

In order to introduce the self-organization model of forage selection that leads to the emergence of collective intelligence of honeybee swarms, first, we need to define three essential components: food sources, unemployed foragers, and employed foragers (Duan et al. 2010, 2011; Xu et al. 2010; Yu and Duan 2012):

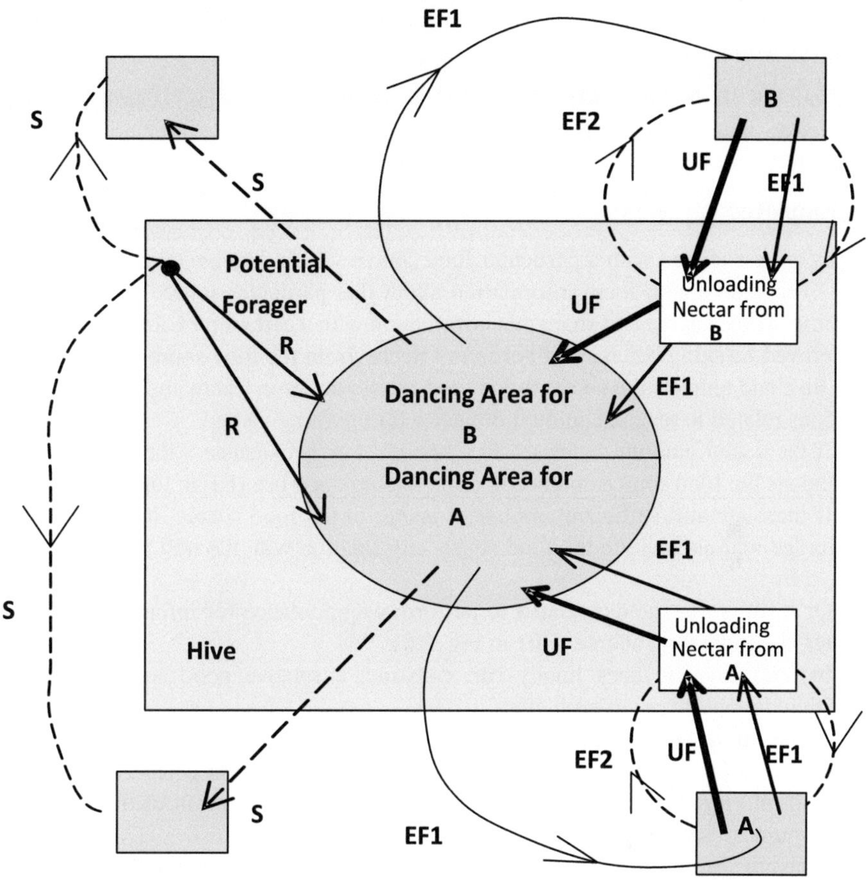

Fig. 2.6 The behavior of honeybee foraging for nectar (Reprinted from Xu and Duan (2010), with kind permission from Elsevier)

1. Food Sources

For the sake of simplicity, the "profitability" of a food source (A and B in Fig. 2.6) can be represented with a single quantity. The position of a food source represents a possible parameter solution to the optimization problem, and the nectar amount of a food source corresponds to the similarity value of the associated solution.

2. Unemployed Foragers

If it is assumed that a bee has no knowledge about the food sources in the search field, the bee initializes its search as an unemployed forager. Unemployed foragers are continually at look out for a food source to exploit. There are two types of unemployed foragers: scouts and onlookers.

Scouts (S in Fig. 2.6): If the bee starts searching spontaneously for new food sources without any knowledge, it will be a scout bee.

Onlookers (R in Fig. 2.6): The onlookers wait in the nest and search the food source through sharing information of the employed foragers, and there is a greater probability of onlookers choosing more profitable sources.

3. Employed Foragers

They are associated with a particular food source which they are currently exploiting. They carry with them information about this particular source and the profitability of the source and share this information with a certain probability. After the employed foraging bee loads a portion of nectar from the food source, it returns to the hive and unloads the nectar to the food area in the hive. There are three possible options related to residual amount of nectar for the foraging bee.

If the nectar amount decreases to a low level or is exhausted, the foraging bee abandons the food source and becomes an unemployed bee (UF in Fig. 2.6).

If there are still sufficient amount of nectar in the food source, it can continue to forage without sharing the food source information with the nest mates (EF2 in Fig. 2.6).

Or it can go to the dance area to perform waggle dance for informing the nest mates about the food source (EF1 in Fig. 2.6).

In this way, the bees finally can construct a relative good solution of the multimodal optimization problems.

At the initial moment, all the bees without any prior knowledge play the role of detecting bees. After a random search for bee sources, the detecting bees can convert into any kind of bees above in accordance with the profit of the searched food sources. The changing rules are described as follows:

When the profit of the food source the bee searched is higher than the threshold, it becomes a leading bee, goes on exploring nectar, and also recruits more bees (EF1) to explore together. When the profit of related food source is relative low, it gives up the food source and again becomes a detecting bee to search for new food source (UF). When the profit is less than certain threshold, it follows leading bees to explore nectar. When searching times around hive exceed a certain limit but still the bees could not find a good resource, it abandons the source and finds a new one.

In ABC algorithm, the position of a food source represents a possible solution to the optimization problem, and the nectar amount of a food source corresponds to the quality (fitness) of the associated solution. The number of the employed bees or the onlooker bees is equal to the number of solutions in the population. At the first step, the ABC generates a randomly distributed initial population, which is corresponding to the food source positions. After initialization, the population of the positions (solutions) is subject to repeated cycles, $T = 1, 2, \ldots, T_{\max}$, of the search processes of the employed bees, the onlooker bees, and the scout bees. An employed bee produces a modification on the position (solution) in her memory depending on the local information (visual information) and tests the nectar amount (fitness value) of the new source (new solution). If the nectar amount of the new one is higher than that of the previous one, the bee memorizes the new position and forgets the old one. Otherwise she keeps the position of the previous one in her memory. After all employed bees complete the search process, they share the nectar information of the food sources and their position information with the onlooker bees. An onlooker bee evaluates the nectar information taken from all employed bees and chooses a food source with a probability related to its nectar amount. As in the case of the employed bee, she produces a modification on the position in her memory and checks the nectar amount of the candidate source. If the nectar is higher than that of the previous one, the bee memorizes the new position and forgets the old one.

2.4.3 Algorithmic Structure of Artificial Bee Colony

In ABC algorithm, each cycle of the search consists of three steps: sending the employed bees onto their food sources and evaluating their nectar amounts; after sharing the nectar information of food sources, the selection of food source regions by the onlookers and evaluating the nectar amount of the food sources; and determining the scout bees and then sending them randomly onto possible new food sources. At the initialization stage, a set of food sources is randomly selected by the bees, and their nectar amounts are determined. At the first step of the cycle, these bees come into the hive and share the nectar information of the sources with the bees waiting on the dance area. A bee waiting on the dance area for making decision to choose a food source is called onlooker, and the bee going to the food source visited by herself just before is named as employed bee. After sharing their information with onlookers, every employed bee goes to the food source area visited by herself at the previous cycle since that food source exists in her memory and then chooses a new food source by means of visual information in the neighborhood of the one in her memory and evaluates its nectar amount. At the second step, an onlooker prefers a food source area depending on the nectar information distributed by the employed bees on the dance area. As the nectar amount of a food source increases, the probability of that food source chosen also increases. After arriving at the selected area, she chooses a new food source in the neighborhood of the one in the memory depending on visual information as in the case of employed bees. The determination of the new food source is carried out by the bees based on the comparison process of food source positions visually. At the third step of the cycle, when the nectar of a food source is abandoned by the bees, a new food source is randomly determined by a scout bee and replaced with the abandoned one. At each cycle at most, one scout goes outside for searching a new food source, and the number of employed and onlooker bees is selected to be equal to each other. These three steps are repeated through a predetermined number of cycles called maximum cycle number $T_{\max}$ or until a termination criterion is satisfied.

Define N_s as the total number of bees, N_e as the colony size of the employed bees, and N_u as the size of unemployed bees, which satisfy the equation $N_s = N_e + N_u$. We usually set N_e equal to N_u. D is the dimension of individual solution vector,

$S = \mathbb{R}^D$ represents individual search space, and S^{N_e} denotes the colony space of employed bees. An employed bee colony can be expressed by N_e dimension vector $\vec{X} = (X_1, \ldots, X_{N_e})$, where $X_i \in S$ and $i \leq N_e$. $\vec{X}(0)$ means the initial employed bee colony, while $\vec{X}(n)$ represents employed bee colony in the nth iteration. Denote $f: S \to R^+$ as the fitness function, and the standard ABC algorithm can be expressed as follows:

Step 1 Randomly initialize a set of feasible solutions $(X_1, \ldots, X_{N_s})$, and the specific solution X_i can be generated by

$$X_i^j = X_{\min}^j + rand\,(0,1)\left(X_{\max}^j - X_{\min}^j\right) \tag{2.14}$$

where $j \in \{1,2, \ldots, D\}$ is the jth dimension of the solution vector. Calculate the fitness value of each solution vector respectively, and set the top N_e best solutions as the initial population of the employed bees $\vec{X}(0)$.

Step 2 For an employed bee in the nth iteration $X_i(n)$, search new solutions in the neighborhood of the current position vector according to the following equation:

$$V_i^j = X_i^j + \varphi_i^j\left(X_i^j - X_k^j\right) \tag{2.15}$$

where $V \in S$, $j \in \{1,2, \ldots, D\}$, $k \in \{1,2, \ldots, N_e\}, k \neq i$, k, and j are randomly generated. φ_i^j is a random number between -1 and 1. It controls the production of neighbor food sources around X_i^j and represents the comparison of two food positions visually by a bee. As can be seen from the above equation, as the difference between the parameters of X_i^j and X_k^j decreases, the perturbation on the position X_i^j gets decreased, too. Thus, as the search approaches the optimum solution in the search space, the step length is adaptively reduced.

Generally, this searching process is actually a random mapping from individual space to individual space, and this process can be denoted with $T_m: S \to S$, and its probability distribution is clearly only related to current position vector $X_i(n)$, and has no relation with past location vectors as well as the iteration number n.

Step 3 Apply the greedy selection operator $T_s: S^2 \to S$ to choose the better solution between searched new vector V_i and the original vector X_i into the next generation. Its probability distribution can be described as follows:

$$P\{T_s(X_i, V_i) = V_i\} = \begin{cases} 1, f(V_i) \geq f(X_i) \\ 0, f(V_i) < f(X_i) \end{cases} \tag{2.16}$$

The greedy selection operator ensures that the population is able to retain the elite individual, and accordingly the evolution will not retreat. Obviously, the distribution of T_s has no relation with the iteration n.

Step 4 Each unemployed bee selects an employed bee from the colony according to their fitness values. The probability distribution of the selection operator T_{s1} : $S^{N_e} \rightarrow S$ can described as follows:

$$P\left\{T_{s1}\left(\vec{X}\right) = X_i\right\} = \frac{f\left(X_i\right)}{\sum_{m=1}^{N_e} f\left(X_m\right)} \tag{2.17}$$

Step 5 The unemployed bee searches in the neighborhood of the selected employed bee's position to find new solutions (see (2.15)). The updated best fitness value can be denoted with f_best, and the best solution parameters can be expressed with $(x_1, x_2, \ldots, x_D)$.

Step 6 If a position cannot be improved further through a predetermined number of cycles, then that food source is assumed to be abandoned. The value of predetermined number of cycles is an important control parameter of the ABC algorithm, which is called "limit" for abandonment. If the searching times surrounding an employed bee *Bas* exceed a certain threshold *Limit*, but still could not find better solutions, then the location vector can be reinitialized randomly according to the following equation:

$$X_i\left(n+1\right) = \begin{cases} X_{\min} + rand\left(0,1\right)\left(X_{\max} - X_{\min}\right), & Bas_i \geq Limit \\ X_i(n), & Bas_i < Limit \end{cases} \tag{2.18}$$

Step 7 If not, go to Step (2). If the iteration value is larger than the maximum number of the iteration (i.e., $T > T_{\max}$), output the optimal fitness value f_best and correlative parameters $(x_1, x_2, \ldots, x_D)$. If not, go to Step (2).

Step (6) is a most prominent aspect making ABC algorithm different from other algorithms, which is designed to enhance the diversity of the population to prevent the population from trapping into the local optimum. Obviously, this step can improve the probability of finding the best solution efficiently and make the ABC algorithm perform much better.

Totally, ABC algorithm employs four different selection processes (Karaboga and Akay 2009):

1. A global probabilistic selection process, in which the probability value is calculated by (2.17) used by the onlooker bees for discovering promising regions.
2. A local probabilistic selection process carried out in a region by the employed bees and the onlookers depending on the visual information such as the color, shape, and fragrance of the flowers (sources) (bees will not be able to identify the type of nectar source until they arrive at the right location and discriminate among sources growing there based on their scent) for determining a food source around the source in the memory in a way described by (2.15).

3. A local selection called greedy selection process carried out by onlooker and employed bees in that if the nectar amount of the candidate source is better than that of the present one, the bee forgets the present one and memorizes the candidate source produced by (2.15). Otherwise, the bee keeps the present one in the memory.
4. A random selection process carried out by scouts as defined in (2.18).

2.5 Differential Evolution

2.5.1 Biological Inspiration

Researchers have been looking into nature for years for inspiration with the purpose of tackling complex computational problems. Optimization is ubiquitous in natural processes. For example, every species had to adapt their physical structures to fit to the environments they were in and to strengthen their survival ability all the time. The underlying relation between optimization and biological evolution led to the development of an important paradigm of computational intelligence, the evolutionary computing techniques for performing very complex search and optimization.

Evolutionary computation uses iterative progress, such as growth or development in a population. This population is then selected in a guided random search using parallel processing to achieve the desired end. The paradigm of evolutionary computing techniques dates back to early 1950s, when the idea to use Darwinian principles for automated problem solving originated. It was not until the 1960s that three distinct interpretations of this idea started to be developed in three different places. Evolutionary programming (EP) was introduced by Lawrence J. Fogel in the United States, while almost simultaneously, I. Rechenberg and H.-P. Schwefel introduced evolution strategies (ESs) in Germany. Almost a decade later, John Henry Holland from the University of Michigan at Ann Arbor devised an independent method of simulating the Darwinian evolution to solve practical optimization problems and called it the genetic algorithm (GA). These areas developed separately for about 15 years. From the early 1990s on, they are unified as different representatives of one technology, called evolutionary computing. Also since the early 1990s, a fourth stream following the same general ideas started to emerge, which is genetic programming (GP). Nowadays, the field of nature-inspired metaheuristics is mostly constituted by the evolutionary algorithms (EA) as well as the swarm intelligence algorithms. Also the field extends in a broader sense to include self-organizing systems, artificial life (digital organism), memetic and cultural algorithms, harmony search, artificial immune systems, and learnable evolution model.

The DE algorithm emerged as a very competitive form of evolutionary computing more than a decade ago. The first written article on DE appeared as a technical report by Storn and Price (1995). One year later, the success of DE was demonstrated at the First International Contest on Evolutionary Optimization in May 1996, which was held in conjunction with the 1996 IEEE International Conference on Evolutionary Computation (CEC). DE finished third at the First International Contest on Evolutionary Optimization (1st ICEO), which was held in Nagoya, Japan. DE turned out to be the best evolutionary algorithm for solving the real-valued test function suite of the 1st ICEO (the first two places were given to non-evolutionary algorithms, which are not universally applicable but solved the test problems faster than DE). Price presented DE at the Second International Contest on Evolutionary Optimization in 1997, and it turned out as one of the best among the competing algorithms. Two journal articles describing the algorithm in sufficient details followed immediately in quick succession. In 2005 CEC competition on real parameter optimization, on 10-*D* problems classical DE secured 2*nd* rank and a self-adaptive DE variant called SaDE secured third rank although they performed poorly over 30-*D* problems.

DE, like most popular EAs, is a population-based tool. DE, unlike other EAs, generates offspring by perturbing the solutions with a scaled difference of two randomly selected population vectors, instead of recombining the solutions under conditions imposed by a probabilistic scheme. In addition, DE employs a one-to-one spawning logic which allows replacement of an individual only if the offspring outperforms its corresponding parent (Duan et al., 2011). DE has been seen as an attractive optimization tool for continuous optimization for the following four reasons: (1) Compared to most other EAs, DE is much more simple and straightforward to implement. (2) As indicated by the recent studies on DE despite its simplicity, DE exhibits much better performance in comparison with several algorithms in solving a wide variety of problems including unimodal, multimodal, separable, non-separable, and so on. (3) The number of control parameters in DE is very few (*Cr*, *F*, and *NP* in classical DE). (4) The space complexity of DE is low as compared to some of the most competitive real parameter optimizers.

2.5.2 *Principle of Differential Evolution*

2.5.2.1 Initialization of the Parameter Vectors

DE algorithm mainly has three evolutionary operations, namely, mutation, recombination, and selection. The positions of individuals are represented as real-coded vectors which are randomly initialized inside the limits of the given search space in the beginning of an optimization process. The individuals are evolved during the optimization process by applying mutation, recombination, and selection to each individual in every generation. A stopping criterion determines after the building of every new generation if the optimization process should be terminated.

Like other evolutionary algorithms, DE also deals with a population of solutions. Suppose that the initial solution population has *NP* individuals and the search space is D dimensional, the solution vector in continuous space can be represented by $x_i = [x_{i1}, x_{i2}, \cdots, x_{iD}], (i = 1, \cdots, NP)$. Let there be some criteria of optimization, usually named fitness or cost function. Then the optimization goal of DE algorithm is to find the values of the variables that minimize the fitness, that is, to find

$$x^* : f\left(x^*\right) = \min_x f(x)$$

2.5.2.2 Mutation with Difference Vectors

Biologically, "mutation" means a sudden change in the gene characteristics of a chromosome. In the context of the evolutionary computing paradigm, mutation is also seen as a change or perturbation with a random element. In DE literature, a parent vector from the current generation is called target vector, a mutant vector obtained through the differential mutation operation is known as donor vector, and finally an offspring formed by recombining the donor with the target vector is called trial vector. In one of the simplest forms of DE mutation, to create the donor vector for each ith target vector from the current population, three other distinct parameter vectors, say $\boldsymbol{x}_{r1}$, $\boldsymbol{x}_{r2}$, and $\boldsymbol{x}_{r3}$, are sampled randomly from the current population. The indices r_1, r_2, and r_3 are mutually exclusive integers randomly chosen from the range $[1, NP]$, which are also different from the base vector index i. These indices are randomly generated once for each mutant vector. Now the difference of any two of these three vectors is scaled by a scalar number F (that typically lies in the interval $[0.4, 1]$), and the scaled difference is added to the third one whence we obtain the donor vector $\boldsymbol{v}_i$. We can express the process as

$$\boldsymbol{v}_i = \boldsymbol{x}_{r_1} + F \times (\boldsymbol{x}_{r_2} - \boldsymbol{x}_{r_3}) \tag{2.19}$$

The process is illustrated on a 2-D parameter space in Fig. 2.7.

2.5.2.3 Crossover

To enhance the potential diversity of the population, a crossover operation comes into play after generating the donor vector through mutation. In this way, the individuals of the population are updated by means of the recombination operation. By coping components from the mutation vector $\boldsymbol{v}_i$ and the target vector $\boldsymbol{x}_i$ in dependence, the trial vector $\boldsymbol{u}_i$ was generated. This process can be written as the following equation:

$$u_{ji} = \begin{cases} v_{ji}, \ if \ randb \leq CR \ or \ j = randr, & j = 1, \cdots, D \\ x_{ji}, \ if \ randb > CR \ or \ j \neq randr, & j = 1, \cdots, D \end{cases} \tag{2.20}$$

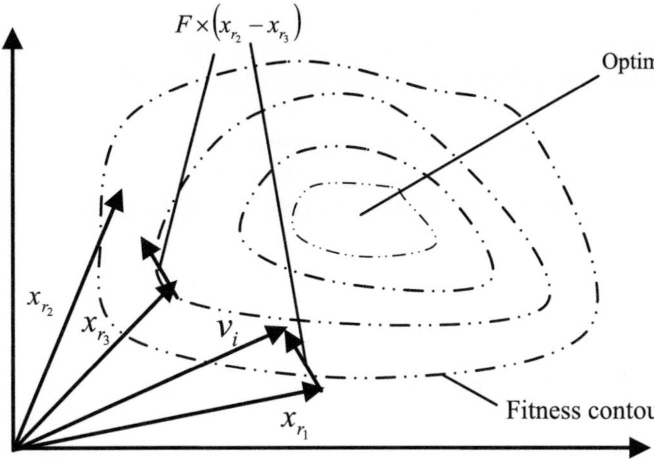

Fig. 2.7 Mutation process of DE

Fig. 2.8 Recombination process of DE

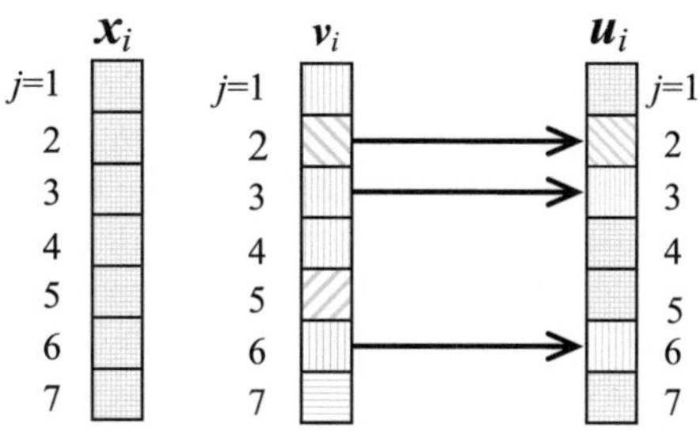

where the random number $randb \in [0,1]$, the recombination control parameter *CR* is a constant in the interval [0,1]. *randr* is an integer randomly chosen from [1,*D*]. The recombination process is described in Fig. 2.8.

2.5.2.4 Selection

To keep the population size constant over subsequent generations, the next step of the algorithm calls for selection to determine whether the target or the trial vector survives to the next generation. Then, selection operation, as a deterministic process in DE algorithm, is implemented to choose the better individuals with lower fitness function value between the target vector and the trial vector, which is inherited by the next generation, expressed as

$$x_i\left(t+1\right)=\begin{cases}u_i, & if\ f\left(u_i\right)\le f\left(x_i\right)\\ x_i, & else\end{cases} \tag{2.21}$$

This selection scheme allows only improvement but not deterioration of the fitness function value; it is called greedy. Selection operator ensures that the best fitness function value cannot get lost when moving from one generation to the next, which usually results in the fast convergence behavior. Therefore, if the new trial vector yields an equal or lower value of the objective function, it replaces the corresponding target vector in the next generation; otherwise the target is retained in the population. Hence, the population either gets better (with respect to the minimization of the objective function) or remains the same in fitness status but never deteriorates.

2.5.3 Control Parameters of Differential Evolution

There are three main control parameters of the DE algorithm: the mutation scale factor F, the crossover constant CR, and the population size NP.

The population size is related to the amount of possible moving vectors. Over all the possible moves given by a population, some moves are beneficial in the search for the optimum, while some others are ineffective and result in a waste of computational effort. Therefore, too small a population size can contain too limited an amount of moves, while too large a population size may contain a high number of ineffective moves which can likely mislead the search. To some extent the population sizing of a DE is analogous to the other EAs, if too small it could cause premature convergence, and if too large it could cause stagnation. A good value can be found by considering the dimensionality of the problem similar to what is commonly performed for the other EAs. A guideline is given in Price and Storn (1997) where a setting of Spop equal to ten times the dimensionality of the problem is proposed. However, this indication is not confirmed by a recent study in Neri and Tirronen (2008) where it is shown that a population size lower than the dimensionality of the problem can be optimal in many cases.

Regarding the scale factor F and the crossover rate CR, these settings may be a difficult task. The setting of these two parameters is neither an intuitive nor a straightforward task but is unfortunately crucial for guaranteeing the algorithmic functioning. Several studies have thus been proposed in literature. The study reported in Lampinen and Zelinka (2000) arrives at the conclusion, after an empirical analysis, that usage of $F = 1$ is not recommended, since according to a conjecture of the authors, it leads to a significant decrease in explorative power. Analogously, the setting $CR = 1$ is also discouraged since it would dramatically decrease the amount of possible offspring solutions. In Price and Storn (1997), the settings $F \in [0.5,1]$ and $CR \in [0.8,1]$ are recommended. The empirical analysis reported in Zielinski et al. (2006) shows that in many cases the setting of $F \geq 0.6$ and $CR \geq 0.6$ leads to results having better performance.

Several studies highlight that an efficient parameter setting is very dependent on problems (e.g., $F = 0.2$ could be a very efficient setting for a certain fitness landscape and completely inadequate for another problem). This result can be seen as a confirmation of the validity of the No Free Lunch Theorem (Wolpert and Macready 1997) with reference to the DE schemes.

The problem of the parameter setting is emphasized when DE is employed for handling difficulties of real-world applications such as high dimensionality and noisy optimization problems. Clearly, the risk of the DE stagnation is higher for larger decision spaces and worsens as the number of dimensions of the problem increases. A large decision space (in terms of dimensions) requires a wide range of possible moves to enhance its capability of detecting new promising solutions. Since, as mentioned before, an enlargement in population size causes an increase in the set of potential ineffective moves, a proper choice of F and CR becomes a crucial aspect in the success of DE.

2.6 Other Algorithms

2.6.1 *Glowworm Swarm Optimization*

Glowworm swarm optimization (GSO) is a novel algorithm designed by Krishnanand and Ghose (2009) for the simultaneous computation of multiple optima of multimodal functions. The algorithm shares a few features with some better known swarm intelligence-based optimization algorithms, such as ACO and PSO, but with several significant differences. The agents in GSO are thought of as glowworms that carry a luminescence quantity called luciferin along with them. The glowworms encode the fitness of their current locations, evaluated using the objective function, into a luciferin value that they broadcast to their neighbors. The glowworm identifies its neighbors and computes its movements by exploiting an adaptive neighborhood, which is bounded above by its sensor range. Each glowworm selects, using a probabilistic mechanism, a neighbor that has a luciferin value higher than its own and moves towards it. These movements—based only on local information and selective neighbor interactions—enable the swarm of glowworms to partition into disjoint subgroups that converge on multiple optima of a given multimodal function.

2.6.2 *Bacteria Foraging Optimization*

Natural selection tends to eliminate animals with poor "foraging strategies" (methods for locating, handling, and ingesting food) and favor the propagation of genes of those animals that have successful foraging strategies since they are more likely to enjoy reproductive success (they obtain enough food to enable them to reproduce). After many generations, poor foraging strategies are either eliminated or shaped into good ones. Such evolutionary principles have led scientists to hypothesize that it is appropriate to model the activity of foraging as an optimization process. Passino (2002) proposed a bacteria foraging optimization by simulating the chemotactic (foraging) behavior of *E. coli* bacteria. There are algorithmic analogies between the genetic algorithm and the above optimization model for foraging. There are analogies between the fitness function and the nutrient concentration function (both a type of "landscape"), selection and bacterial reproduction (bacteria in the most favorable environments gain a selective advantage for reproduction), crossover and bacterial splitting (the children are at the same concentration, whereas with crossover they generally end up in a region around their parents on the fitness landscape), and mutation and elimination and dispersal. However, the algorithms are not equivalent, and neither is a special case of the other. Each has its own distinguishing features. The fitness function and nutrient concentration functions are not the same (one represents likelihood of survival for given phenotypic characteristics, whereas the other represents nutrient/noxious substance concentrations or for other foragers predator/prey characteristics). Crossover represents mating and resulting differences in offspring, something we ignore in the bacterial foraging algorithm (we could, however, have made less than perfect copies of the bacteria to represent their splitting). Moreover, mutation represents gene mutation and the resulting phenotypical changes, not physical dispersal in an environment.

2.6.3 Bat-Inspired Algorithm

Most microbats are insectivores. Microbats use a type of sonar, called echolocation, to detect prey, avoid obstacles, and locate their roosting crevices in the dark. These bats emit a very loud sound pulse and listen for the echo that bounces back from the surrounding objects. Their pulses vary in properties and can be correlated with their hunting strategies, depending on the species. Most bats use short, frequency-modulated signals to sweep through about an octave, while others more often use constant-frequency signals for echolocation. Their signal bandwidth varies and depends on the species and often increased by using more harmonics. Yang (2010) formulated a new Bat Algorithm for continuous constrained optimization problems. Though the implementation is more complicated than many other metaheuristic algorithms, however, it does show that it utilizes a balanced combination of the advantages of existing successful algorithms with innovative feature based on the echolocation behavior of bats. New solutions are generated by adjusting frequencies, loudness, and pulse emission rates, while the proposed solution is accepted or does not depend on the quality of the solutions controlled or characterized by loudness and pulse rate which are in turn related to the closeness or the fitness of the locations/solution to the global optimal solution. Moreover, the author argues that PSO and harmony search are the special cases of the Bat Algorithm under appropriate simplifications.

2.7 Conclusions

In this chapter, we focus on four popular bio-inspired computation algorithms, which are, respectively, ACO, PSO, ABC, and DE. PSO and ACO are currently the most popular algorithms in the swarm intelligence domain. Dorigo and his colleagues introduced the first ACO algorithms in the early 1990s, which is a metaheuristic suitable for solving hard combinatorial optimization problems. The inspiring source of ACO is the pheromone trail laying and following behavior of real ants, which use pheromones as a communication medium. Ants are social insects, being interested mainly in the colony survival rather than individual survival. Of interests is ants' ability to find the shortest path from their nest to food. This idea was the source of the algorithms inspired from ants' behavior. The initial ideas on particle swarms of Kennedy and Eberhart were essentially aimed at producing computational intelligence by exploiting simple analogues of social interaction, which soon developed into a powerful optimization method-PSO. Unlike in the other evolutionary computation techniques, each particle in PSO is also associated with a velocity. Particles fly through the search space with velocities, which are dynamically adjusted according to their historical behaviors. Therefore, the particles have the tendency to fly towards the better and better search area over the course of search process. ABC was originally presented by Karaboga and Basturk, under the inspiration of collective behavior on honeybees, and it has been proved to possess a better performance in function optimization problem. Compared with the usual algorithms, the major advantage of ABC algorithm lies in that it conducts both global search and local search in each iteration, and as a result the probability of finding the optimal parameters is significantly increased, which efficiently avoid local optimum to a large extent. The DE algorithm emerged as a very competitive form of evolutionary computing more than a decade ago. DE has been seen as an attractive optimization tool for continuous optimization for the following reasons: (1) simple and straightforward for implementation, (2) better performance (compared with several other algorithms in solving a variety of problems including unimodal, multimodal, separable, non-separable, and so on), (3) few control parameters, and (4) less space complexity. Besides, we have also given a brief introduction of other bio-inspired computation algorithms such as GSM, BFO, and BA.

References

Bavelas A (1950) Communication patterns in task-oriented groups. J Acoust Soc Am 22(6):725–730

Bonabeau E, Dorigo M, Theraulaz G (1999) Swarm intelligence: from natural to artificial systems. Oxford University Press, New York

Bullnheimer B, Hartl RF, Strauss C (1999) A new rank based version of the Ant System: a computational study. Central European J Operations Res Econom 7(1):25–38

Clerc M, Kennedy J (2002) The particle swarm-explosion, stability, and convergence in a multidimensional complex space. IEEE Trans Evolut Comput 6(1):58–73
Colorni A, Dorigo M, Maniezzo V (1991) Positive feedback as a search strategy. Techn Rep, politecnico di milano
Crina G, Ajith A (2006) Stigmergic optimization: inspiration, technologies and perspectives. In: Stigmergic optimization. Springer Berlin Heidelberg, pp 1–24
Deneubourg J-L, Aron S, Goss S, Pasteels JM (1990) The self-organizing exploratory pattern of the argentine ant. J Insect Behav 3(2):159–168
Dorigo M (1992) Optimization, learning and natural algorithms. PhD Thesis, Politecnico di Milano, Italy
Dorigo M, Blum C (2005) Ant colony optimization theory: a survey. Theor Comput Sci 344(2):243–278
Dorigo M, Gambardella LM (1997) Ant colony system: a cooperative learning approach to the traveling salesman problem. IEEE Trans Evolut Comput 1(1):53–66
Dorigo M, Stützle T (2003) The ant colony optimization metaheuristic: algorithms, applications, and advances. In: Glover F, Kochenberger GA (eds) Handbook of Metaheuristics. Springer, Boston, MA, pp 250–285
Dorigo M, Maniezzo V, Colorni A (1996) Ant system: optimization by a colony of cooperating agents. IEEE Trans Syst Man Cybern B Cybern 26(1):29–41
Dornhaus A, Klügl F, Puppe F, Tautz J (1998) Task selection in honeybees-experiments using multi-agent simulation. In: Proceedings of The Third German Workshop on Artificial Life, Bochum. Verlag Harry Deutsch, pp 171–183
Duan H (2005) Ant colony algorithms: theory and applications. Science Press, Beijing, China
Duan H (2010) Ant colony optimization: principle, convergence and application. In: Bijaya Ketan Panigrahi, Yuhui Shi, Lim M-H (eds) Handbook of Swarm Intelligence. Springer Berlin Heidelberg, pp 373–388
Duan H, Liu S (2010) Non-linear dual-mode receding horizon control for multiple unmanned air vehicles formation flight based on chaotic particle swarm optimisation. IET Control Theory Appl 4(11):2565–2578
Duan H, Xing Z (2009) Improved quantum evolutionary computation based on particle swarm optimization and two-crossovers. Chin Phys Lett 26(12):120304
Duan H, Xu C, Xing Z (2010) A hybrid artificial bee colony optimization and quantum evolutionary algorithm for continuous optimization problems. Int J Neural Syst 20(01):39–50
Duan H, Zhang X, Xu C (2011) Bio-inspired computing. Science Press, Beijing, China
Eberhart R, Kennedy J (1995) A new optimizer using particle swarm theory. In: Proceedings of the Sixth International Symposium on Micro Machine and Human Science, Nagoya. IEEE, pp 39–43
Grosan C, Abraham A (2011) Swarm intelligence. In: Intelligent systems: a modern approach. Springer, Berlin, Heidelberg, pp 409–422
Heppner F, Grenander U (1990) A stochastic nonlinear model for coordinated bird flocks. In: Krasner S (ed) The ubiquity of chaos. AAAS Publications, Washington, DC, pp 233–238
Karaboga D, Akay B (2009) A comparative study of artificial bee colony algorithm. Appl Math Comput 214(1):108–132
Karaboga D, Basturk B (2007) A powerful and efficient algorithm for numerical function optimization: artificial bee colony (ABC) algorithm. J Global Optim 39(3):459–471
Kennedy J (1998) The behavior of particles. Evolutionary programming VII. In: David Hutchison, Takeo Kanade, Josef Kittler (eds) Lecture Notes in Computer Science. Springer Berlin Heidelberg, pp 579–589
Kennedy J (1999) Small worlds and mega-minds: effects of neighborhood topology on particle swarm performance. In: Proceedings of the 1999 Congress on Evolutionary Computation, Washington, DC. IEEE, pp 1931–1938
Kennedy J, Eberhart R (1995) Particle swarm optimization. In: Proceedings of IEEE International Conference on Neural Networks, Piscataway, NJ. IEEE, pp 1942–1948

Kennedy J, Mendes R (2002) Population structure and particle swarm performance. In: Proceedings of the 2002 Congress on Evolutionary Computation (CEC'02), Honolulu, HI. IEEE, pp 1671–1676

Krishnanand K, Ghose D (2009) Glowworm swarm optimization for simultaneous capture of multiple local optima of multimodal functions. Swarm Intell 3(2):87–124

Lampinen J, Zelinka I (2000) On stagnation of the differential evolution algorithm. In: Proceedings of 6th International Mendel Conference Soft Computing, Brno, Czech Republic. pp 76–83

Menzel R, De Marco RJ, Greggers U (2006) Spatial memory, navigation and dance behaviour in Apis mellifera. J Comp Physiol A 192(9):889–903

Millonas MM (1994) Swarms, phase transitions, and collective intelligence. In: Artificial life III. Reading, MA. Addison-Wesley, pp 417–445

Neri F, Tirronen V (2008) On memetic differential evolution frameworks: a study of advantages and limitations in hybridization. In: Proceedings of IEEE Congress on Evolutionary Computation, 2008 (IEEE World Congress on Computational Intelligence), Hong Kong. IEEE, pp 2135–2142

Passino KM (2002) Biomimicry of bacterial foraging for distributed optimization and control. IEEE Contr Syst 22(3):52–67

Poli R, Kennedy J, Blackwell T (2007) Particle swarm optimization: an overview. Swarm Intell 1(1):33–57

Price K, Storn R (1997) Differential evolution–a simple evolution strategy for fast optimization. Dr Dobb's J 22(4):18–24

Reynolds CW (1987) Flocks, herds and schools: a distributed behavioral model. Comput Graphics 21(4):25–34

Seeley TD (1985) Honeybee ecology: a study of adaptation in social life. Princeton University Press, Princeton

Shi Y, Eberhart R. (1998) A modified particle swarm optimizer. In: Proceedings of The 1998 IEEE International Conference on Evolutionary Computation (IEEE World Congress on Computational Intelligence), Anchorage, AK. IEEE, pp 69–73

Storn R, Price K (1995) Differential Evolution-a simple and efficient adaptive scheme for global optimization over continuous spaces. Techn Rep International Computer Science Institute, Berkeley, CA

Stutzle T, Hoos H (1997) MAX-MIN ant system and local search for the traveling salesman problem. In: Proceeding of IEEE Conference on Evolutionary Computation, Indianapolis, IN. IEEE, pp 309–314

Wolpert DH, Macready WG (1997) No free lunch theorems for optimization. IEEE Trans Evolut Comput 1(1):67–82

Xu C, Duan H, Liu F (2010) Chaotic artificial bee colony approach to Uninhabited Combat Air Vehicle (UCAV) path planning. Aerosp Sci Technol 14(8):535–541

Yang X-S (2010) A new metaheuristic bat-inspired algorithm. In: Gonzalez JR et al. (eds) Nature inspired cooperative strategies for optimization (NISCO 2010). Studies in computational intelligence, vol 284. Springer, Berlin, pp 65–74

Yu J, Duan H (2012) Artificial Bee Colony approach to information granulation-based fuzzy radial basis function neural networks for image fusion. Optik 124(17):3103–3111

Zielinski K, Weitkemper P, Laur R, Kammeyer K-D (2006) Parameter study for differential evolution using a power allocation problem including interference cancellation. In: Proceedings of IEEE Congress on Evolutionary Computation, Vancouver, BC. IEEE, pp 1857–1864

Chapter 3
UAV Modeling and Controller Design

Haibin Duan

Abstract As a complicated multi-input, multi-output, and time-varying nonlinear system, flight control system of unmanned aerial vehicle (UAV), which determines the whole system's performance directly, is crucial for the simulation training system design. This chapter mainly focuses on parameter identification of flight control system based on the modeling of UAVs and a specific controller design for the pendulum-like oscillation in micro aerial vehicle (MAV). A predator–prey particle swarm optimization (PSO) algorithm for identifying parameters of UAV flight control system is presented, with the aim of reducing the workload of the designers during the process of designing complicated UAV control system. Besides, a software environment for UAV controller design was developed based on the UAV model and the proposed method. Then a specific kind of controller design involving the pendulum-like oscillation in the hover and stare state for a MAV in the presence of external disturbances is discussed in detail, since the pendulum-like oscillation caused by uncertainty and external disturbances badly jeopardize the performance of hover and stare, resulting in blurred images or even MAV's overturning. A novel type of PSO-based linear-quadratic regulator (LQR) controller for stabilizing the pendulum-like oscillation is developed, which can enhance the MAV's performance efficiently.

3.1 Introduction

Flight control system, which is a complicated multi-input, multi-output, and time-varying nonlinear system, is the core of the simulation training system design of unmanned aerial vehicle (UAV), which also determines the whole system's performance directly (McLean 1990). For the existence of strong coupling among the inputs and the nonexistence of mapping relationship between the performance index and the controller parameter, the selection of the controller parameters is a very tough problem in the design of the flight control system. Presently, cut and

---

The original version of this chapter was revised. A correction to this chapter is available at https://doi.org/10.1007/978-3-642-41196-0_9

H. Duan and P. Li, *Bio-inspired Computation in Unmanned Aerial Vehicles*,
DOI 10.1007/978-3-642-41196-0_3, © Springer-Verlag Berlin Heidelberg 2014

try method is commonly used to identify all the control loop parameters of flight control system. But this design method is low efficient and, to a great extent, much depends on the experience of the designer, while the flight control system will be more complex with the improvement of the aircraft performances, and these are becoming the bottleneck of the flight control system design (Zhang and An 2008). In the first part of this chapter, we proposed a parameter identification method for UAV control system based on predator–prey particle swarm optimization (PSO). While using PSO to optimize flight control systems, there are two problems to be solved. Firstly, the UAV model selected is very important. Secondly, it is very crucial for PSO algorithm to choose the fitness function, because there is no obvious mapping relationship between the property index and the UAV controller. Therefore, how to evaluate each particle's quality turns out to be a key issue which must be resolved.

Micro aerial vehicles (MAVs), essentially small-scale flying robots, became an area of interest in the aerospace community with the initiation of the micro UAV (MUAV) program by Defense Advanced Research Projects Agency (DARPA). The technological feasibility of MAVs as one possible solution to new challenging reconnaissance mission scenarios in urban warfare (local, close-up range, hidden reconnaissance, operation between obstacles and maybe even inside buildings) is depending on a bunch of questions, which are only partly answered so far (Johnson and Turbe 2006; Bloss 2009). One of the challenging problems for MAVs is to design a robust flight control system for such a miniaturized "bird," which is generally one order of magnitude smaller than any today's operational UAV. Hover and stare is a key issue to the performance of MAVs and similar kinds of unmanned vehicles, which are designed to perform surveillance and reconnaissance missions. However, pendulum-like oscillation triggered by external disturbances and other uncertain factors will badly impair its performance, thus resulting in blurred images or even overturn of the vehicle (Pflimlin et al. 2010). As a result, control techniques of such a vehicle are becoming more and more important for their wide applications in civil and military fields. The second part of this chapter designed a novel type of pendulum-like oscillation controller for MAV hover and stare state in the presence of external disturbances, which is based on linear-quadratic regulator (LQR) and PSO (Duan and Sun 2013). A linear mathematical model of pendulum phenomenon based upon actual wind tunnel test data representing the hover mode is established, and a hybrid LQR and PSO approach is proposed to stabilize oscillation. PSO is applied for parameter optimization of the designed LQR controller.

3.2 Parameter Identification for UAVs Based on Predator–Prey PSO

3.2.1 Mathematical Model of UAVs

The 6-DOF nonlinear model of UAVs is illustrated in this section, which is the prerequisite for simplifying and linearizing the mathematical model (Zhang 2004).

3.2.1.1 Nonlinear Equations of 6-DOF Modeling

UAV nonlinear equations of 6-DOF can be deduced by the aerodynamic and kinematical equations as follows:

$$\begin{cases} \dot{V} = \frac{1}{m}\left(P_x \cos\alpha\cos\beta - P_y \sin\alpha\cos\beta + P_z \sin\beta + Z\sin\beta - Q\right) \\ \quad - g\Big(\cos\alpha\cos\beta\sin\vartheta - \sin\alpha\cos\beta\cos\vartheta\cos\gamma - \sin\beta\cos\vartheta\sin\gamma \\ \dot{\alpha} = -\frac{1}{mV\cos\beta}\left(P_x\sin\alpha + P_y\cos\alpha + Y\right) + \omega_z - \tan\beta\left(\omega_x\cos\alpha - \omega_y\sin\alpha\right)\frac{\delta y}{\delta x} \\ \quad + \frac{g}{V\cos\beta}\left(\sin\alpha\sin\vartheta + \cos\alpha\cos\vartheta\cos\gamma\right) \\ \dot{\beta} = \frac{1}{mV}\left(-P_x\cos\alpha\sin\beta + P_y\sin\alpha\sin\beta + P_z\cos\beta + Z\right) + \omega_x\sin\alpha \\ \quad + \omega_y\cos\alpha + \frac{g}{V}\left(\cos\alpha\sin\beta\sin\vartheta - \sin\alpha\sin\beta\cos\vartheta\cos\gamma\right. \\ \quad \left. + \cos\beta\sin\gamma\cos\vartheta\right) \end{cases} \tag{3.1}$$

where m is the mass of the UAV; α is attack angle; β is sideslip angle; ϑ is pitch angle; γ is roll angle; P is engine thrust; X, Y, Z are the projections of aerodynamic force in body axis; and ω_x, ω_y, ω_z denote the coordinate components of palstance. These three equations already contain three forces in the body axis which are generated by the thrust vector:

$$\begin{cases} \dot{\omega}_x = b_{11}\omega_y\omega_z + b_{12}\omega_x\omega_z + \frac{I_y\left(M_x + M_{px}\right) + I_{xy}\left(M_y + M_{py}\right)}{I_x I_y - {I_{xy}}^2} \\ \dot{\omega}_y = b_{21}\omega_y\omega_z + b_{22}\omega_x\omega_z + \frac{I_{xy}\left(M_x + M_{px}\right) + I_x\left(M_y + M_{py}\right)}{I_x I_y - {I_{xy}}^2} \\ \dot{\omega}_z = \frac{I_x - I_y}{I_z}\omega_x\omega_y + \frac{I_{xy}}{I_z}\left({\omega_x}^2 - {\omega_y}^2\right) + \frac{\left(M_z + M_{pz}\right)}{I_z} \end{cases} \tag{3.2}$$

$$b_{11} = \frac{{I_y}^2 - I_y I_z - {I_{xy}}^2}{I_x I_y - {I_{xy}}^2}, \quad b_{22} = \frac{I_x I_z - {I_x}^2 - {I_{xy}}^2}{I_x I_y - {I_{xy}}^2},$$

$$b_{12} = \frac{I_{xy}\left(I_z - I_y - I_x\right)}{I_x I_y - {I_{xy}}^2}, \quad b_{21} = \frac{I_{xy}\left(I_y - I_z - I_x\right)}{I_x I_y - {I_{xy}}^2}.$$

where I_x, I_y, I_z and M_x, M_y, M_z denote the coordinate components of inertia moment and resultant moment, respectively.

In the body axis, we have

$$\begin{cases} \dot{\gamma} = \omega_x - \tan\vartheta\left(\omega_y \cos\gamma - \omega_z \sin\gamma\right) \\ \dot{\vartheta} = \omega_y \sin\gamma + \omega_z \cos\gamma \\ \dot{\psi} = \frac{1}{\cos\vartheta}\left(\omega_y \cos\gamma - \omega_z \sin\gamma\right) \end{cases} \tag{3.3}$$

where ψ is drift angle. Aerodynamic equations can be described as

$$Y = C_y qS, C_y = C_y(\alpha, \delta_z), Z = \sum C_z qS, \sum C_z = C_z(\alpha, \delta_x) + C_z(\alpha, \delta_y)$$
$$Q = C_x qS, C_x = C_x(\alpha, \delta_z)$$

where δ_x, δ_y, δ_z are the coordinate components of deflection angles of the controlling surface. Aerodynamic moments can be given by

$$M_x = \sum m_x qsl, \sum m_x = m_x{}^{\beta}\beta + m_x(\alpha, \delta_x) + m_x(\alpha, \delta_y) + m_x{}^{\omega x}\omega_x \frac{l}{2V} + m_x{}^{\omega y}\omega_y \frac{l}{2V}$$

$$M_y = \sum m_y qsl, \sum m_y = m_y{}^{\beta}\beta + m_y(\alpha, \delta_x) + m_y(\alpha, \delta_y) + m_y{}^{\omega x}\omega_x \frac{l}{2V} + m_y{}^{\omega y}\omega_y \frac{l}{2V}$$

$$M_z = \sum m_z qsb_A, \sum m_z = m_z(\alpha, \delta_z) + m_z{}^{\omega z}\omega_z \frac{b_A}{V} + m_z{}^{\dot{\alpha}}\dot{\alpha}\frac{b_A}{V}.$$

When the height and mach are fixed, aerodynamic coefficients $C_y(\alpha,\delta_z)$, $C_x(\alpha,\delta_z)$, $C_z(\alpha,\delta_x)$, $C_z(\alpha,\delta_y)$, $m_x(\alpha,\delta_x)$, $m_x(\alpha,\delta_y)$, $m_y(\alpha,\delta_x)$, $m_y(\alpha,\delta_y)$, $m_z(\alpha,\delta_z)$ are the functions of the height, mach, attack angle, and control surface. Aerodynamic derivatives $m_z{}^{\omega z}$, $m_z{}^{\dot{\alpha}}$, $m_x{}^{\beta}$, $m_x{}^{\omega x}$, $m_x{}^{\omega y}$, $m_y{}^{\beta}$, $m_y{}^{\omega x}$, $m_y{}^{\omega y}$ are specified values.

3.2.1.2 Nonlinear Equations of 5-DOF Modeling

Suppose that UAV equations can be simplified into nonlinear equations of 5-DOF if the thrust and the resistance of the aircraft are the same. Without consideration of the thrust vector, we have $\dot{V} = 0, P = Q, P_x = P \ll G, P_y = P_z = 0$. On this condition, the equations of the UAV speed level off (Zhang 2004). At the same time, it is assumed that $I_{xy} \ll I_x I_y - I_{xy}{}^2$, state variable $x = (\alpha,\beta,\omega_x,\omega_y,\omega_z)^T$. The UAV nonlinear equations of 5-DOF are represented as follows:

$$
\begin{cases}
\dot{\alpha} = -\dfrac{1}{mV\cos\beta}(P\sin\alpha + Y) + \omega_z - \tan\beta\left(\omega_x\cos\alpha - \omega_y\sin\alpha\right) \\
\quad + \dfrac{g}{V\cos\beta}(\sin\alpha\sin\vartheta + \cos\alpha\cos\vartheta\cos\gamma) \\
\dot{\beta} = \dfrac{1}{mV}(-P\cos\alpha\sin\beta + Z) + \omega_x\sin\alpha \\
\quad + \omega_y\cos\alpha + \dfrac{g}{V}(\cos\alpha\sin\beta\sin\vartheta - \sin\alpha\sin\beta\cos\vartheta\cos\gamma \\
\quad + \cos\beta\sin\gamma\cos\vartheta) \\
\dot{\omega}_x = \dfrac{{I_y}^2 - I_yI_z - {I_{xy}}^2}{I_xI_y - {I_{xy}}^2}\omega_y\omega_z + \dfrac{I_{xy}\left(I_z - I_y - I_x\right)}{I_xI_y - {I_{xy}}^2}\omega_x\omega_z + \dfrac{I_y\sum M_x}{I_xI_y - {I_{xy}}^2} \\
\quad b_{11}\omega_y\omega_z + b_{12}\omega_x\omega_z + \dfrac{I_yM_x}{I_xI_y - {I_{xy}}^2} \\
\dot{\omega}_y = \dfrac{I_{xy}\left(I_y - I_z - I_x\right)}{I_xI_y - {I_{xy}}^2}\omega_y\omega_z + \dfrac{I_xI_z - {I_x}^2 - {I_{xy}}^2}{I_xI_y - {I_{xy}}^2}\omega_x\omega_z + \dfrac{I_x\sum M_y}{I_xI_y - {I_{xy}}^2} \\
\quad b_{21}\omega_y\omega_z + b_{22}\omega_x\omega_z + \dfrac{I_xM_y}{I_xI_y - {I_{xy}}^2} \\
\dot{\omega}_z = \dfrac{I_x - I_y}{I_z}\omega_x\omega_y + \dfrac{I_{xy}}{I_z}\left({\omega_x}^2 - {\omega_y}^2\right) + \dfrac{\sum M_z}{I_z}\dfrac{I_x - I_y}{I_z}\omega_x\omega_y \\
\quad + \dfrac{I_{xy}}{I_z}\left({\omega_x}^2 - {\omega_y}^2\right) + \dfrac{M_z}{I_z}
\end{cases}
\tag{3.4}
$$

3.2.1.3 Linearization Modeling

In most cases, the UAV maintains steady straight level flight, and (3.4) can be modeled as linear time invariant state-space perturbation models, with the nominal trajectory being steady-level trimmed flight. The UAV's linear equations are as follows:

$$
\begin{cases}
\Delta\dot{\alpha} = \Delta\omega_z - (Y^\alpha\Delta\alpha + \Delta\delta_Z) \\
\Delta\dot{\omega}_z = {M_z}^\alpha\Delta\alpha + {M_z}^{\omega_z}\Delta\omega_z + {M_z}^{\delta_z}\Delta\delta_z \\
\Delta\dot{\beta} = \Delta\omega_y + Z^{\delta_x}\Delta\delta_x + Z^{\delta_y}\Delta\delta_y \\
\Delta\dot{\omega}_x = {M_x}^\beta\Delta\beta + {M_x}^{\omega_x}\Delta\omega_x + {M_x}^{\omega_y}\Delta\omega_y + {M_x}^{\delta_x}\Delta\delta_x + {M_x}^{\delta_y}\Delta\delta_y \\
\Delta\dot{\omega}_y = {M_y}^\beta\Delta\beta + {M_y}^{\omega_x}\Delta\omega_x + {M_y}^{\omega_y}\Delta\omega_y + {M_y}^{\delta_x}\Delta\delta_x + {M_y}^{\delta_y}\Delta\delta_y
\end{cases}
\tag{3.5}
$$

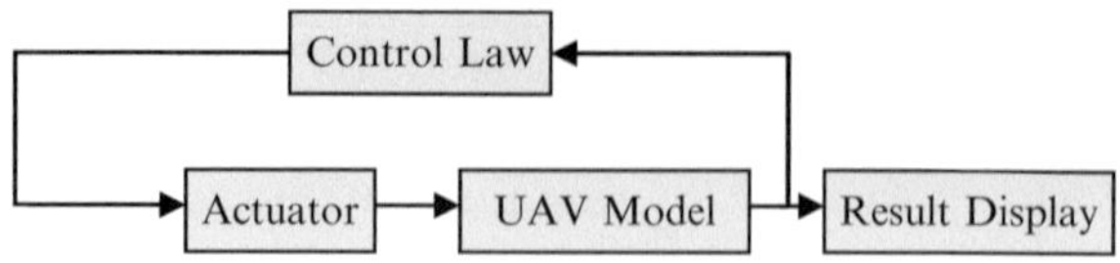

Fig. 3.1 UAV system model (Reprinted from Duan et al. (2013a), with kind permission from Springer Science+Business Media)

We have the state equations.

$$\begin{aligned} \dot{x} &= Ax + Bu \\ y &= Cx \end{aligned} \tag{3.6}$$

where the state variable $x = (\alpha,\omega_z,\beta,\omega_x,\omega_y)^T$ and the control surface $u = (\delta_z,\delta_x,\delta_y)^T$ and A, B, C can be denoted by

$$A = \begin{bmatrix} -Y^{\alpha} & 1 & 0 & 0 & 0 \\ M_z^{\alpha} & M_z^{\omega_z} & 0 & 0 & 0 \\ 0 & 0 & Z^{\beta} & 0 & 1 \\ 0 & 0 & M_x^{\beta} & M_x^{\omega_x} & M_x^{\omega_y} \\ 0 & 0 & M_y^{\beta} & M_y^{\omega_x} & M_y^{\omega_y} \end{bmatrix}, \quad B = \begin{bmatrix} -Y\delta_z & 0 & 0 \\ M_z^{\delta_z} & 0 & 0 \\ 0 & Z^{\delta_x} & Z^{\delta_y} \\ 0 & M_x^{\delta_x} & M_x^{\delta_y} \\ 0 & M_y^{\delta_x} & M_y^{\delta_y} \end{bmatrix},$$

$$C = \begin{bmatrix} 1 & 0 & 0 & 0 & 0 \\ 0 & 1 & 0 & 0 & 0 \\ 0 & 0 & 1 & 0 & 0 \\ 0 & 0 & 0 & 1 & 0 \\ 0 & 0 & 0 & 0 & 1 \end{bmatrix}$$

3.2.2 *Predator–Prey PSO for Parameter Identification*

3.2.2.1 UAV Flight Control System Design in Matlab Environment

Based on control augmentation system of UAVs, the aircraft linear equations are generally obtained by a series of equilibrium points. The flight envelope of this UAV must satisfy $0 \le H \le 18km$ and $0.6 \le M \le 2.2$ (Zhang 2004). Figure 3.1 shows the schematic diagram of UAV system.

As is obvious in Fig. 3.1, the UAV system is comprised of four subsystems: control law, actuator, mathematical model, and simulation result display.

Matrix K for the function of control law can be obtained from Matlab main program. Actuator module is responsible for control adjusting, which can also limit the amplitudes of control surface deflection angle. The actuator module can be shown with Fig. 3.2.

The actuator module requires that deflection angle of the elevator must be less than that of aileron and rudder. In order to prevent the deflection angles of the control

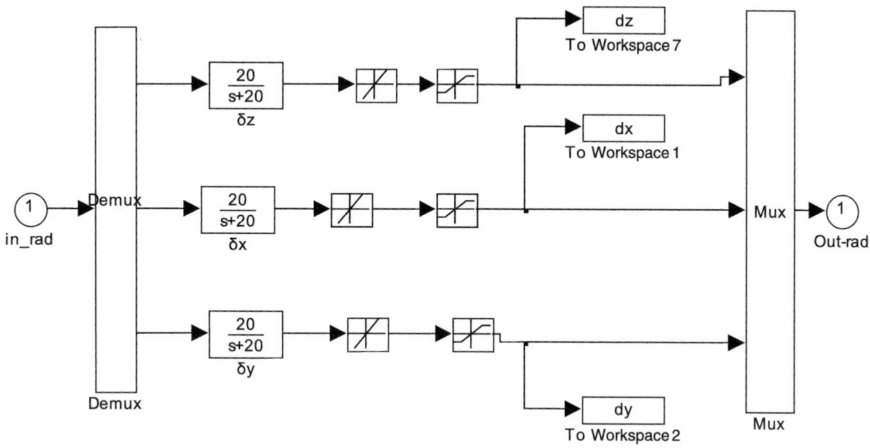

Fig. 3.2 Actuator module of UAVs in Matlab environment (Reprinted from Duan et al. (2013a), with kind permission from Springer Science+Business Media)

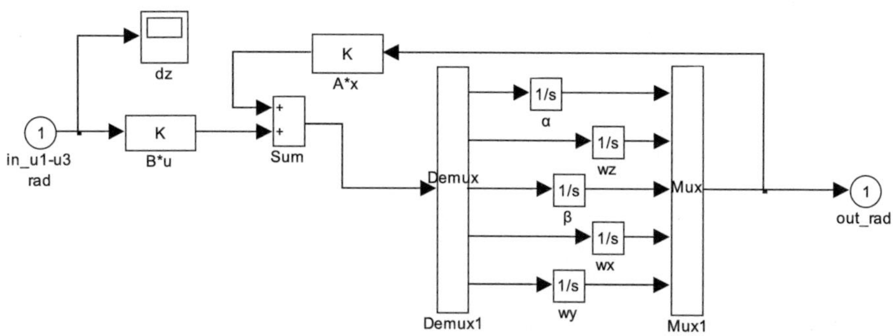

Fig. 3.3 UAV model module in Matlab environment (Reprinted from Duan et al. (2013a), with kind permission from Springer Science+Business Media)

surface from deflecting too fast, the angle authority of the elevator is set at $-18°$ to $12°$, aileron and rudder at $-25°$ to $25°$, and the angle rate authority of elevator and aileron at $50°/s$, and rudder is at $80°/s$. UAV model module is displayed in Fig. 3.3. where matrix A and B are both obtained from the workspace of Matlab. Figure 3.4 shows the simulation result display module.

The control law $u = -Kx$ is used. K denotes the state-feedback gain, and it can be illustrated with the following matrix:

$$K = \begin{bmatrix} k_1 & k_2 & 0 & 0 & 0 \\ 0 & 0 & k_3 & k_4 & k_5 \\ 0 & 0 & k_6 & k_7 & k_8 \end{bmatrix}$$

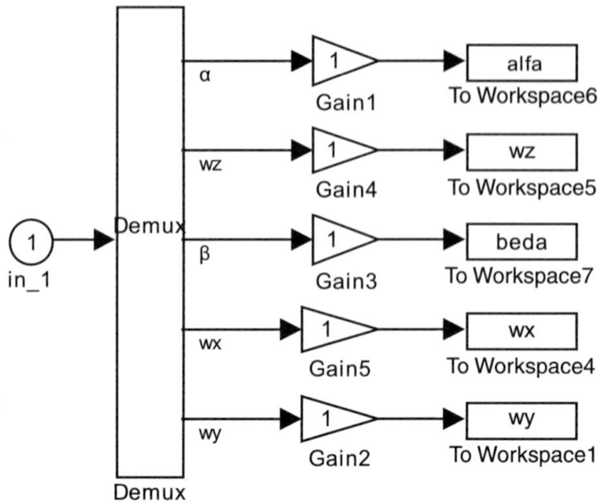

Fig. 3.4 Simulation result display module in Matlab environment (Reprinted from Duan et al. (2013a), with kind permission from Springer Science+Business Media)

3.2.2.2 Predator–Prey PSO for Identifying Controller Parameters

In the Gbest model of PSO, each particle has its current position and velocity in a space of solution. The best solution found so far are Pbest and Gbest. Each particle aims to get a global optimal solution by current velocity, Pbest, and Gbest. Gbest model can be expressed as (Shi and Eberhart 1998a)

$$v_{ij}(k+1) = \omega v_{ij}(k) + c_1 r_1 \left[p_i(k) - x_{ij}(k)\right] + c_2 r_2 \left[g_i(k) - x_{ij}(k)\right] \quad (3.7)$$

$$x_{ij}(k+1) = x_{ij}(k) + v_{ij}(k+1) \quad (3.8)$$

where $v_i(k)$ and $x_i(k)$ respectively denote the velocity and position of the i th particle at step k, j is the dimension of particle i, c_1 and c_2 are weight coefficients, r_1 and r_2 are random numbers between 0 and 1, p_i is the best position of the i th particle, and g_i is the best position which particles have ever found.

Generally, the basic PSO algorithm mentioned above is easily falling to local optimal solutions. In this case, the concept of predator–prey behavior is proposed to improve the basic PSO. Predator–prey PSO is a method which takes a cue from the behavior of schools of sardines and pods of killer whales (Higashitani et al. 2006; Wang and Duan 2013). In this model, particles are divided into two categories, *predator* and *prey*. Predators show the behavior of chasing the center of preys' swarm; they look like chasing preys (Duan et al. 2011), and preys escape from predators in multidimensional solution space. After taking a trade-off between predation risk and their energy, escaping particles would take different escaping behaviors. This helps the particles avoid the local optimal solutions and find the global optimal solution.

The velocities of the predator and the prey in the improved PSO can be defined by (Higashitani et al. 2006)

$$v_{dij}(k+1) = \omega_d v_{dij}(k) + c_1 r_1 \left[p_{dij}(k) - x_{dij}(k)\right] + c_2 r_2 \left[g_{dj}(k) - x_{dij}(k)\right] + c_3 r_3 \left[g_j(k) - x_{dij}(k)\right] \tag{3.9}$$

$$\begin{aligned} v_{rij}(k+1) = {} & \omega_r v_{rij}(k) + c_4 r_4 \left[p_{rij}(k) - x_{rij}(k)\right] \\ & + c_5 r_5 \left[g_{rj}(k) - x_{rij}(k)\right] + c_6 r_6 \left[g_j(k) - x_{rij}(k)\right] \\ & - Pasign\left[x_{dIj}(k) - x_{rij}(k)\right] \exp\left[-b| x_{dij}(k) - x_{rij}(k) |\right] \end{aligned} \tag{3.10}$$

where d and r denote predator and prey, respectively, p_{di} is the best position of predators, g_d is the best position which predators have ever found, p_{ri} is the best position of preys, g_r is the best position which preys have ever found, g is the best position which all the particles have ever found, and ω_d and ω_r can be defined as

$$\omega_d = 0.2 \exp\left(-10 \frac{iteration}{iteration_{\max}}\right) + 0.4 \tag{3.11}$$

$$\omega_r = \omega_{\max} - \frac{\omega_{\max} - \omega_{\min}}{iteration_{\max}} iteration \tag{3.12}$$

PSO can be also improved by a modification of the inertia weight ω_r in (3.12). The inertia weight, whose value is between 0 and 1[Adaptive Particle Swarm Optimization], can be used to balance the local and global search during the optimization process. If the inertia weight is big, it is possible to enhance global search. Otherwise, smaller inertia weight will enhance the local search. In (3.12) $iteration_{\max}$ is maximum iteration, $\omega_{\max}$ and $\omega_{\min}$ are, respectively, maximum and minimum of ω_r. In our experiment, $\omega_{\max}$ and $\omega_{\min}$ are 0.9 and 0.2, respectively. And the definition of I is given by

$$I = \left\{ k \,\middle|\, \min_k \left(|x_{dk} - x_{ri}|\right) \right\} \tag{3.13}$$

Then I denotes the number of the ith prey's nearest predator. In (3.10), P is used to decide if the prey escapes or not ($P = 0$ or $P = 1$), and a, b are the parameters which decide the difficulty of the preys escaping from the predators. The closer the prey and the predator, the harder the prey escapes from the predator. a, b are denoted by

$$a = x_{span}, \qquad b = \frac{100}{x_{span}} \tag{3.14}$$

where x_{span} is the span of the variable.

The parameter identification of the conventional flight controller can be treated as the typical continual spatial optimization problem. PSO is a novel way for solving the problem. PSO, which is a bio-inspired computation algorithm, can be applied to flight system control to reduce the workload of conventional designer. The bounds of the control gain parameters are set, and PSO searches for the corresponding space automatically to find the optimal parameters. The process in conventional design is conducted manually; now it can be done automatically. Bio-inspired computation can be applied to promote the automation of conventional controller design.

Using the proposed predator–prey PSO algorithm to obtain the optimal parameter combination for the UAV flight control system here, the fitness function is given by

$$J = \frac{1}{2}\int \left(x'Qx + u'Ru \right) dt \tag{3.15}$$

where x and u are, respectively, the state vector and the control vector. Q and R are diagonal positive matrix. Here the weighting matrices are chosen as Q =diag(50,10,20,30,30) and R =diag(100,100,100). The smaller J, the better the particle.

The position vector of the predator and the prey is defined by

$$\begin{aligned} x_d &= \left(k_1\ k_2\ k_3\ k_4\ k_5\ k_6\ k_7\ k_8 \right) \\ x_r &= \left(k_1\ k_2\ k_3\ k_4\ k_5\ k_6\ k_7\ k_8 \right) \end{aligned}$$

where x_d and x_r have the constraint of ± 10, which is set according to exact experience.

The process of proposed predator–prey PSO algorithm for solving UAV controller parameter identification can be described with Fig. 3.5 (Duan and Sun 2013).

The above mentioned flow chart of the predator–prey PSO algorithm process can also be illustrated with Fig. 3.6.

The complexity of predator–prey PSO algorithm can be computed, and Table 3.1 shows a comparison of the complexity analysis between basic PSO and predator–prey PSO.

In Table 3.1, $m = m_d + m_r$, and n is the dimension of particle's position. The total complexity of basic PSO and predator–prey PSO can be expressed as

$$T(n)_{basic\,PSO} = O\left(14N_{\max}\, mn\right) \tag{3.16}$$

$$T(n)_{improved\,PSO} = O\left(N_{\max}\, m_r\, n^2\right) \tag{3.17}$$

3.2.3 Experiments

In order to investigate the feasibility and effectiveness of the proposed predator–prey PSO approach for identification of UAV controller parameters, a series of experiments are conducted under some constrained conditions.

PROCEDURE Optimization of UAV controller parameter based on the predator-prey PSO
BEGIN
Step 1: Initialization
Set the maximum iteration number $N_{\max}$, the number of the predators m_d and the number of the preys m_r. Initialize randomly the positions and velocities of the predators x_d and v_d respectively, both of which have the same dimensions m_d by 8. So are x_r and v_r. And run simulation module of Simulink to compute fitness of each particle. After that, find out the minimum fitness value of the predators as $pbest_d(0)$, that of the preys as $pbest_r(0)$, and that of all the particles as $gbest(0)$.
Step 2: (1) Let *N*=1;
(2) Calculate the fitness value of all the particles in iteration *k* through running simulation module, and then find out the minimum fitness value of the predators as $pbest_d(N)$, that of the preys as $pbest_r(N)$, and that of all the particles as $gbest(N)$.
Step 3: (1) Let $N \leftarrow N+1$;
(2)Update all the positions and the velocities according to (3.8)-(3.10). Then repeat (2) in **Step 2**.
Step 4: $N \geq N_{\max}$?
(1)Yes: stop and output results;
(2)No: go to **Step 3**.
End

Fig. 3.5 The pseudocode of predator–prey PSO algorithm for UAV flight controller (Reprinted from Duan et al. (2013a), with kind permission from Springer Science+Business Media)

The predator–prey PSO algorithm is implemented in a Matlab 2008a programming environment with an Intel Core 2 PC running Windows XP SP2. No commercial PSO tools are used in these experiments.

Case 1: In this case, the parameter values of predator–prey PSO are set to $\alpha = 10°, \beta = 10°$, $N_{\max} = 100$, $t = 20s$, mach=0.8, $H = 8000m$, $m_d = 10$, $m_r = 20$ where t is the simulating time of the controller. Comparison of the experiment results between the improved PSO and the LQR method which could directly compute the state-feedback gain K is illustrated from Fig. 3.7 (a–e). Figure 3.7f shows the evolution curve of the proposed PSO.
The final optimal result is K = [−0.1471,−0.4012, 1.2393, −0.4847, 0.1373, 1.1180, −0.3967, 4.9621], the minimum fitness value $J_{\min}$ = 223.0907, and the best iteration *bestN* = 100. As indicated in Fig. 3.7, as expected, the proposed algorithm can guarantee that the achieved states are almost the same as the ones obtained by LQR. And the realization of the proposed method is simpler

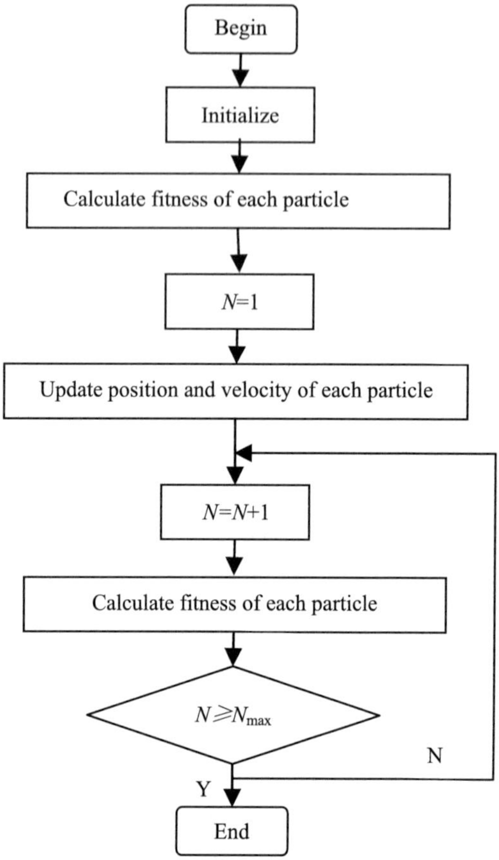

Fig. 3.6 Flow chart of the predator–prey PSO (Reprinted from Duan et al. (2013a), with kind permission from Springer Science+Business Media)

Table 3.1 Comparison of complexity analysis between basic PSO and predator–prey PSO

		Complexity	
Step	Operation	Basic PSO	Predator–prey PSO
1	Initialize	$O(2mn)$	$O[2(m_d+m_r)n]$
2	Calculate the fitness value of all the particles	$O(7m)$	$O(7(m_d+m_r))$
3	Update all the positions and the velocities	$O(14mn)$	$O(14m_dn+(31+n)m_rn)$
4	Stop and output result	$O(1)$	$O(1)$

than LQR. Figure 3.7f also demonstrates that the algorithm can converge to the optimal solution quickly.

Case 2: In this case, the parameter values are $\alpha = 10°, \beta = 10°$, $N_{\max} = 100$, $t = 20s$, $mach = 1.2$, $H = 15000m$, $m_d = 10$, $m_r = 20$ where t is the simulating time of the controller. Comparisons of the experiment results between the improved PSO and the LQR method are illustrated from Fig. 3.8a–e. Figure 3.8f shows the evolution curve of the proposed PSO.

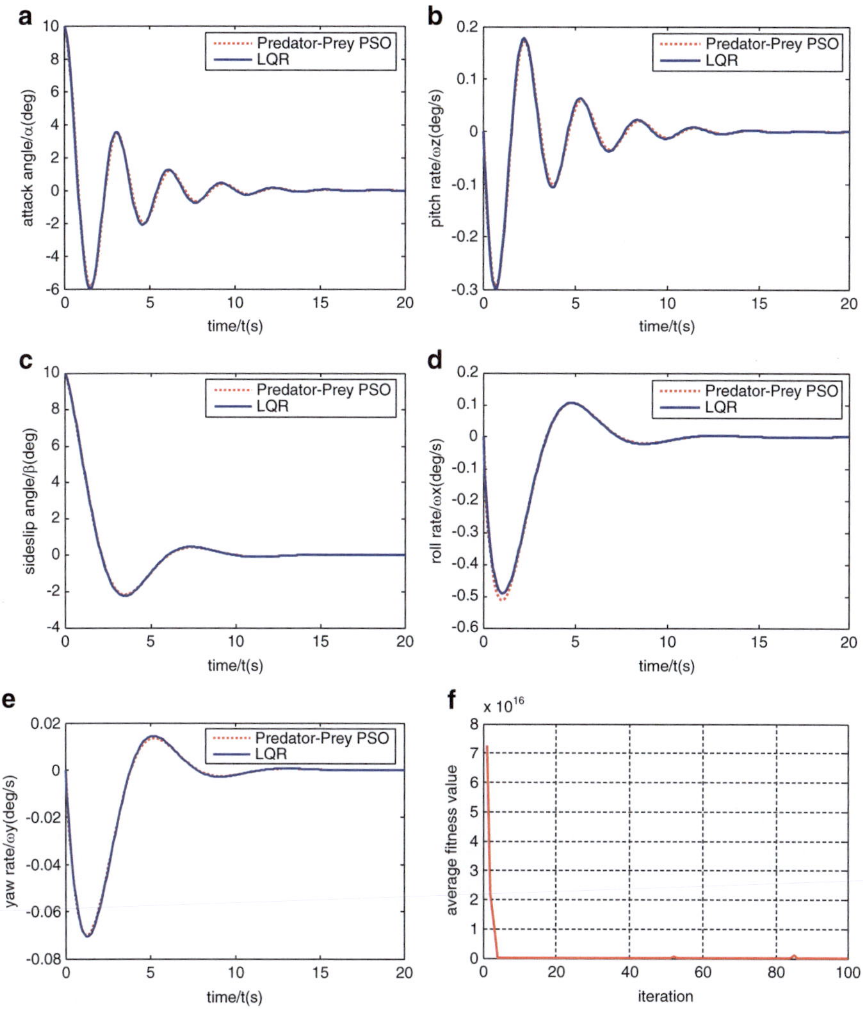

Fig. 3.7 Results of identifying controller parameters for UAVs based on predator–prey PSO in Case 1 (**a**) Comparison of attack angle responses. (**b**) Comparison of pitch rate responses. (**c**) Comparison of sideslip angle responses. (**d**) Comparison of roll rate responses. (**e**) Comparison of yaw rate responses. (**f**) Evolution curve of predator–prey PSO (Reprinted from Duan et al. (2013a), with kind permission from Springer Science+Business Media)

The final optimal results are $K = [0.0416, 0.8298, -2.7208, 1.4825, -1.0683, 1.9002, -0.2273, 2.6742]$, the minimum fitness value $J_{\min} = 68.3361$, and the best iteration $bestN = 99$. Although the initial conditions are much different from those of experiment 1, the improved algorithm can find the optimal solution.

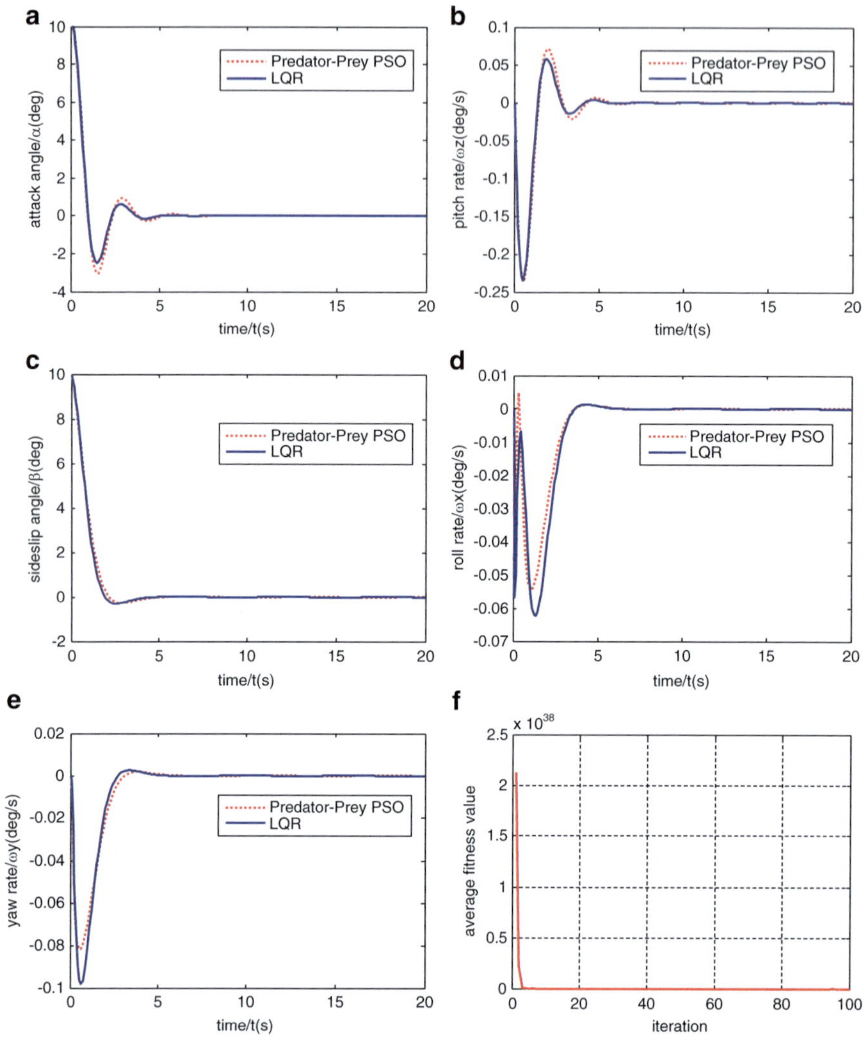

Fig. 3.8 Results of identifying controller parameters for UAVs based on predator–prey PSO in Case 2 (**a**) Comparison of attack angle responses. (**b**) Comparison of pitch rate responses. (**c**) Comparison of sideslip angle responses. (**d**) Comparison of roll rate responses. (**e**) Comparison of yaw rate responses. (**f**) Evolution curve of predator–prey PSO (Reprinted from Duan et al. (2013a), with kind permission from Springer Science+Business Media)

From these two experimental results, it is obvious that the proposed predator–prey PSO approach could make the UAV controlling system obtain better performance than the conventional LQR method. The state-feedback gain obtained according to the predator–prey PSO can guarantee fast response, precise control, and strong robustness.

Based on the UAV model and the proposed predator–prey PSO algorithm, we developed a software platform of UAV controller design. The graphical user interfaces (GUI) of this platform are shown in Fig. 3.9.

3.3 PSO Optimized Controller for Unmanned Rotorcraft Pendulum

MAV offer several advantages as autonomous UAVs (Pflimlin et al. 2010; Sun and Duan 2013; Duan and Sun 2013). They can be very small with a compact layout. Many are capable of high-speed flight in addition to the normal hover and vertical take-off and landing capabilities. These features make them well suited for a variety of missions, especially in urban environments. A recent announcement by the US Government of plans to greatly increase the number of unmanned aircraft on station in Iraq and Afghanistan helps drive the surge in interest for unmanned vehicles. US Defense Secretary Robert Gates (Bloss 2009) commented that unmanned aircraft is essential to provide real-time video of insurgent activity and argued that the need is growing at 300 % per year. Nowadays, many companies are functioning towards research and development of MAVs, and the representative achievements are Cypher series by Sikorsky and MAVs by Honeywell etc.

Hover and stare is of the paramount importance to the performance of MAVs, as they are designed to perform surveillance and reconnaissance missions. However, pendulum-like oscillation (see Fig. 3.10) triggered by external disturbances and other uncertain factors will badly impair its performance, resulting in blurred images or even MAVs' overturning. As a result, control techniques of MAVs are becoming more and more important for their wide applications in civil and military fields, with special regard to the hover and stare state for better performances of surveillance and reconnaissance missions. This section focuses on a particular kind of MAV, which is driven by a rotor and a ducted fan, and proposes a combination control law design approach to stabilize the hover and stare pendulum-like oscillation based on LQR, in which an improved PSO algorithm is utilized for parameter optimization of matrix Q and R in the linear-quadratic regulator. In this way, the MAV's dynamic properties can be ameliorated efficiently while executing surveillance missions that requires perfect stability and rapid responses.

3.3.1 *Mathematical Model of Pendulum Oscillation for MAVs*

The MAV in this section adopts the axial symmetrical layout. Owing to the fact that the suspension center moves freely with rotor wings in the plane, pendulum-like oscillation is nonlinear, strongly coupled, and of high order (Pflimlin et al. 2010). In this section, a mathematical model representing the pendulum-like oscillation

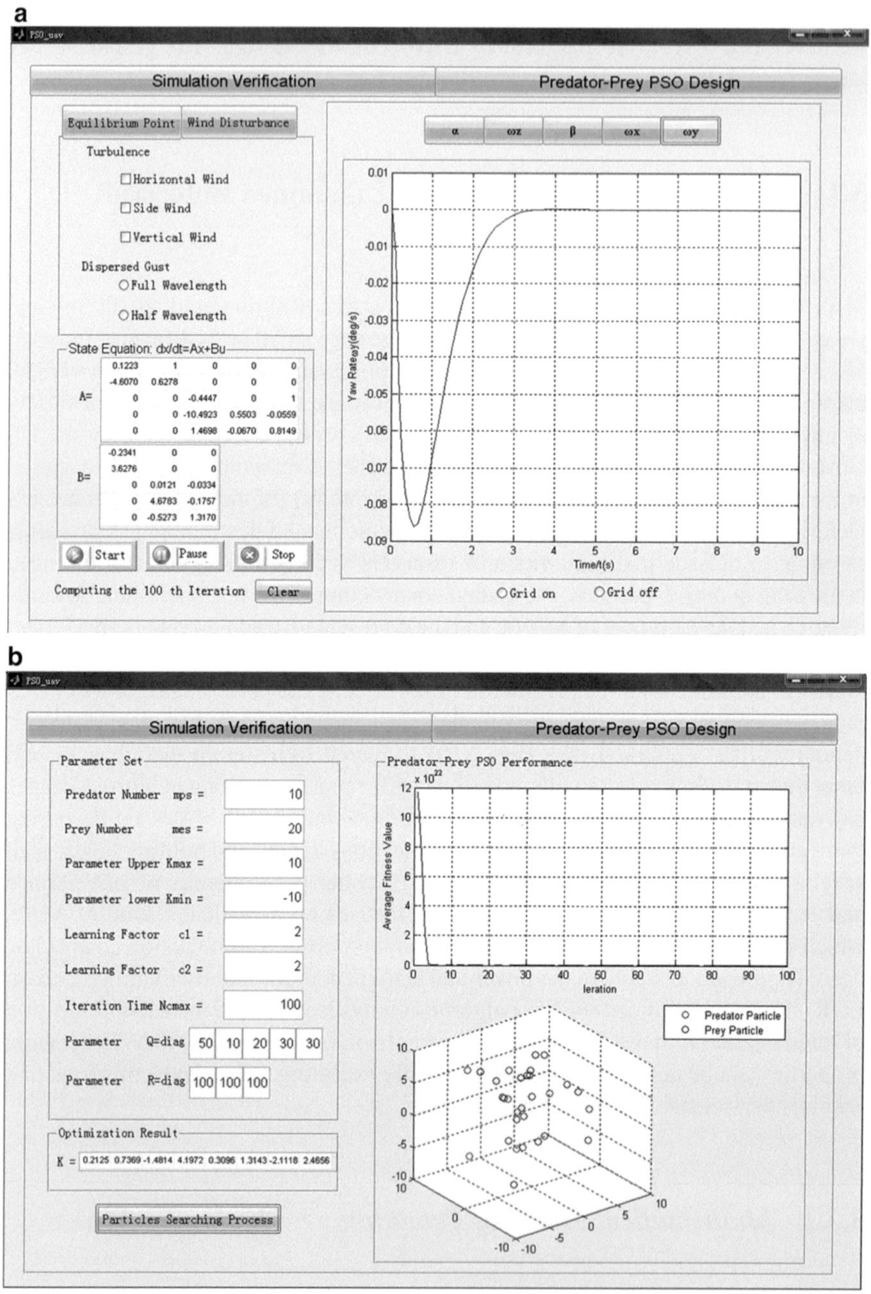

Fig. 3.9 UAV controller design software platform based on predator–prey PSO. (**a**) Yaw rate response interface. (**b**) Predator–prey PSO convergence process interface (Reprinted from Duan et al. (2013a), with kind permission from Springer Science+Business Media)

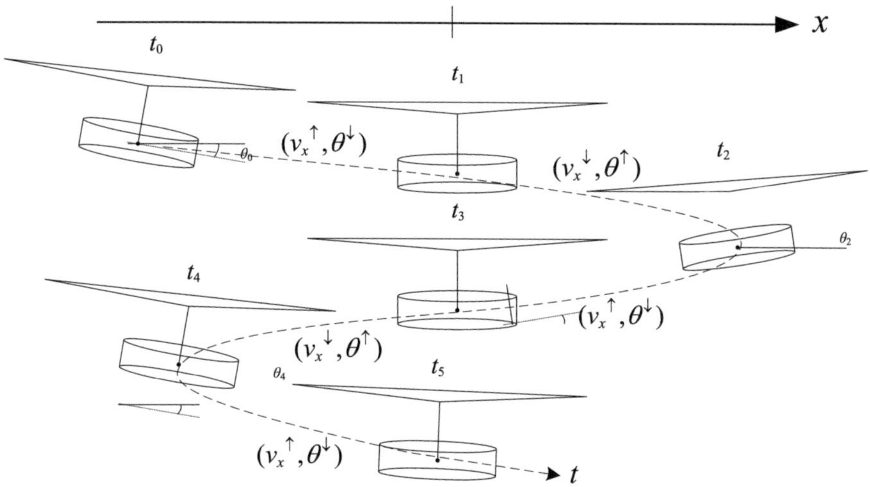

Fig. 3.10 Pendulum-like oscillation in actual flight (x-axis) (Reprinted from Duan and Sun (2013), with kind permission from Springer Science+Business Media)

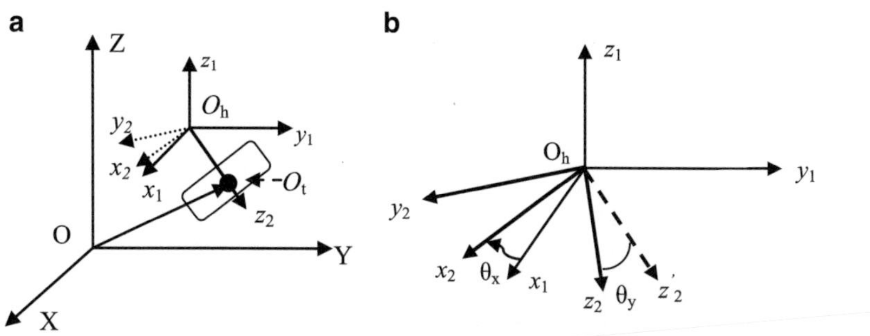

Fig. 3.11 Coordinates of the pendulum model and decomposition of oscillation angle. (**a**) Coordinates of the pendulum model. (**b**) Decomposition of oscillation angle (Reprinted from Duan and Sun (2013), with kind permission from Springer Science+Business Media)

in the hover and stare state is obtained by using the Lagrangian method, and the corresponding linearized model is obtained in the neighborhood of the hover and stare equilibrium.

3.3.1.1 Nonlinear Mathematical Model of MAV Pendulum-Like Oscillation

The pendulum can be abstracted as a system consisting of the suspension point O_h, the pendulum rod O_hO_t, and the pendulum mass O_t (see Fig. 3.11).

In Fig. 3.11, *OXYZ* is the ground-fixed coordinate system, $O_h x_1 y_1 z_1$ represents a mobile ground coordinate system, with the point O_h as the origin and parallel to *OXYZ*, while $O_h x_2 y_2 z_2$ is defined as the pendulum-body coordinate frame, which originates from O_h, and axis z_2 points downward along the pendulum rod $O_h O_t$. Decompose the two-dimensional pendulum-like oscillation to one in the direction of axis *X* and *Y*, and the transition matrix *R* from $O_h x_2 y_2 z_2$ to $O_h x_1 y_1 z_1$ is obtained as follows:

$$R = \begin{bmatrix} \cos\theta_y & \sin\theta_x \sin\theta_y & \cos\theta_x \sin\theta_y \\ 0 & \cos\theta_x & -\sin\theta_x \\ -\sin\theta_y & \sin\theta_x \cos\theta_y & \cos\theta_x \cos\theta_y \end{bmatrix} \tag{3.18}$$

Suppose the position vector of the suspension point O_h in the coordinate frame of *OXYZ* and the pendulum mass O_h in the coordinate frame of $O_h x_2 y_2 z_2$ are presented with $r_h = [x,y,z]^T$, $r'_t = [0,0,l]^T$, respectively. Then, the coordinate of point O_t in $O_h x_1 y_1 z_1$ can be calculated according to the following equation:

$$r_{ht} = R \cdot r'_t = \left[l\cos\theta_x \sin\theta_y, -l\sin\theta_x, l\cos\theta_x\cos\theta_y\right]^T \tag{3.19}$$

where *l* denotes the distance from O_h to O_t, i.e., the length of the pendulum.

Finally, we have the coordinate of O_t in coordinate frame *OXYZ* presented as in (3.20):

$$r_t = r_h + R \cdot r' \tag{3.20}$$

The Lagrange function of the pendulum system:

$$\begin{aligned} L &= T_0 - V = T_m + T'_M + T''_M - V \\ &= \frac{1}{2} m \dot{r}_h^T \dot{r}_h + \frac{1}{2} m_t V_t^T V + \frac{1}{2} J_{x2}\omega_{x2}^2 + \frac{1}{2} J_{y2}\omega_{y2}^2 \\ &\quad - \left(-m_t g l \cos\theta_x \cos\theta_y + (m + m_t)(z - z_0)g\right) \\ &= \frac{1}{2}(m + m_t)\left(\dot{x}^2 + \dot{y}^2 + \dot{z}^2\right) + \frac{1}{6} m_t l \left[\left(3 + \cos^2\theta_y\right) l\dot{\theta}_x^2 + \left(1 + 3\cos^2\theta_x\right) l\dot{\theta}_y^2\right] \\ &\quad + m_t l \Big(\dot{x}\dot{\theta}_y \cos\theta_x \cos\theta_y - \dot{x}\dot{\theta}_x \sin\theta_x \sin\theta_y - \dot{y}\dot{\theta}_x \cos\theta_x - \dot{z}\dot{\theta}_x \sin\theta_x \cos\theta_y \\ &\quad - \dot{z}\dot{\theta}_y \cos\theta_x \sin\theta_y\Big) + m_t g l \cos\theta_x \cos\theta_y - (m + m_t)(z - z_0) g \end{aligned} \tag{3.21}$$

where T_m, T'_M, T''_M represent, respectively, kinetic energy of the suspension center O_h, translation kinetic energy of the pendulum mass O_t, and rotational kinetic energy of the pendulum rod around the centroid of O_t and *V* is potential energy of the system, choosing the initial state of the pendulum-like oscillation as the zero-potential energy surface.

Table 3.2 Main parameters of the MAV system structure

Symbol	Value	Physical meaning
m	10 kg	Suspension center quality
m_t	20 kg	Pendulum quality
l	0.86 m	Pendulum length
g	9.8 m/s^2	Gravity acceleration
θ_x, θ_y	–	Pendulum angle around axis x, y
ϕ_x, ϕ_y	–	Pitch and roll angle
a_x, a_y	–	Suspension center acceleration
u_x, u_y	–	Control force on O_h

Therefore, from the Lagrange equation it can be deduced:

$$\begin{cases} \dfrac{d}{dt}\left(\dfrac{\partial L}{\partial \dot{\theta}_x}\right) - \dfrac{\partial L}{\partial \theta_x} = 0 \\ \dfrac{d}{dt}\left(\dfrac{\partial L}{\partial \dot{\theta}_y}\right) - \dfrac{\partial L}{\partial \theta_y} = 0 \end{cases} \tag{3.22}$$

Considering all the above equations (3.18, 3.19, 3.20, 3.21, and3.22), the nonlinear mathematical model of the MAV pendulum-like oscillation is finally obtained as follows:

$$\begin{cases} \ddot{\theta}_x = \Big(3\ddot{x}\sin\theta_x\sin\theta_y + 3\ddot{y}\cos\theta_x + 3\ddot{z}\sin\theta_x\cos\theta_y + 2l\dot{\theta}_x\dot{\theta}_y\cos\theta_x\sin\theta_y \\ \qquad - 3l\dot{\theta}_y^2\sin\theta_x\cos\theta_x - 3g\sin\theta_x\cos\theta_x\Big)/\left(\cos^2\theta_y + 3\right)l \\ \ddot{\theta}_y = \Big(-3\ddot{x}\cos\theta_x\cos\theta_y + 3\ddot{z}\cos\theta_x\sin\theta_y + 6l\dot{\theta}_x\dot{\theta}_y\sin\theta_x\cos\theta_x \\ \qquad - l\dot{\theta}_x^2\sin\theta_y\cos\theta_y - 3g\cos\theta_x\sin\theta_y\Big)/\left(\cos^2\theta_x + 3\right)l \end{cases} \tag{3.23}$$

Table 3.2 gives the main parameters of the MAV system structure in this section.

3.3.1.2 Linearization of Mathematical Model

Step 1: Change the pendulum angle around axis x, y (θ_x,θ_y) to the pitch and roll angle (ϕ_x, ϕ_y) according to $\phi_x = \theta_y$, $\phi_y = -\theta_x$, and then (3.23) can be described as

$$\begin{cases} \ddot{\phi}_y = \Big(3\ddot{x}\sin\phi_x\sin\phi_y - 3\ddot{y}\cos\phi_y + 3\ddot{z}\sin\phi_y\cos\phi_x + 2l\dot{\phi}_x\dot{\phi}_y\cos\phi_y\sin\phi_x \\ \qquad - 3l\dot{\phi}_x^2\sin\phi_y\cos\phi_y - 3g\sin\phi_y\cos\phi_y\Big)\Big/\left(\cos^2\phi_x + 3\right)l \\ \ddot{\phi}_x = \Big(-3\ddot{x}\cos\phi_y\cos\phi_x + 3\ddot{z}\cos\phi_y\sin\phi_x + 6l\dot{\phi}_x\dot{\phi}_y\sin\phi_y\cos\phi_y \\ \qquad - l\dot{\phi}_y^2\sin\phi_x\cos\phi_x - 3g\cos\phi_y\sin\phi_x\Big)\Big/\left(\cos^2\phi_y + 3\right)l \end{cases} \tag{3.24}$$

Step 2: Choose the state variables $X_e = \left[x, \phi_x, \dot{x}, \dot{\phi}_x, y, \phi_y, \dot{y}, \dot{\phi}_y, z, \dot{z}, \ddot{x}, \ddot{y}, \ddot{z}\right]^T$, the input of the system $u = [\ddot{x}, \ddot{y}]^T$. Expand (3.24) into Taylor series in the vicinity of the equilibrium point, and the linear model of the pendulum-like oscillation is finally obtained as follows:

$$\begin{bmatrix} \dot{X}_x \\ \dot{X}_y \end{bmatrix} = \begin{bmatrix} A_x & 0 \\ 0 & A_y \end{bmatrix} \begin{bmatrix} X_x \\ X_y \end{bmatrix} + \begin{bmatrix} B_x & 0 \\ 0 & B_y \end{bmatrix} \begin{bmatrix} u_x \\ u_y \end{bmatrix} \tag{3.25}$$

where the state vector $X_x = \left[x, \phi_x, \dot{x}, \dot{\phi}_x\right]^T$, $X_y = \left[y, \phi_y, \dot{y}, \dot{\phi}_y\right]^T$, and
$$A_x = \begin{bmatrix} 0 & 0 & 1 & 0 \\ 0 & 0 & 0 & 1 \\ 0 & 0 & 0 & 0 \\ 0 & -\frac{3g}{4l} & 0 & 0 \end{bmatrix}, A_y = \begin{bmatrix} 0 & 0 & 1 & 0 \\ 0 & 0 & 0 & 1 \\ 0 & 0 & 0 & 0 \\ 0 & -\frac{3g}{4l} & 0 & 0 \end{bmatrix}, B_x = \begin{bmatrix} 0 \\ 0 \\ 1 \\ -\frac{3}{4l} \end{bmatrix}, B_y = \begin{bmatrix} 0 \\ 0 \\ 1 \\ -\frac{3}{4l} \end{bmatrix}.$$

3.3.2 *Oscillation Controller Design Based on LQR and PSO*

3.3.2.1 Characteristics of Pendulum-Like Oscillation

From the linear model, the MAV pendulum-like oscillation in the direction of *X* and *Y* is no longer coupled with each other. The pendulum system can be reduced to two four-order subsystems, which are independent of one another. Due to similarity between matrix A_x and A_y, we choose either of the two subsystems to analyze characteristics of the MAV pendulum-like oscillation.

Considering pendulum motion only in the direction of axis *X*, the state-space equation of the linearized *X* subsystem is presented as follows:

$$\begin{cases} \dot{X} = AX + BU \\ Y = CX \end{cases} \tag{3.26}$$

where $X = [x_1, x_2, x_3, x_4]^T = \left[x, \phi_x, \dot{x}, \dot{\phi}_x\right]^T$, $A = \begin{bmatrix} 0 & 0 & 1 & 0 \\ 0 & 0 & 0 & 1 \\ 0 & 0 & -0.6500 & 0.6500 \\ 0 & -8.547 & 0.5669 & -0.3924 \end{bmatrix}$,

$$B = \begin{bmatrix} 0 \\ 0 \\ 1 \\ -0.872 \end{bmatrix}, C = \begin{bmatrix} 1 & 0 & 0 & 0 \\ 0 & 1 & 0 & 0 \\ 0 & 0 & 1 & 0 \\ 0 & 0 & 0 & 1 \end{bmatrix}.$$

In the actual flight, external disturbances such as crosswinds jeopardize stability of hover and stare state and lead to a considerate degradation of the surveillance performance and result in even false intelligence information. Assume there is a

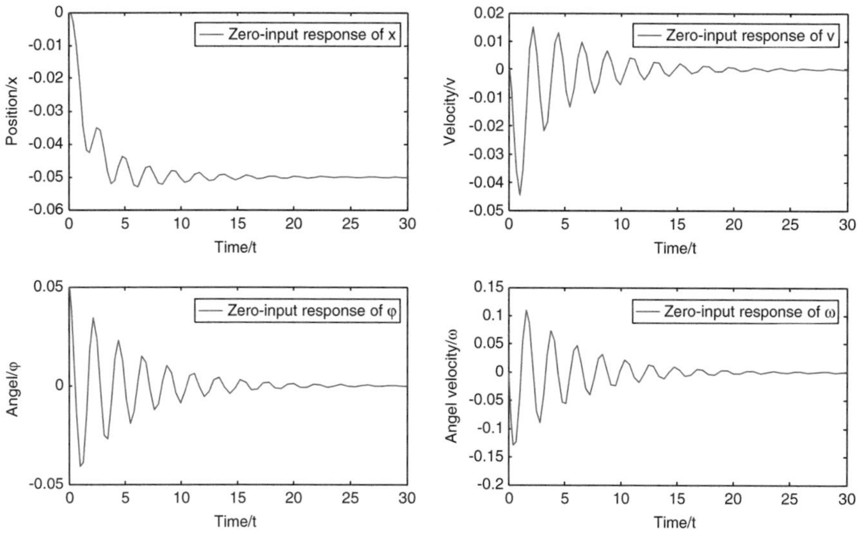

Fig. 3.12 Responses of the MAV pendulum-like oscillation system (Reprinted from Duan and Sun (2013), with kind permission from Springer Science+Business Media)

sudden gust, the pendulum system gets an initial state $\phi_x(0) = 0.05$rad. The inherent characteristics of the MAV pendulum-like oscillation system and the responses are given in Fig. 3.12.

As shown in Fig. 3.12, due to external disturbances of crosswinds, which result in an initial state $\phi_x(0) = 0.05$rad, the pendulum-like oscillation occurs in MAV that is required to be stable enough to carry out hover and stare missions. However, the actual fact shown by the analysis results states clearly that the position of the suspension, denoted by x in the Fig. 3.12, does not remain in the original place but moves to another site in 20 s, which may bring about deviation from the ideal monitoring precision. Furthermore, the MAV's body swings back and forth just in the way as a pendulum does and eventually converges to the equilibrium point after 25 s.

3.3.2.2 Control Law Design Based on LQR

As mentioned above, the hover and stare state is inherently unstable, and external disturbances would give rise to pendulum-like oscillation depicted as in Fig. 3.12. A novel type of optimal control law based on LQR and PSO is designed in this section, which is used to eliminate pendulum-like oscillation with the shortest duration. In this controller, the position of the suspension point x and the pendulum angle ϕ_x are the two main state variables to be controlled to the desired value. In order to ensure the robustness and optimality of the close loop, the LQR design technique is applied due to the fact that it has a very nice robustness property and has been widely used in many applications. The key issue of LQR controller design is how to select an appropriate control vector $u(t)$ so that the given quadratic performance index (see (3.27)) obtains the minimum value. It is proved that the performance index (10) can reach its minimum by the designed linear control law in (3.28).

$$J = \int_0^{\infty} \left(X^T Q X + u^T R u \right) dt \tag{3.27}$$

$$u(t) = -KX(t) = R^{-1} B^T P X(t) \tag{3.28}$$

and the optimal matrix P can be calculated from the following Algebraic Riccati Equation:

$$A^T P + PA - PBR^{-1} B^T P + Q = 0 \tag{3.29}$$

Taking all factors into account, including the performance of the control system and restrictions on the total energy consumed, matrix Q and R can be defined in the form of $Q = \mathrm{diag}(q_{11}, q_{22}, 0, 0)$, $R = 1$, in which parameters q_{11} and q_{22} are crucial for a splendid dynamic response. As a result, the proposed algorithm takes advantage of PSO's high operating efficiency, fast convergence speed, and model simplicity, which is used to search the appropriate parameter setting of the LQR control law design approach.

Let the input u be the control force acting on the suspension center, and the corresponding pendulum-holding back control law from the LQR is described as follows:

$$u = -KX = -\left(k_1 x_1 + k_2 x_2 + k_3 x_3 + k_4 x_4 \right) \tag{3.30}$$

where K denotes the feedback parameters obtained for the LQR and X denotes the states of the system, i.e., the position of the suspension center, the angle of the pendulum, the speed of the suspension center, and the angular velocity of the pendulum, respectively.

The PSO-based LQR controller for prohibiting MAV pendulum-like oscillation of the hover and stare state in presence of external disturbances can be illustrated with Fig. 3.13.

3.3.2.3 Key Settings for PSO

1. *Fitness Function f*

   The fitness function f chosen in LQR controller can be described as follows:

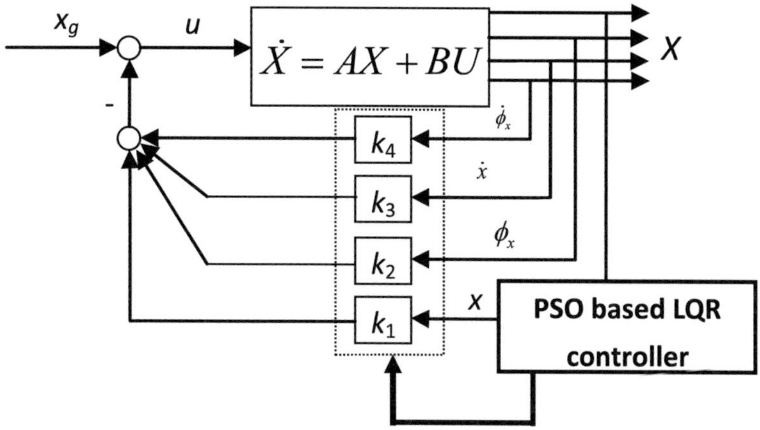

Fig. 3.13 Structure of pendulum-like oscillation control system (Reprinted from Duan and Sun (2013), with kind permission from Springer Science+Business Media)

$$f = \frac{1}{\int_0^{\infty} \left(X^T Q X + u^T R u \right) dt} \tag{3.31}$$

where $u = -(k_1x_1 + k_2x_2 + k_3x_3 + k_4x_4)$, $K = [k_1,k_2,k_3,k_4] = lqr(A,B,Q,R)$, $Q = \text{diag}(q_{11},q_{22},0,0)$.

2. *Inertia Weight* ω

   The inertia weight ω controls the exploration properties of the algorithm, with larger values facilitating a more global behavior and smaller values facilitating a more local behavior; thus results of the algorithm depend largely on ω selection (Duan and Liu 2010; Liu et al. 2012; Duan and Sun 2013). Generally, there are two ways to choose the inertia weight, namely, constant ω and time-variant ω.Shi suggested using $0.8 < \omega < 1.4$, which starts with bigger ω values (a more global search behavior) that is dynamically reduced (a more local search behavior) during the optimization. In this section, ω can be declined linearly from 1.4 to 0.8 in the former 75 % phylogenetic scale and keep constant in the rest time.

3. *Population Size m*

   According to the scale of the exact optimization problem, m is set between 40 and 150. Here we choose $m = 100$.

4. *Acceleration Constants* c_1 *and* c_2

   Shi and Eberhart (1998b) suggests $c_1 = c_2 = 2$. Related work showed that having each particle put slightly more trust in the swarm (larger c_2 value) and slightly less trust in itself (smaller c_1 value), which seems to act better for the structural design problems. According to experiences, we choose $c_1 = 1.8$ and $c_2 = 1.3$.

Table 3.3 Control parameters of the PSO-based LQR approach

Parameter	Optimized value
Matrix Q	Diag(245.6,250.3 0,0)
Feedback vector K	[15.6709, −17.1806, 8.6616, 2.2921]

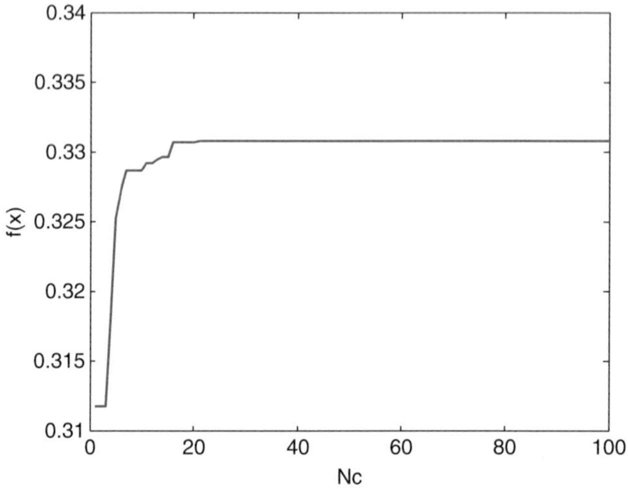

Fig. 3.14 The average evolution curves of the PSO (Reprinted from Duan and Sun (2013), with kind permission from Springer Science+Business Media)

3.3.3 Experiments

Assume the designed MAV executes a surveillance mission using the hover and stare mode in the actual flight. Considering external disturbances, we take crosswinds, for example, the system gets an initial state $x = [0, 0.05, 0, 0]$, which gives rise to pendulum oscillation without an appropriate control.

Using the proposed controller described in Fig. 3.1 and experimental settings given above, the control parameters which include the optimal weight matrix Q and the resulting feedback vector K are obtained, which are shown in Table 3.3.

Fig. 3.14 shows the evolution curve of the PSO algorithm for optimizing the weight parameters of the designed LQR controller.

The zero-input responses of the MAV pendulum-like oscillation with an initial pendulum angle of 0.05rad, which is brought about by crosswinds in the actual flight environment (See Fig. 3.15).

Compared with the responses without control in Fig. 3.13, the dynamic behaviors of x and ϕ_x state obviously that the closed loop eliminates the pendulum-like oscillation and the state converges to the equilibrium point with an average time of 6 s.

Furthermore, considering a constant interference force acting upon the pendulum system besides the initial state, the resulting responses are given in Fig. 3.16. The interference value is taken as 0.2m/s^2, and the control framework and relative parameters remain the same as mentioned above in Fig. 3.13 and Table 3.3.

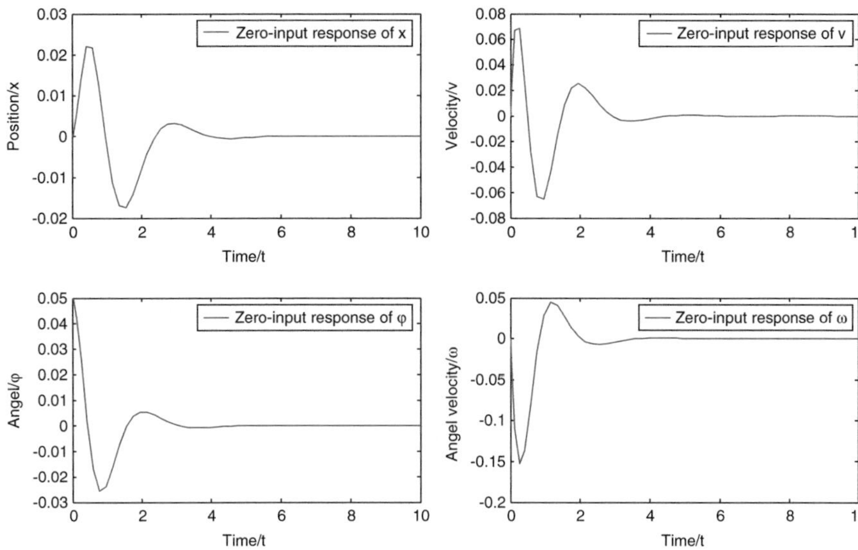

Fig. 3.15 The zero-input response of the MAV pendulum-like oscillation system (Reprinted from Duan and Sun (2013), with kind permission from Springer Science+Business Media)

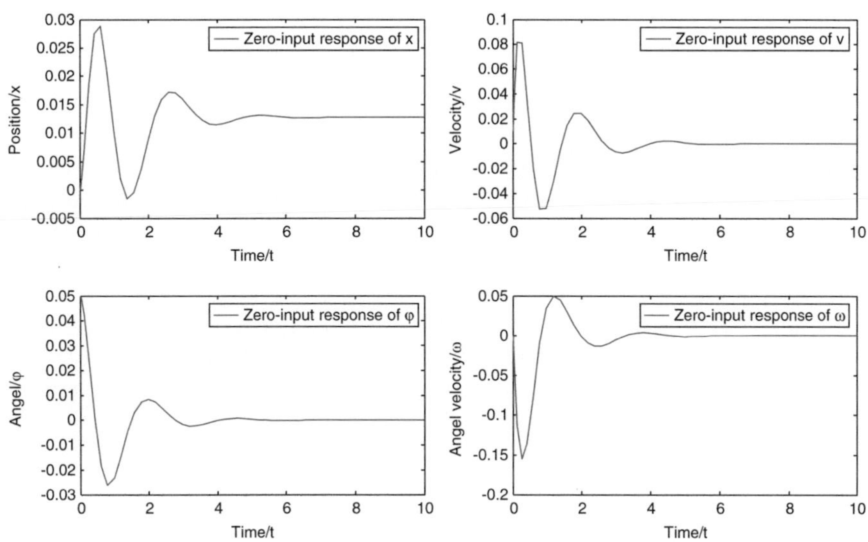

Fig. 3.16 Responses of the MAV pendulum system with a constant disturbance (Reprinted from Duan and Sun (2013), with kind permission from Springer Science+Business Media)

As presented in Fig. 3.16, the PSO-based LQR controller can eliminate the interference of the constant disturbance and the initial state in 6 s. The MAV is stabilized to a state of $x = [0.0128, 0, 0, 0]$, which means that the LQR control design technique based on PSO is robust for external disturbances.

3.4 Conclusions

As a key component of UAV system, controller acts as the brain of UAVs. So the selection of controller parameters is crucial in the design of the flight control system. However, for the existence of strong coupling among the inputs and the nonexistence of mapping relationship between the performance index and the controller parameter, it is a tough work. The 6-DOF nonlinear model of UAVs is illustrated, which is the prerequisite for simplifying and linearizing the mathematical model. Then UAV equations can be simplified into nonlinear equations of 5-DOF with the assumption that the thrust and the resistance of the aircraft maintains the same. And it can be modeled as linear time invariant state-space perturbation models, with the nominal trajectory being steady-level trimmed flight. To reduce the workload of the designers during the process of designing complicated UAV control system, a predator–prey PSO algorithm for identifying parameters of UAV flight control system is presented.

The performance of hover and stare is the key issue to MAVs when carrying out new challenging reconnaissance missions in urban warfare (local, close-up range, hidden reconnaissance, operation between obstacles, and maybe even inside buildings). However, pendulum-like oscillation caused by uncertainty will badly impair the performance of hover and stare, resulting in blurred images or even overturn. However, pendulum-like oscillation caused by uncertainty and external disturbances badly jeopardize the performance of hover and stare, resulting in blurred images or even MAV's overturning. So the second part of this chapter mainly deals with control issue of pendulum-like oscillation in an MAV's hover and stare state in the presence of external disturbances; a novel type of PSO-based LQR controller for stabilizing the pendulum-like oscillation is developed, which can enhance the MAV's performance efficiently. Simulation results verify the feasibility, effectiveness, and robustness of our proposed approach, which provides a more effective way for control law design.

References

Bloss R (2009) Latest unmanned vehicle show features both innovative new vehicles and miniaturization. Ind Robot Int J 36(1):13–18

Duan H, Liu S (2010) Non-linear dual-mode receding horizon control for multiple unmanned air vehicles formation flight based on chaotic particle swarm optimisation. IET Control Theory Appl 4(11):2565–2578

Duan H, Sun C (2013) Pendulum-like oscillation controller for micro aerial vehicle with ducted fan based on LQR and PSO. Sci China Technol Sci 56(2):423–429

Duan H, Zhang X, Xu C (2011) Bio-inspired computing. Science Press, Beijing

Duan H, Yu Y, Zhao Z (2013a) Parameters identification of UCAV flight control system based on predator–prey particle swarm optimization. Sci China Inf Sci 56(1):012202

Duan H, Luo Q, Ma G, Shi Y (2013b) Hybrid particle swarm optimization and genetic algorithm for multi-UAVs formation reconfiguration. IEEE Comput Intell Mag 8(3):16–27

Higashitani M, Ishigame A, Yasuda K (2006) Particle swarm optimization considering the concept of predator–prey behavior. In: Proceedings of IEEE Congress on Evolutionary Computation, Vancouver, BC, Canada. IEEE, pp 434–437

Johnson EN, Turbe MA (2006) Modeling, control, and flight testing of a small-ducted fan aircraft. J Guid Control Dyn 29(4):769–779

Liu F, Duan H, Deng Y (2012) A chaotic quantum-behaved particle swarm optimization based on lateral inhibition for image matching. Optik 123(21):1955–1960

McLean D (1990) Automatic flight control systems. Prentice Hall, New York

Pflimlin J-M, Binetti P, Soueres P, Hamel T, Trouchet D (2010) Modeling and attitude control analysis of a ducted-fan micro aerial vehicle. Control Eng Pract 18(3):209–218

Shi Y, Eberhart R (1998a) A modified particle swarm optimizer. In: Proceedings of the 1998 IEEE International Conference on Evolutionary Computation (IEEE World Congress on Computational Intelligence), Anchorage, AK. IEEE, pp 69–73

Shi Y, Eberhart RC (1998b) Parameter selection in particle swarm optimization. In: Proceedings of 7th International Conference on Evolutionary Programming, California. Springer, Berlin/Heidelberg, pp 591–600

Sun C, Duan H (2013) Artificial Bee colony optimized controller for MAV pendulum. Aircr Eng Aerosp Technol 85(2):104–114

Wang X, Duan H (2013) Predator–prey biogeography-based optimization for bio-inspired visual attention. Int J Comput Intell Syst 6(6):1151–1162

Zhang WY (2004) Study of intelligent robust design approach of large flight envelope flight control system. Master thesis, Beihang University, Beijing

Zhang M, An J (2008) Application of intelligent computation to promote the automation of flight control design. In: Proceedings of International Conference on Intelligent Computation Technology and Automation (ICICTA), Hunan. IEEE, pp 991–994

Chapter 4
UAV Path Planning

Haibin Duan and Pei Li

Abstract Path planning of unmanned aerial vehicle (UAV) is a rather complicated global optimum problem which is about seeking a superior flight route considering the different kinds of constrains under complex dynamic field environment. Several significant considerations for an ideal path planner includes optimality, completeness, and computational complexity, last one of which is the most important requirement since path planning has to be executed quickly due to fast vehicle dynamics. This chapter mainly focuses on path-planning problem for UAVs, from 2-D path planning to 3-D path planning, from path planning for a single UAV to coordinated path replanning for multiple UAVs. Under the assumption that the UAV maintains constant flight altitude and speed when on a mission, a chaotic artificial bee colony (ABC) approach to 2-D path planning is proposed. Besides, a new hybrid meta-heuristic ant colony optimization (ACO) and differential evolution (DE) algorithm is proposed to solve the UAV path-planning problem in three-dimensional scenario. Then path-smoothing strategies are adopted to make the generated path feasible and flyable. Finally, based on the construction of the basic model of multiple UAV coordinated path replanning, which includes problem description, threat modeling, constraint conditions, coordinated function, and coordination mechanism, a novel Max–Min adaptive ACO approach to multiple UAV coordinated path replanning is presented.

4.1 Introduction

Unmanned aerial vehicle (UAV) is one of the inevitable trends of the modern aerial weapon equipments owing to its potential to perform dangerous, repetitive tasks in remote and hazardous environments (Beard et al. 2002). Research on UAV can

---

The original version of this chapter was revised. A correction to this chapter is available at https://doi.org/10.1007/978-3-642-41196-0_9

H. Duan and P. Li, *Bio-inspired Computation in Unmanned Aerial Vehicles*,
DOI 10.1007/978-3-642-41196-0_4, © Springer-Verlag Berlin Heidelberg 2014

directly affect battle effectiveness of the air force, therefore is crucial to safeness of a nation. Path planning is an imperative task required in the design of UAVs, which is to search out an optimal or near-optimal flight path between an initial location and the desired destination under specific constraint conditions. There are several considerations for an ideal path planner including optimality, completeness, and computational complexity, last one of which is the most important requirement since path planning has to occur quickly due to fast vehicle dynamics (Yang and Kapila 2002). The flight path planning in a large mission area is a typical large-scale optimization problem; a series of algorithms have been proposed to solve this complicated multi-constrained optimization problem, such as the A* algorithm, evolutionary computation, genetic algorithm, and ant colony algorithm. However, those methods can hardly solve the contradiction between the global optimization and excessive information.

4.1.1 Characteristic of Path Planning for UAVs

Compared with the path-planning problem in other applications, path planning for UAVs has the following attributes (Zheng et al. 2005):

- ***Stealth:*** The air vehicles are usually required to carry out missions in threatened environments. In such a circumstance, stealth means safety. It is very important to minimize the probability of detection by hostile radar. There are two ways to achieve this. One is absorbing incoming radar radiation as much as possible and/or reflecting it in a direction different than the ambient direction, so that little is reflected back to the original radar site. The other is flying along a path which keeps away from perceived threats and/or has a lower altitude to avoid radar detection utilizing the masking of terrain.
- ***Physical feasibility:*** The physical feasibility of a path refers to the physical limitations from the use of UAVs. They include the following constraints:
  - *Maximum path distance:* It determines the elapsing time between the start and goal points and finally it depends on the fuel supply of aircraft.
  - *Minimum path leg length:* Due to inertia of motion, aircraft will fly straightly along a path for a certain distance before initiating a turn. We call it path length. In order to decrease navigational error, we must make the path leg length larger than a minimum threshold.
- ***Performance of mission:*** Each flight has its special mission. This depends on the application. In order to complete the mission, some requirements must be met when we design a path. These requirements include:
  - *Maximum turning angle:* This constrains the path allowing only to turn an angle less than or equal to a prespecified threshold. Aircraft usually does not wish to make severe turns in some flight scenarios. For example, aircraft in tight formation cannot make severe turns without a greater risk of collision.
  - *Maximum climbing/diving angle:* This has the same definition as maximum turning angle but in altitude direction. It can be either positive or negative,

which depends on its moving direction (it is positive if it is climbing). In order to decrease the risk of collision, a sharp climb or dive should be avoided.
  - *Minimum flying height:* As mentioned before, in order to reduce the probability of being detected by the hostile radar, the aircraft should fly along a low penetration path so as to enhance terrain masking effect. But on the other hand, this definitely increases the probability of crashing into the ground. For this reason, we must maintain a minimum flying height for the entire path.
  - Specific approaching angle to goal point: In some applications such as attacking operation, an optimal direction is fixed a priori. Thus the aircraft should go along a specific direction to approach the goal point.

- ***Cooperation:*** The path-planning algorithm must be compatible with the cooperative nature envisioned for the use of UAVs. In the future operation of UAVs, a flight mission might involve multiple UAVs. In such an application, the ability to coordinate the arriving time of each UAV will be vital in many missions.
- ***Real-time implementation:*** The flight environments of the UAVs are usually constantly changing. Therefore, our path-planning algorithm must be computationally efficient. The replanning ability of path is critical for adapting to unforeseen threats.

4.1.2 Main Features of Path Replanning for Multiple UAVs

The current combating environments are not static; they are constantly changing with many uncertain factors. In the air battlefield, there are not only a number of static threats which have been known a priori, but also other "pop up" or some threats become known only when one UAV maneuvers into their proximity. Furthermore, even those static threats whose locations have been detected ahead of time, their threat grade or threat scope may be changing frequently, which also makes them uncertain. Considering these uncertain factors, the preplanned trajectories often are not adapted to the practice changing complicated air battlefield. In order to increase the survival chance of multiple UAVs, the path replanning is essential while encountering the pop-up threats (Zucker et al. 2007; Duan et al. 2009). Suppose maximizing the probability that the mission in dynamic and uncertain environments will succeed, it is desirable to assign multiple UAVs to conduct all the missions together (Beard and McLain 2003), and thus the problem of multiple UAV coordinated path replanning in dynamic and uncertain environments is put forward. The coordination of multiple UAVs is considered mainly from two aspects (Duan et al. 2009):

- ***Simultaneous arrival***, which requires determining an estimated time of arrival (ETA) at the specific destinations. Each UAV selects candidate path and adjusts the flight velocity corresponding to the ETA. Of course, the multiple UAV trajectories and velocities will change coordinately when some pop-up threats occur.

- ***Collision avoidance***, which requires that the trajectories of multiple UAVs should have no overlaps or crosses between each other.

4.2 Modeling for Path Planning

4.2.1 Environment Representation

Modeling of the threat sources is the key task in single UAV optimal path planning. The typical UAV combating field model in three-dimensional scenario can be depicted with Fig. 4.1 (Duan and Huang 2013).

In order to simplify the single UAV path-planning problem, the UAV task region can be divided into two-dimensional mesh, thus forming a two-dimensional network diagram connecting the starting point and goal point, which can be shown in Fig. 4.2. In this way, the problem of single UAV path planning is the general path optimization problem in essence.

In Fig. 4.3, suppose the flight task for the single UAV is from node B to node A. There are some threatening areas in the task region. Let OA be the x -axis and OB be the y -axis; a coordinate system is then established. We divide OA into m subsections and divide OB and OC into n subsections equally. There are $(m-1)$ vertical lines between node B and node A, which can be labeled with L_1, $L_2, \ldots, L_{m-1}$. The $(m-1)$ vertical lines and the $(2n+1)$ horizontal lines cross-constitute $(m-1)(2n+1)$ nodes. We named these nodes as $L_1(x_1,y_1)$, $L_2(x_2,y_1)$, $\ldots$, $L_{m-1}(x_{m-1},y_1), \ldots, L_1(x_1,y_{2n+1}), \ldots, L_{m-1}(x_{m-1},y_{2n+1})$. Where $L_i(x_i,y_i)$ is the ith node in the vertical line L_i. In this way, the path from the starting node (A) to the target node (B) can be described as follows (Xu et al. 2010; Li and Duan 2012):

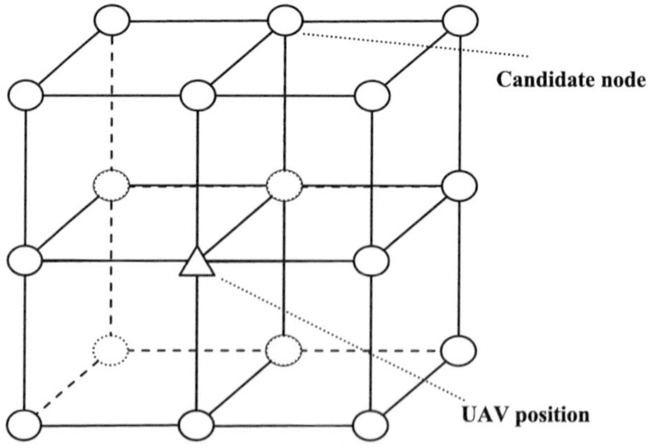

Fig. 4.1 Typical UAV combating field model in three-dimensional scenario

$$Path = \{o, L_1(x_1, y_{k1}), L_2(x_2, y_{k2}), \cdots, L_{m-1}(x_{m-1}, y_{k(m-1)}), A\} \tag{4.1}$$

where $k_i = 1, 2, \cdots, 2n+1$.

To accelerate the search speed of the algorithm, we can let line ST be the x -axis and take the coordinate transformation on each discrete point $(x(k), y(k))$ according to (4.2), where θ is the angle that the original x -axis contrarotates to parallel segment ST, while (x_s, y_s) represents the coordinates in the original coordinate system:

$$\begin{bmatrix} x'(k) \\ y'(k) \end{bmatrix} = \begin{bmatrix} \cos\theta & \sin\theta \\ -\sin\theta & \cos\theta \end{bmatrix} \begin{bmatrix} x(k) - x_s \\ y(k) - y_s \end{bmatrix} \tag{4.2}$$

Thus, the x coordinate of each point can be obtained by a simple equation $x'(k) = \frac{|ST|}{D+1} \cdot k$; therefore the points collection C can be simplified into $C' = \{0, L_1(y'(1)), L_2(y'(2)), \ldots, L_k(y'(k)), \ldots, L_D(y'(D)), 0\}$, which can greatly reduce the computational cost.

4.2.2 Evaluation Function

The performance indicators of the UAV route mainly include the threat cost J_t and the fuel cost J_f, the calculating equation of which are presented as follows:

$$J_t = \int_0^L w_t dl \tag{4.3}$$

$$J_f = \int_0^L w_f dl \tag{4.4}$$

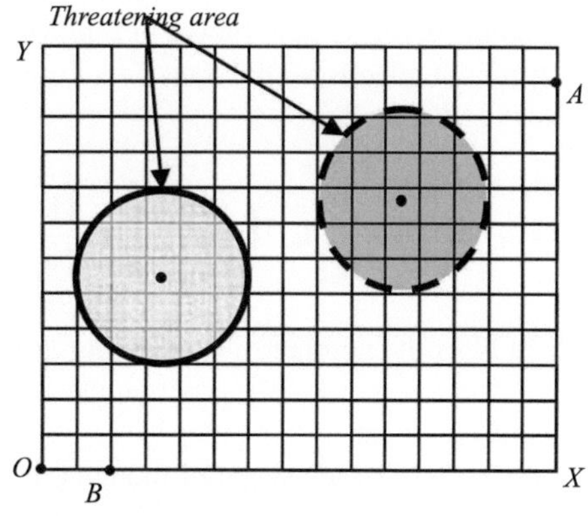

Fig. 4.2 Typical UAV mission area (Reprinted from Duan et al. (2009), with kind permission from Elsevier)

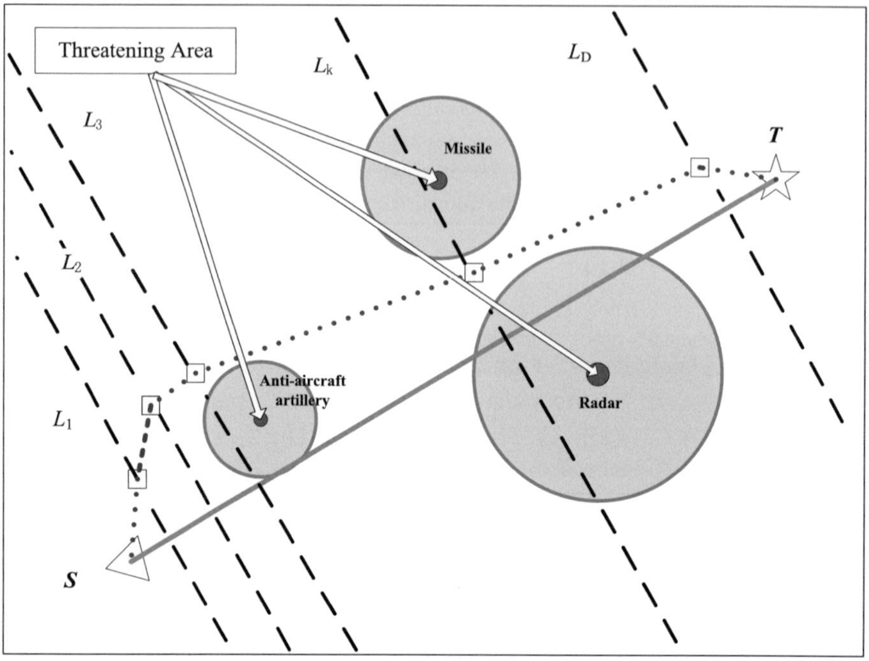

Fig. 4.3 Typical UAV battle field model (Reprinted from Xu et al. (2010), with kind permission from Elsevier)

where w_t and w_f are variables closely related with the current path point and changing along with "l," which respectively present the threat cost and fuel cost of each line segment on the route, while L is the total length of the generated path.

In order to simplify the calculations, a computationally more efficient and acceptably accurate approximation to the exact solution is adopted (Duan et al. 2011). The threat cost of each edge connecting two discrete points was calculated at five points along it, as is shown in Fig. 4.4.

If the ith edge is within the effect range, the threat cost can be computed by the following expression:

$$w_{t,L_i} = \frac{L_i}{5} \cdot \sum_{k=1}^{N_t} t_k \cdot \left(\frac{1}{d_{0.1,i,k}^4} + \frac{1}{d_{0.3,i,k}^4} + \frac{1}{d_{0.5,i,k}^4} + \frac{1}{d_{0.7,i,k}^4} + \frac{1}{d_{0.9,i,k}^4} \right) \tag{4.5}$$

where N_t is the number of threatening areas, L_i is the ith sub-path length, $d_{0.1,i,k}$ is the distance from the 1/10 point on the ith edge to the kth threat, and t_k is the threat level of kth threat.

Assuming that the speed of the UAV is a constant, then the fuel cost of the path J_f can be considered equal to L, the total length of path.

The total cost for traveling along the path comes from a weighted sum of the threat and fuel costs, which can defined as follows:

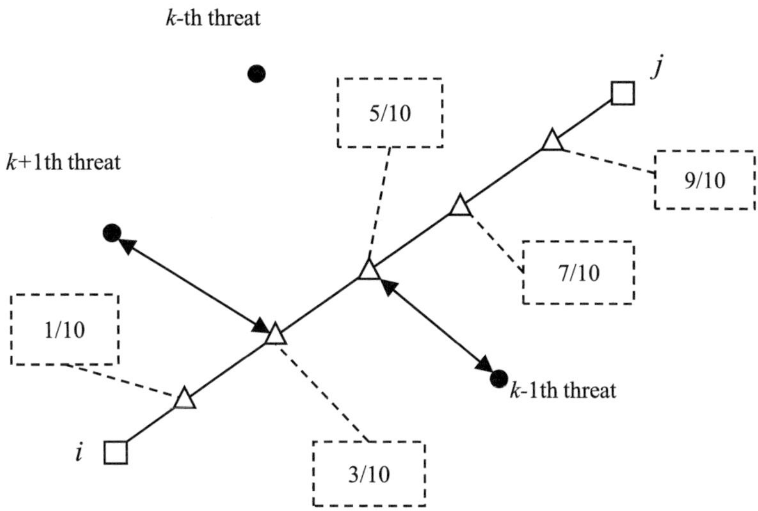

Fig. 4.4 Computation of threat cost (Reprinted from Xu et al. (2010), with kind permission from Elsevier)

$$J = kJ_t + (1 - k)\,J_f \tag{4.6}$$

where k is a variable between 0 and 1 (0.5 in our algorithm), which gives the designer certain flexibility to dispose relations between the threat exposition degree and the fuel consumption. When k is more approaching 1, a shorter path is needed to be planned, and less attention is paid to the radar's exposed threat. Otherwise, when k is more approaching 0, it requires avoiding the threat as far as possible on the cost of sacrifice the path length. The optimized path is founded only when function J reaches its minimal value.

4.3 Chaotic ABC Approach to UAV Path Planning

4.3.1 Brief Introduction to Chaos Theory

Chaos is the highly unstable motion of deterministic systems in finite phase space which often exists in nonlinear systems. Chaos theory is epitomized by the so-called butterfly effect detailed by Lorenz (1963). Attempting to simulate numerically a global weather system, Lorenz discovered that minute changes in initial conditions steered subsequent simulations toward radically different final stales, rendering long-term prediction impossible in general. Until now, chaotic behavior has already been observed in the laboratory in a variety of systems including electrical circuits, lasers, oscillating chemical reactions, fluid dynamics, as well as computer models of chaotic processes. Chaos theory has been applied to a number of fields, among which, one of the most applications was in ecology, where dynamic systems have been used to show how population growth under density dependence can lead to chaotic dynamics.

Sensitive dependence on initial conditions is observed not only in complex systems but even in the simplest logistic equation (Duan et al. 2010a; Xu et al. 2010; Liu et al. 2012). In the following well-known logistic equation:

$$x_{n+1} = 4x_n\left(1 - x_n\right) \tag{4.7}$$

where $0 < x_n < 1$, a very small difference in the initial value of x would give rise to large difference in its long-time behavior, which is the basic characteristic of chaos. The track of chaotic variable can travel ergodically over the whole space of interest. The variation of the chaotic variable has a delicate inherent rule in spite of the fact that its variation looks like in disorder. Therefore, after each search round, we can conduct the chaotic search in the neighborhood of the current optimal parameters by listing a certain number of new generated parameters through chaotic process. In this way, we can make use of the ergodicity and irregularity of the chaotic variable to help the algorithm to jump out of the local optimum as well as finding the optimal parameters.

4.3.2 Procedures of Path Planning Using Chaotic ABC Approach

Due to the flexibility, versatility, and robustness in solving optimization problems, ABC algorithm has already aroused intense interest (Karaboga and Basturk 2007; Duan et al. 2010b; Yu and Duan 2013). However, there still exist some flaws on this algorithm, such as the large number of iterations to reach the global optimal solution and the tendency to converge prematurely. In order to overcome these flaws of ABC and upon the merits of chaotic variable, chaotic ABC algorithm, which integrates ABC with chaotic variable, was proposed. After the search process of each bee, conduct the chaotic search in the neighborhood of current best solution in order to choose one better solution into next generation. In this way, our proposed algorithm takes the advantage of the characteristics of the chaotic variable to make the individuals of subgenerations distributed ergodically in the defined space and thus to avoid from the premature of the individuals, as well as to increase the speed of reaching the optimal solution.

The implementation procedure of our proposed chaotic ABC approach to UAV path planning can be described as follows:

Step 1: According to the environmental modeling method, initialize the detailed information about the path-planning task, as well as the threaten information including the coordinates of threat centers, threat radiuses, and threat levels. In order to simplify the calculation, conduct the coordinate transformation on discrete points related with the task according to (4.2).

Step 2: Initialize the parameters of artificial bee colony optimization algorithm, such as the population of the bee colony N_s, the number of employed bees N_e, and the number of the unemployed bees N_u, which satisfy the condition shown as follows:

$$N_s = N_e + N_u \tag{4.8}$$

Obviously, a larger N_s will contribute to a larger possibility of finding the best solution of the problem; however, it also means an increased computing complexity of the algorithm. In general, we define $N_e = N_u$, and according to our special problem, we set $N_s = 60$. Denote the largest searching times with *Limit* (30 in our experiments), current iterations with T, and the largest iterations with $T_{\max}$. Initialize within the bound of the battlefield the employed bee population of D -dimensional parameters $C' = \{y'(1), y'(2), \ldots, y'(k), \ldots, y'(D)\}$, which represent the y coordinates of each discrete point as we have discussed, while the corresponding x coordinates could be easily obtained by the equation $x'(k) = \frac{|ST|}{D+1} \cdot k$. Each group of parameters can engender a path that is leading the UAV from the starting point S to the target point T, and the goal is to find the optimal combination of parameters that can provide relative satisfactory performance. Initialize the search time of each bee $Bas = 0$ and the starting iteration $T = 1$.

Step 3: According to the parameters of the employed bees, calculate the cost of each path formed by relative parameters based on (4.3), (4.4), (4.5), and (4.6). The smaller the cost value is, the better performance the path maintains.

Step 4: The employed bees search around their current positions (parameters) to find new solutions and update their positions if the new cost value is lower than the original value. The search strategy can be described as follows: for the ith employed bee, first engender a random integer j between 1 and D and a random integer k between 1 and N_e, and then the jth parameter of the ith employed bee could be updated by

$$y_i''(j) = y_i'(j) + \left(y_i'(j) - y_k'(j)\right) \cdot (rand - 0.5) \cdot 2 \tag{4.9}$$

where *rand* represents a random value between 0 and 1. Calculate the new cost value of the updated parameters and choose the one that possesses a lower cost as the new employed bee.

Step 5: The unemployed bees apply the roulette selection method to choose the bee individual that maintains a relatively low cost value as the leading bee according to the calculated cost results of employed bees. Each recruited unemployed bee continues to search new solutions just around the leading bee's solution space similar with Step 4 and calculate their cost values. If the capability of the new solution is better than the original one, the unemployed bee converts into an employed bee, which means that update the positions of the employed bees and continue exploring with *Bas* re-initialized as 0, or else keep searching around, and its *Bas* value plus one.

Step 6: If the search times *Bas* is larger than certain threshold *Limit*, the employed bee gives up the solution and re-search the new food resources, which is realized by re-initializing the parameters and calculating the cost value.

Step 7: Store the best solution parameters and the best cost value.

Step 8: Conduct the chaotic search around the best solution parameters based on (4.7) after transforming the parameters ranges into (0–1). Among the engendered series of solutions, select the best one and use it to replace a random employed bee.

Step 9: If $T < T_{\max}$, go to Step 4. Otherwise, output the optimal parameters and optimal cost value.

The detailed procedure can also be shown with Fig. 4.5.

4.3.3 Experiments

In order to investigate the feasibility and effectiveness of the proposed method, series of experiments are conducted, and further comparative experimental results with the standard ABC algorithm were also given.

Set the coordinates of the starting point as (11, 11) and the target point as (75, 75), while the initial parameters of ABC algorithm were set as $N_s = 60$, $N_e = 30$, $N_u = 30$, $T_{\max} = 100$, $Limit = 30$.

Respectively assume D as 10, 20, and 30 to carry our experiments, the results of which are shown in Figs. 4.6, 4.7, 4.8, 4.9, 4.10. Figure 4.11 depicts the path-planning result of chaotic ABC algorithm when D is set as 10, and the achieved path shown in the figure obviously maintains a favorable performance, hence proves the feasibility of our algorithm without any doubt.

When $D = 10$, the experimental results of standard ABC and chaotic ABC have slightly differences due to the calculating complexity. However, when the value of D is increased to 20, even to 30, we can clearly see the superiority of our proposed method over the standard ABC algorithm in the comparative experimental results shown in Figs. 4.7, 4.8, 4.9, 4.10, and 4.11.

To further prove the performance of our proposed method against standard ABC algorithm, we run the program for 100 times to obtain the average cost of our generated best path. It turns out that the average cost of our algorithm is 50.9346, while the average cost of standard ABC algorithm is 53.9071, apparently showing that our method can find the feasible and optimal path for UAVs more stable than basic ABC algorithm and can effectively solve the path-planning problem of UAVs in complex combat field environment.

When the threats move, we can recalculate the path according to current threat positions. The simulation results can be shown in Fig. 4.11, which show the feasibility of chaotic ABC algorithm under moving threatens.

From the above experimental results, we can clearly see that using standard ABC algorithm could possibly lead to a path that does not satisfy the requirements,

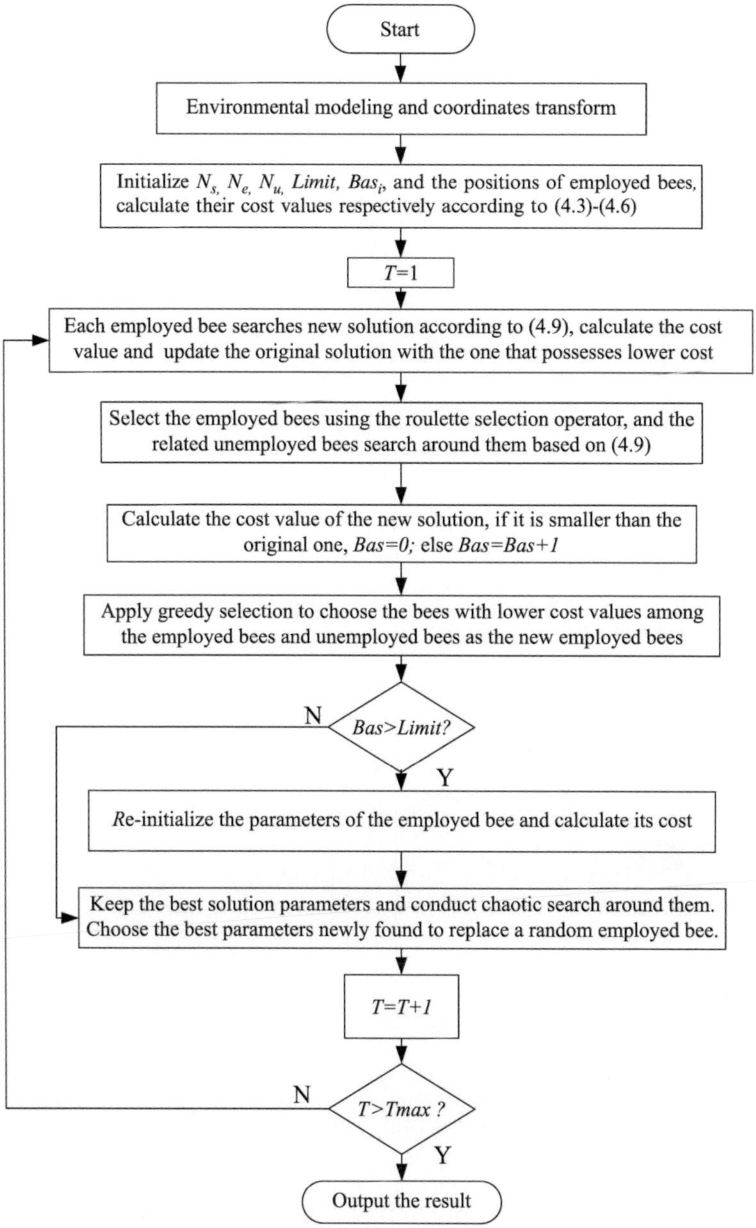

Fig. 4.5 The procedure of our proposed method (Reprinted from Xu et al. (2010), with kind permission from Elsevier)

especially when the optimized dimension increases. Therefore, we make use of the ergodicity of chaotic variable to help the basic ABC algorithm to jump out of the local best and obtain a favorable path.

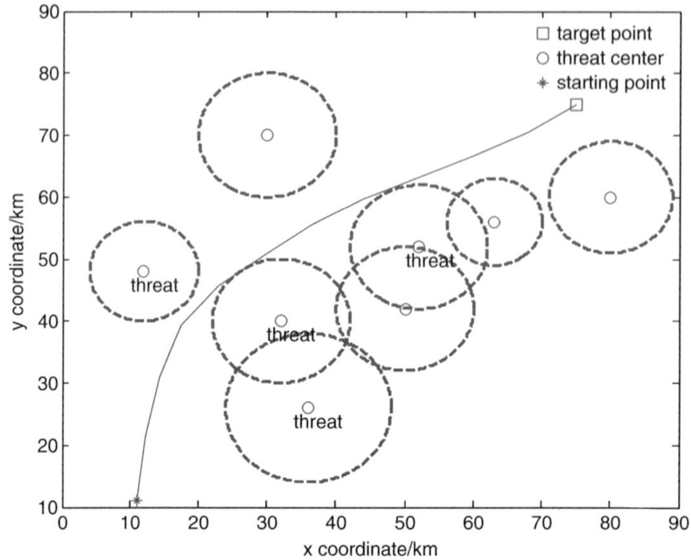

Fig. 4.6 The path-planning result using chaotic ABC algorithm, $D=10$ (Reprinted from Xu et al. (2010), with kind permission from Elsevier)

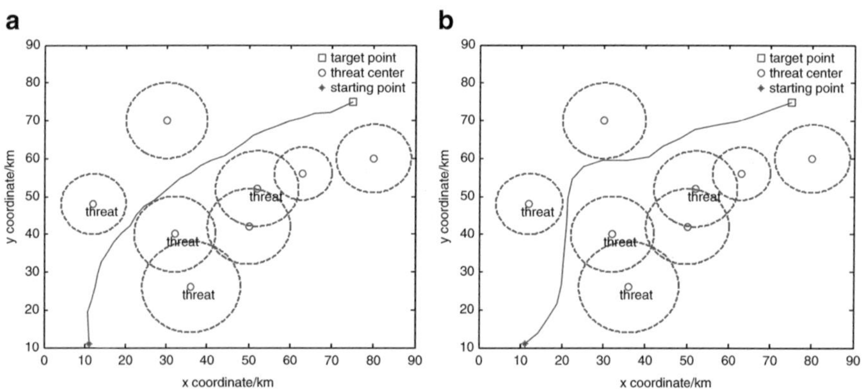

Fig. 4.7 The comparative path-planning results, $D=20$: (**a**) chaotic ABC algorithm (**b**) standard ABC algorithm (Reprinted from Xu et al. (2010), with kind permission from Elsevier)

4.4 Hybrid ACO-DE Approach to Three-Dimensional Path Planning for UAVs

4.4.1 Hybrid Meta-heuristic ACO-DE Algorithm

The ACO pheromone plays a very important role in the path exploration and exploitation. A reasonable distribution of the pheromone trial can directly affect ants

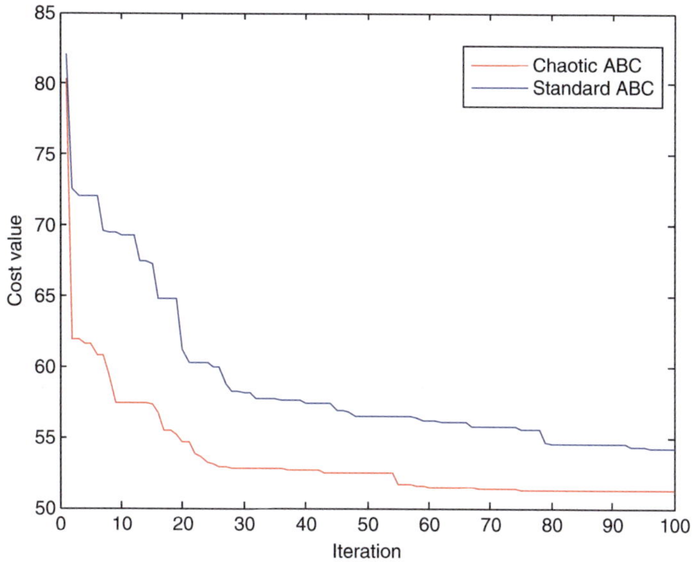

Fig. 4.8 The evolution curves of two algorithms, $D=20$ (Reprinted from Xu et al. (2010), with kind permission from Elsevier)

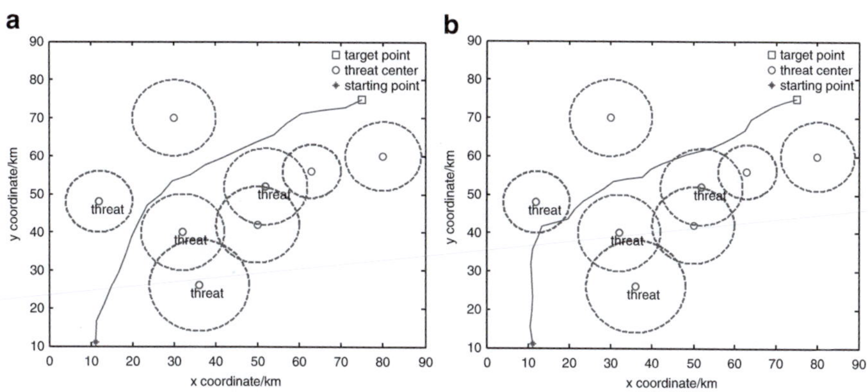

Fig. 4.9 The comparative path-planning results, $D=30$ (Reprinted from Xu et al. (2010), with kind permission from Elsevier)

to explore their optimal paths. In view of this, we propose a hybrid meta-heuristic ACO and DE model. DE (Price and Storn 1997) is used to make some random deviation disturbance in the ACO pheromone trail. Through this kind of random disturbance, we intend to realize that the pheromone trail between two neighboring nodes left by ant colony can reach a more reasonable distribution, which can lead ants to find out the optimal path.

In our proposed hybrid meta-heuristic ACO and DE algorithm model, we set the pheromone on the path left by ants in ACO as the object of the mutation, crossover,

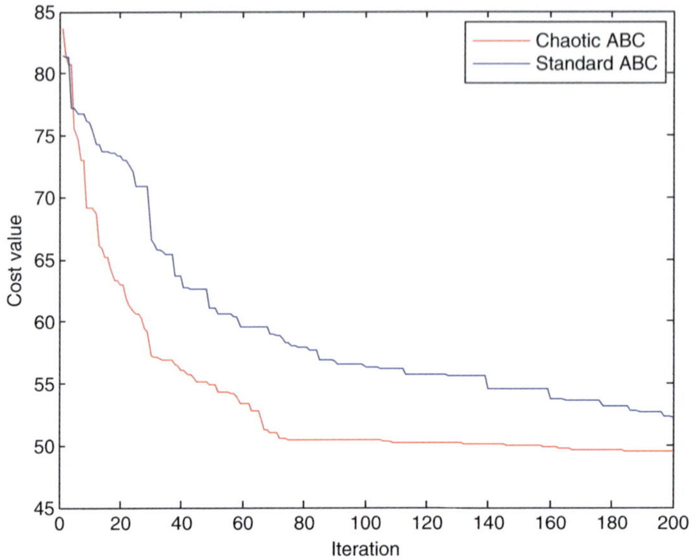

Fig. 4.10 The evolution curves of two algorithms, D=30 (Reprinted from Xu et al. (2010), with kind permission from Elsevier)

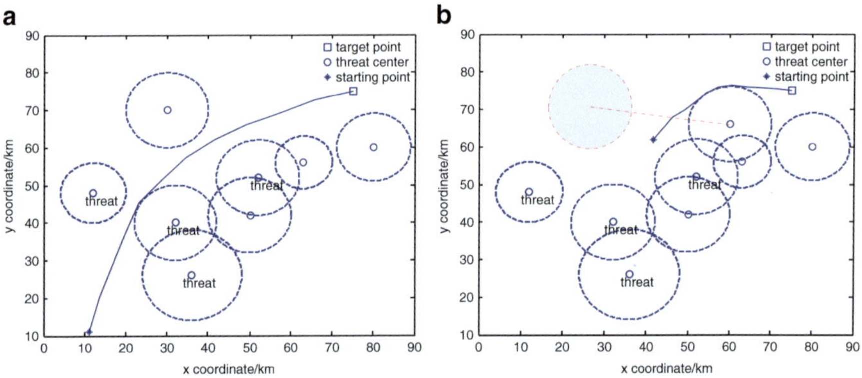

Fig. 4.11 Experimental results on moving threatens: (**a**) initial optimized path, (**b**) recalculated path after the threaten moves (Reprinted from Xu et al. (2010), with kind permission from Elsevier)

and selection in DE (Duan et al. 2010c). In solving the UAV three-dimensional path-planning problem, the objective function of the pheromone on all sub-paths between two neighboring nodes is the length of the best tour found by ants, which can be obtained according to the pheromone trail.

Some slightly adjustments are added to the basic ACO model (Dorigo et al. 1996; Dorigo and Di Caro 1999): we divide the entire ant colony into several independent ant teams, and the team number is denoted with *Team*, which is a restriction of the total ant number *m*. For each ant team, the pheromone amount left on the links between each two neighboring nodes are named as $\boldsymbol{\tau} = \{\boldsymbol{\tau}_i\}, i = 1, \ldots, Team$. Obviously, $\boldsymbol{\tau}_i$ is a $n \times n$ matrix. As to the current pheromone of each ant team, DE mutation operation takes effect, and the new trial pheromone trail distribution is generated by the following equation:

$$\tau_{1_i} = \tau_{r_1} + F \times (\tau_{r_2} - \tau_{r_3}), i = 1, 2, \cdots, Team \tag{4.10}$$

where r_1, r_2, and r_3 are integers, which can be chosen randomly from the interval [1, *Team*]; τ_{r_1}, τ_{r_2}, and τ_{r_3} are three pheromone trail individuals, which are selected randomly among all ant-team units and $r_1 \neq r_2 \neq r_3 \neq i$. F is a real and constant factor between [0, 2], which is named *Constant of Mutation*, and this constant factor can control the amplification of the differential variation $(\tau_{r_2} - \tau_{r_3})$. Obviously, the smaller the differential variation between two individuals, the weaker the disturbance which it brings about. It signifies that when the pheromone of each ant team converges to the vicinity of a kind of reasonable pheromone distribution, the disturbance generated through mutation will become weaker automatically.

In our proposed hybrid meta-heuristic ACO and DE algorithm, in order to improve the diversification of pheromone trail between nodes, we can take advantage of the DE crossover operation to make the new trial pheromone trail τ_{1_i}, which is generated through mutation, combined with the current target pheromone τ_i. The proposed hybrid meta-heuristic ACO and DE algorithm generates a new pheromone matrix $\tau_{2_i} = \begin{bmatrix} \tau_{2_i}^{1,1} & \cdot\ \cdot & \tau_{2_i}^{1,n} \\ \cdot & \cdot\ \cdot\ \cdot & \\ \cdot & \cdot\ \cdot\ \cdot & \\ \tau_{2_i}^{n,1} & \cdot\ \cdot & \tau_{2_i}^{n,n} \end{bmatrix}$, $i = 1, \ldots, Team$, which can be expressed as follows:

$$\tau_{2_i}{}^{j,k} = \begin{cases} \tau_i{}^{j,k}, & if \;\; randb \leq CR \;\; or \;\; rand_k = k, \\ \tau_{1_i}{}^{j,k}, & if \;\; randb > CR \;\; or \;\; rand_k \neq k, \end{cases} \tag{4.11}$$

where $\tau_i^{j,k}$ denotes the amount of pheromone between city j and k of ith ant team, $\tau_{1i}^{j,k}$ denotes the trial pheromone trail between city j and k of the ith ant team after the mutation operation, and $\tau_{2i}^{j,k}$ denotes the ith ant-team pheromone trail between city j and k, after the crossover operation toward $\tau_i^{j,k}$ and $\tau_{1i}^{j,k}$. *randb* is a random positive number between [0, 1]. *CR* is a constant between [0, 1], which is known as *Constant of Crossover*; the larger it is, the greater possibility the crossover operation happens; $CR = 0$ represents that no DE crossover occurs. $rand_k$ is an integer number selected randomly between [1, n]. It is obvious that the new generated pheromone matrix τ_{2_i} will surely get at least one element from that mutation trial pheromone τ_{1_i}. Otherwise, it is possible that the pheromone trail will not change at all, which can weaken the pheromone exchange between different ant teams.

In the UAV three-dimensional path planning, ants in each team construct their paths by the transition probability $p_{j,k}$, which can be calculated by their pheromone matrix τ_i. L_best_i denotes the length of the shortest path among all paths obtained by ants, which is the objective function of the pheromone trail τ_i at the same time. Toward the newly generated pheromone trails and the path explorations of ant colony based on them, should we accept them or not? We need to compare the objective function value of both the original target pheromone τ_i and the new τ_{2_i}. After that, we select one solution by the so-called *Greedy* selection model. If and only if the new pheromone trail individual τ_{2_i} has a better objective function value than the original one, it can be accepted and reserved into the pheromone trail matrix of the next generation; otherwise, the original target pheromone τ_i will remain in the pheromone trail between nodes of each ant team. Thus, we can express the crossover operation to pheromone trail as follows:

$$\tau'_{i,t} = \begin{cases} \tau_{2_{i},t}\,,\ if\ \ L\_best_{2_i} < L\_best_{0_i} \\ \tau_{i,t}\,,\ if\ \ L\_best_{2_i} \geq L\_best_{0_i} \end{cases} \tag{4.12}$$

where $\tau_{i,t}$ denotes the original pheromone trail left by the ith ant team, when the number of iteration is t; $\tau_{2i,t}$ denotes, at the tth iteration, the new pheromone trail of the ith ant team after DE mutation and crossover operation; and $\tau'_{i,t}$ is equal to the pheromone matrix which has high objective function value between $\tau_{i,t}$ and $\tau_{2i,t}$. $L\_best_{0_i}$ represents the length of the optimal route gained by $\tau_{i,t}$, which is the objective value of the original pheromone $\tau_{i,t}$ ant team, while $L\_best_{2_i}$ represents the length of the optimal route gained by $\tau_{2i,t}$, which is also the objective value of the new pheromone $\tau_{2i,t}$ of the ith ant team.

After the selection operation, the ith ant teams, which have generated their tours by the pheromone trail $\tau'_{i,t}$ or $\tau_{2i,t}$, release their own pheromone regarding the length of tours they each covered and update the selected pheromone trail $\tau'_{i,t}$ to gain the new pheromone trail $\tau_{i,t+1}$. Then pass the new pheromone of each ant team on to next iteration to continue path exploration and exploitation or stop.

4.4.2 Procedures of Three-Dimensional Path Planning Using Hybrid ACO-DE

The process of our proposed hybrid meta-heuristic ACO and DE algorithm for solving UAV three-dimensional path planning can be described as follows:

Step 1: Initialization of parameters – set the current number of iteration $Nc = 1$. Set the maximum number of iteration as $Nc\max$; set the number of ants as m and the number of ant team as *Team*; set the initial amount of pheromone trail on each link between two path nodes $\tau^{j,k} = const$, where *const* is a positive constant number.

Step 2: Initialization of the ant colony – divide the whole ant colony into different ant teams; the numbers of ant in each ant team are recorded in the matrix T_m, $(1\times\ Team)$; for the ith ant team, the number of ant individuals is $T\_m(i)$; then put ants in each ant teams on the starting node. Set other parameters of ACO and DE: α, β, ρ, Q, F, e, CR.

Step 3: Set $Nc = 1$, $i = 1$, the $T\_m(i)$ ants in the ith ant team select the path nodes k and go forward as the transition probability $p_{j,k}$, until the whole ant colony arrives at the target point; then update the pheromone trail left on the paths to generate $\tau_{i,2}$; set $i = i + 1$, and return to Step 3 until $i > Team$.

Step 4: $Nc = Nc + 1$, take mutation and crossover operation to the original pheromone trail τ_i of each ant team passed from the former iteration and generate the new pheromone trail τ_{2_i}; set $i = i + 1$, and return to Step 4 until $i > Team$.

Step 5: Set $i = 1$.

Step 6: Each individual ant of the ith ant team finally arrives at the target point to construct their tours according to the pheromone trail τ_i by the following equation:

$$p_{j,k} = \begin{cases} \dfrac{\left[\tau_i{}^{j,k}\right]^{\alpha}\left[\eta_{j,k}\right]^{\beta}}{\sum\limits_{s\in allowed_{t-m(i)}} \left[\tau_i{}^{j,k}\right]^{\alpha}\left[\eta_{j,k}\right]^{\beta}} & if\ \ k \in allowed_{T-m(i)} \\ 0 \ otherwise & \end{cases} \tag{4.13}$$

Then calculate length of tours gained by each ant, choose the shortest one, and record it as $L\_best_{0_i}$.

Step 7: Each individual ant of the ith ant team visits the whole nodes in the three-dimensional paths to gain their tours by the pheromone trail τ_{2i} as follows:

$$p_{j,k} = \begin{cases} \dfrac{\left[\tau_{2i}{}^{j,k}\right]^{\alpha}\left[\eta_{j,k}\right]^{\beta}}{\sum\limits_{s\in allowed_{t-m(i)}} \left[\tau_{2i}{}^{j,k}\right]^{\alpha}\left[\eta_{j,k}\right]^{\beta}} & if\ \ k \in allowed_{T-m(i)} \\ 0 \ otherwise & \end{cases} \tag{4.14}$$

Then calculate the tour length gained by each ant, choose the shortest one, and denote it with L_best_{2i}

Step 8: Compare L_best_{0i} and L_best_{2i}, take the DE selection operation by (4.12), and set the $\tau_i{}'$ as τ_i or τ_{2i}.

Step 9: Update the current pheromone $\tau_i{}'$ to gain τ_i of next iteration as follows:

$$\Delta\tau_{T\_m(i)}^{j,k} = \begin{cases} \dfrac{Q}{W_{k\{T-m(i)\}}}, \ if\ \ T\_m(i) - th\ ant\ \ passed\ \ (j,k) \\ 0, \ otherwise \end{cases} \tag{4.15}$$

$$\Delta\tau_i = \sum_{s=1}^{T\_m(i)} \Delta\tau_s^{j,k} \tag{4.16}$$

$$\tau_i = (1-\rho)\cdot\tau'_i + \rho\cdot\Delta\tau_i \tag{4.17}$$

If you select τ_i as $\tau_i{}'$ in Step 8, use the tours gained by ants in Step 6 to update the pheromone; if $\tau_i{}' = \tau_{2i}$ after selection operation, choose the tours gained in Step 7 to update the pheromone.

Step 10: Set $i = i + 1$; return to Step 6, until $i > Team$.

Step 11: Return to Step 4 until $Nc \geq Nc\max$.

Step 12: The proposed hybrid meta-heuristic ACO and DE algorithm terminates and outputs the best path in three-dimensional space.

The abovementioned flowchart of the hybrid meta-heuristic ACO and DE algorithm process can also be described in the Fig. 4.12.

4.4.3 Path-Smoothing Strategies

The generated UAV optimal path using the proposed hybrid meta-heuristic method is usually hard for exact flying. Because there are some turning points on the optimized path (Kito et al. 2003), we adopt a class of dynamically feasible path smooth strategy called κ -trajectories (Anderson et al. 2005). Consider the waypoint path defined by the three waypoints w_{i-1}, w_i, and w_{i+1}, and let

$$q_i = (w_i - w_{i-1}) / \| w_i - w_{i-1} \| \tag{4.18}$$

$$q_{i+1} = (w_{i+1} - w_i) / \| w_{i+1} - w_i \| \tag{4.19}$$

Denote the unit vectors along the corresponding path segments as shown in Fig. 4.13.

Let β represent the angle between q_i and q_{i+1} so we can get $\beta = \arccos(-q_{i+1} \cdot q_i)$. As shown in Fig. 4.13, let $\overline{C}$ be a circle of radius

$$R = 0.5 \min \{ \|w_i - w_{i-1}\|, \|w_{i+1} - w_i\| \} \tan \frac{\beta}{2} \tag{4.20}$$

where center C_i lies on the bisector of the angle formed by the three waypoints, such that the circle intersects both the lines $\overline{w_{i-1}w_i}$ and $\overline{w_i w_{i+1}}$ at exactly one point each. The bisector of β will intersect $\overline{C}$ at two locations. So the center C_i is given by

$$C_i = w_i + \left(R / \sin \frac{\beta}{2} \right) (q_{i+1} - q_i) / \| q_{i+1} - q_i \| \tag{4.21}$$

After this process, the original path $\overline{w_{i-1}w_i} \rightarrow \overline{w_i w_{i+1}}$ could be replaced by $\overline{w_{i-1}A} \rightarrow \overset{\frown}{A}\overset{\frown}{C}\overset{\frown}{B} \rightarrow \overline{Bw_{i+1}}$. In this way, the optimized path can be smoothed for feasible flying. This path-smoothing algorithm has a small computational load and can be run in real time.

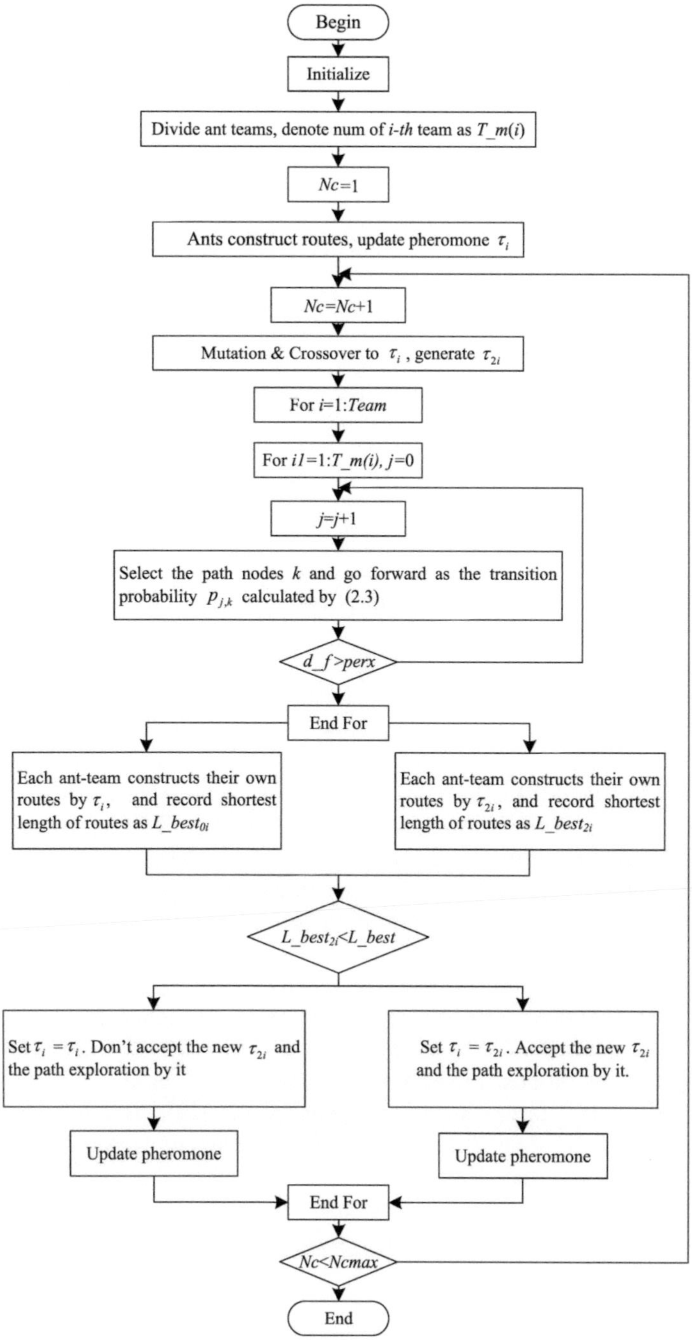

Fig. 4.12 Flowchart of the proposed hybrid meta-heuristic ACO and DE algorithm (Reprinted from Duan et al. (2010c), with kind permission from Elsevier)

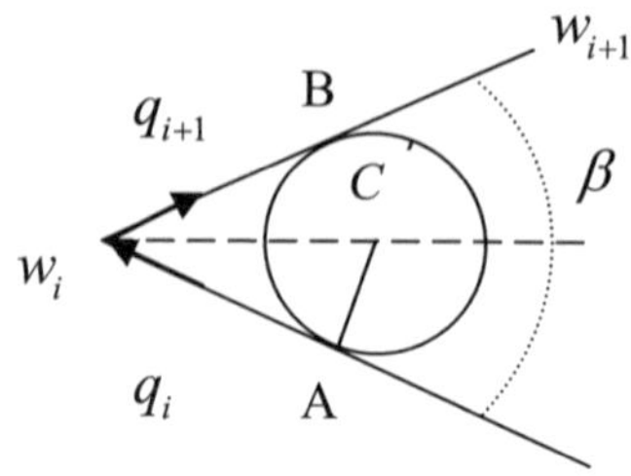

Fig. 4.13 Feasible sub-path (Reprinted from Duan et al. (2010c), with kind permission from Elsevier)

4.4.4 Experiments

In order to investigate the feasibility and effectiveness of the proposed hybrid meta-heuristic ACO and DE approach to UAV three-dimensional path planning, a series of experiments have been conducted under complex combat field environment.

The hybrid meta-heuristic ACO and DE algorithm was implemented in a Matlab 7.2 programming environment on an Intel Core 2 PC running Windows XP SP2. No commercial ACO tools or DE tools were used. In all experiments, the same set of ACO algorithm parameter values were $\alpha = 2$, $\beta = 3$, $\rho = 0.7$, $Q = 100$, $Nc\max = 15$.

Figure 4.14 shows the UAV path-planning results comparison between basic ACO and the proposed hybrid meta-heuristic ACO and DE algorithm in three-dimensional space with $m = 10$ and $Team = 5$, and the curve path comparison by the smooth algorithm, and also the evolution curves comparison. Figure 4.15 shows the UAV path-planning results comparison and the evolution curves comparison between basic ACO and improved ACO with $m = 20$ and $Team = 5$. The symbol "Δ" denotes the starting point, the sphere denotes the threaten area, while the symbol "◊" denotes the target point. And the thin line is the path generated by the basic ACO, while the thick one is generated by the improved ACO.

The values of each optimal solution searched by the different algorithm could be given by the value of the "shortest length," which can be shown in Table 4.1.

From the experimental results presented in Figs. 4.14, 4.15 and Table 4.1, it is obvious that the proposed hybrid meta-heuristic ACO and DE method can find feasible and optimal three-dimensional path for the UAV very quickly and can effectively solve the three-dimensional path planning of UAVs in complicated combating environments. The results also show that the more different ant teams dividing the whole ant colony, the better the solution is. This method provides a new way for three-dimensional path planning of UAVs in exact application.

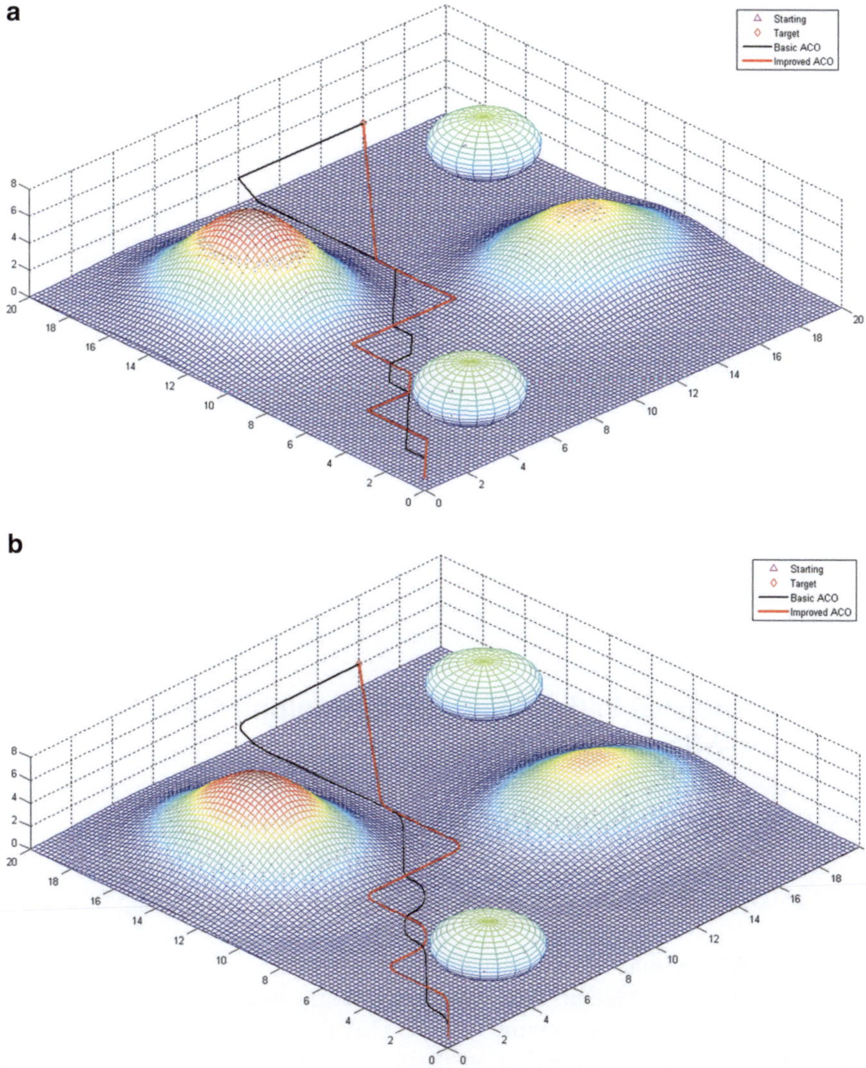

Fig. 4.14 Parameter values were m=10, $Team$=5. (**a**) Path-planning original results comparison between basic ACO and improved ACO. (**b**) Route comparison after using the smoothing strategy. (**c**) Evolution curves comparison between the basic ACO and the improved ACO (Reprinted from Duan et al. (2010c), with kind permission from Elsevier)

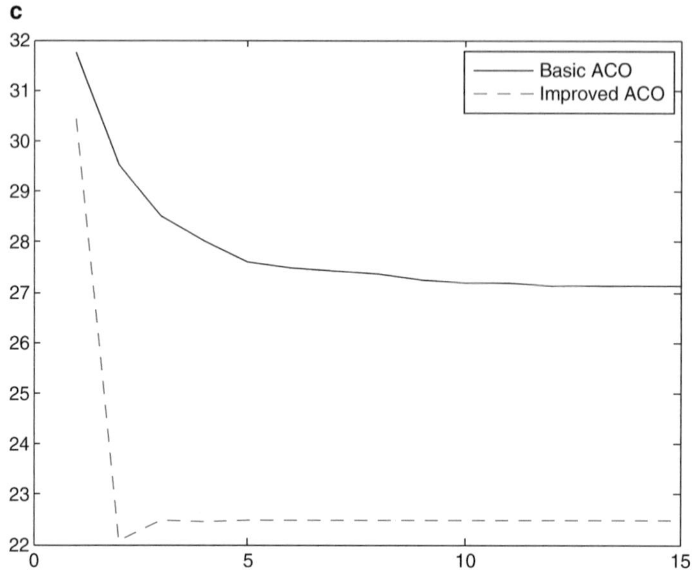

Fig. 4.14 (continued)

4.5 Coordinated Path Replanning for Multiple UAVs Using Max–Min Adaptive ACO

4.5.1 Model of Multiple UAV Coordinated Path Replanning

4.5.1.1 Description of Multiple UAV Coordinated Path Replanning

Multiple UAV coordinated path planning is to generate a safe and short path for each UAV (Beard and McLain 2003; Zheng et al. 2002). In addition, the path should satisfy the requirements concerning multiple UAVs' coordinateness. Therefore, in the issue of multiple UAV coordination, the planned trajectories may be not optimal for any individual vehicle, but they are required to be optimal or near optimal for the whole team.

Suppose that a formation of multiple UAVs is required to fly through the enemy territory and to attack same or different known target locations. There are a number of threats in the flight environment; some of them are known a priori, whereas others pop up or become known only when a UAV maneuvers into its proximity. We assume that each UAV is equipped with sensing capability so that they can detect the pop-up threats in their surroundings. We also assume that the multiple UAVs are equipped with a communication network, so they can inform other UAVs of the pop-up threats' information that they just detected. As is shown in Fig. 4.16, the UAVs' formation is commanded to attack an enemy objective. UAVs fly along

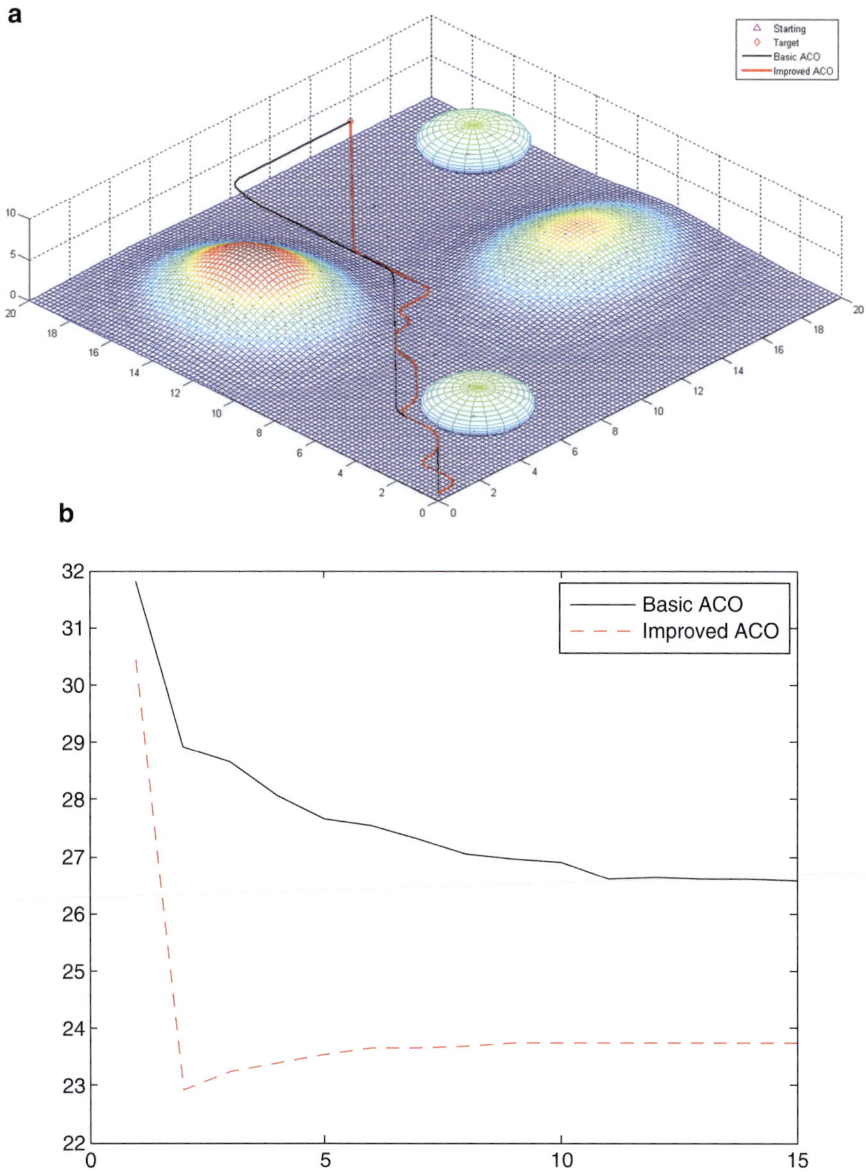

Fig. 4.15 Parameter values were $m=20$, *Team*=5. (**a**) Path-planning results comparison between basic ACO and improved ACO. (**b**) Evolution curves comparison between the basic ACO and the improved ACO (Reprinted from Duan et al. (2010c), with kind permission from Elsevier)

their preplanned trajectories with respective flight velocity, and they are required to arrive at the same time given via planning. When one pop-up threat appears just on one UAV's flight route and pose a threat to it, the current path is not feasible but even dangerous. At this moment, the UAV has to find other new path. Meanwhile, in

Table 4.1 Shortest length comparison between the basic ACO and the improved ACO

	Basic ACO	Improved ACO
Optimal length (m=10)	27.1421	21.2426
Optimal length (m=20)	26.5563	21.6569

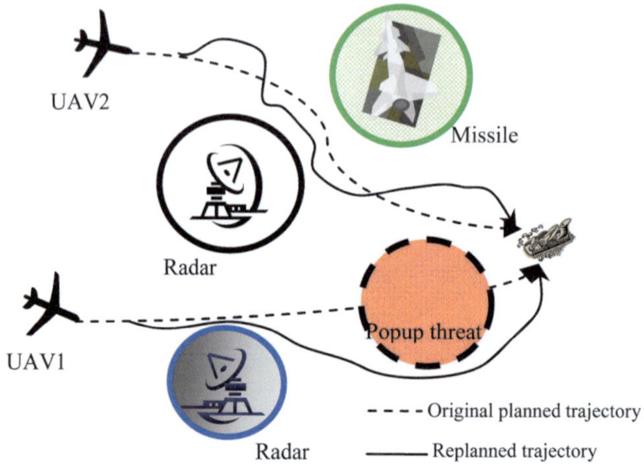

Fig. 4.16 Description of multiple UAV coordinated path replanning (Reprinted from Duan et al. (2009), with kind permission from Elsevier)

order to ensure that the multiple UAV formation launches an attack simultaneously and avoids collision, it's also necessary for other UAVs in the formation to adjust their respective trajectories, and thus, the coordinated path replanning for the whole formation is inevitable. Moreover, along with the change of multiple UAVs' path, the arrival time ETA is also confirmed again accordingly, as well as the flight velocity of each UAV (Duan et al. 2009).

4.5.1.2 Constraint Conditions of Multiple UAV Coordinated Path Replanning

Comparing with the path planning of a single UAV, the difference of multiple UAV coordination path planning also lies in the constraint conditions that ought to be taken into account. Besides the physical properties and mission demands of single individual vehicle, the coordination and cooperation among various UAVs brings several extra co-constraints, including timing constraint that UAVs should reach objectives simultaneously and collision avoidance. Constraints involved in the process of coordinated path replanning are mainly the following three aspects:

Minimum flight turning radius constraint: Considering the maneuverability of UAV, the turning radius in the generated path must be larger than minimum turning

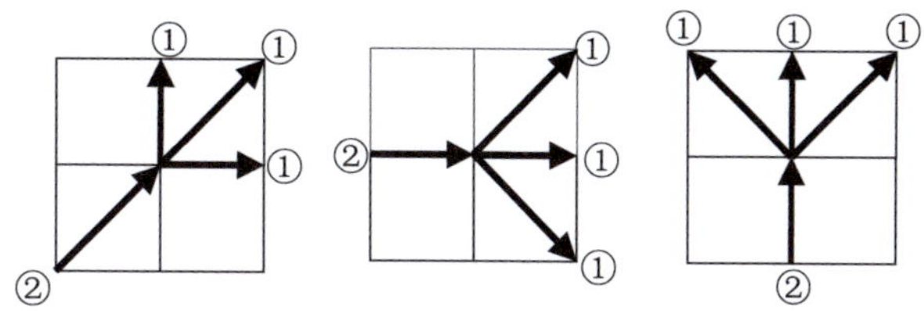

Fig. 4.17 Constraint on the node selection (Reprinted from Duan et al. (2009), with kind permission from Elsevier)

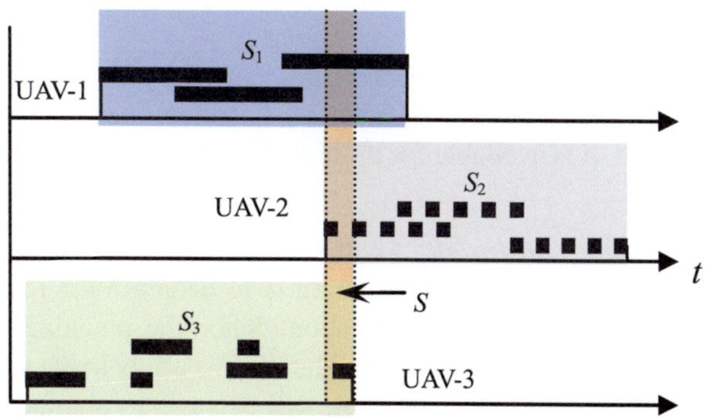

Fig. 4.18 Feasible region of arrival time (Reprinted from Duan et al. (2009), with kind permission from Elsevier)

radius of each UAV, which means the planned path should avoid larger turning. We set the constraint on the choice of waypoint (node) as shown in Fig. 4.17.

In Fig. 4.17, the center node in the square is the current site of the UAV, and ② is the last waypoint just before current one. Considering that UAV cannot take too big turning, next waypoint can only be selected from those points labeled with ①.

Timing coordination constraint: Under this kind of coordination constraint, the goal is to decide the coordinated arrival time for multiple UAVs. The path generated for multiple UAVs should ensure the multiple UAVs arrive at their respective targets simultaneously (as shown with Fig. 4.18). The multiple UAVs must minimize their exposure to threats under the constraint of simultaneous arrival. Therefore, we should comprehensively consider both the length of path and UAVs' flight velocity to assign the team-optimal ETA for the multiple UAV formation.

Suppose that there are N UAVs participating in the flying mission. Each vehicle flies along its route with the velocity constraints $v \in [V_{\min}, V_{\max}]$. For the jth path planned for the ith UAV, of which the length is labeled as $L_{i,j}$, we determine the range of its ETA as follows:

$$T_{i,j} \in \left[\frac{L_{i,j}}{V_{\max}}, \frac{L_{i,j}}{V_{\min}} \right] \tag{4.22}$$

We assume the ith UAV has generated num candidate trajectories, and thus, its estimated time for arrival is a union S_i determined as the following equation:

$$S_i = \left[\frac{L_{i,1}}{V_{\max}}, \frac{L_{i,1}}{V_{\min}}\right] \cup \left[\frac{L_{i,2}}{V_{\max}}, \frac{L_{i,2}}{V_{\min}}\right] \cup \cdots \cup \left[\frac{L_{i,num}}{V_{\max}}, \frac{L_{i,num}}{V_{\min}}\right] \tag{4.23}$$

As for the UAVs formation with N vehicles, the arrival time must be contained in the time intersection S as follows:

$$S = S_1 \cap S_2 \cap \cdots \cap S_N \tag{4.24}$$

If S is not a void, assume such a time $T_a \in S$, and every UAV must have at least one path satisfying the arrival time T_a, and thus T_a can just be regarded as the ETA. Through this method, it is available for multiple UAVs to satisfy the requirement of simultaneous arrival.

Air collision avoidance constraint: Another coordination requirement concerned in multiple UAV coordinated path replanning is to decrease the risk of collision, which is the so-called air–space coordination. Since the planning space is two-dimensional with the assumption that individual UAVs all fly at the same altitude, a proper approach to ensure that no collision will occur is to eliminate any overlap between two UAVs' trajectories. It is clear that if the proportion of overlaps in the entire path is bigger, then the probability of collision will be greater as well. Therefore, multiple UAVs' coordinated path should make it non-overlap as far as possible to implement air–space coordination.

4.5.1.3 Coordination Function

As for the multiple UAV coordinated path replanning problem, the essential idea is that if every vehicle knows the coordination variable and responds appropriately, the coordinated behavior will be achieved. For the aforementioned timing coordinated constraint of simultaneous arrival, the key coordination variable is the arrival time. That is to say, the key work of the multiple UAV coordination is to select a proper factor from the time intersection S as the value of the coordination variable. To make it, it's necessary to construct the coordination function J_{co} to determine the coordination variable:

$$J_{co,i,j}\left(T_{i,j}\right) = f_1 \cdot J_{i,j} + f_2 \cdot T_{i,j} \tag{4.25}$$

where f_1 and f_2 are two constants, and these two factors of various UAVs may be the same or different. The variable $J_{i,j}$ denotes the cost of the jth path planned by the ith UAV, which is determined by the equation $J_{i,j} = f \cdot J_{i,j,threat} + (1-f) \cdot J_{i,j,fuel}$. As for a specific path, it's definite. Thus, the arrival time $T_{i,j}$ is the only independent variable determining $J_{co,i,j}$. The entire cost of the multiple UAVs is

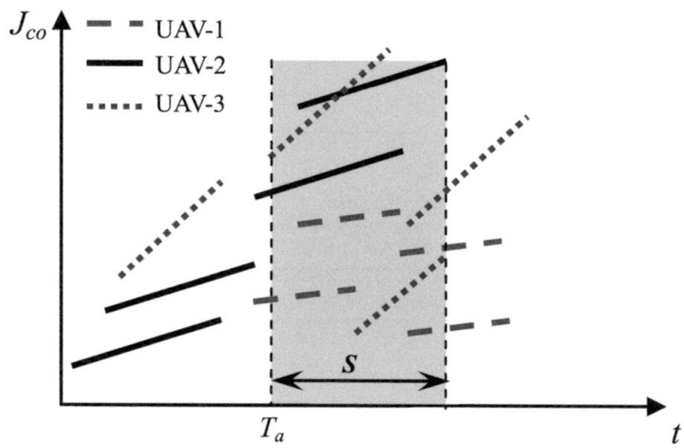

Fig. 4.19 Determination of the coordination variable (Reprinted from Duan et al. (2009), with kind permission from Elsevier)

$$J_{co} = \sum_{i=1}^{N} \sum_{j=1}^{Num} J_{co,i,j}\left(T_{i,j}\right) \tag{4.26}$$

Figure 4.19 shows the relation between $J_{co,i,j}$ and $T_{i,j}$ for each path. As shown in Fig. 4.19, in order to minimize the entire coordination function, the coordinated arrival time T_a often selects the minimum of the feasible region, that is, $T_a = \min S$.

4.5.2 *Coordination Mechanism of Multiple UAV Path Replanning*

The framework of the path planning for a single UAV consists of three main layers: the coordination decider, the path planner, and the path smoother. The path planner quickly calculates a series of safe-enough straight-line trajectories. Communication between the path planner and the coordination decider can help to generate candidate trajectories avoiding overlaps between those of other UAVs'. These candidate trajectories are used by the coordination decider to determine coordination information such as the coordinated time. Because the straight-line trajectories produced by the path planner are not dynamically feasible for the UAV to fly, the path smoother is employed to generate flyable trajectories and send commands to the UAV autopilot. The function of the dynamic path smoother is to smooth junctions in the path with a sequence of radial arcs that can be flown by the UAV. It is essential for timing-critical missions that the length of the original straight-line path must be preserved in the smoothing process. The following Fig. 4.20 displays this mechanism.

The following Fig. 4.21 describes the coordinated mechanism of multiple UAV coordinated path replanning. The framework is distributed, enabling each UAV to

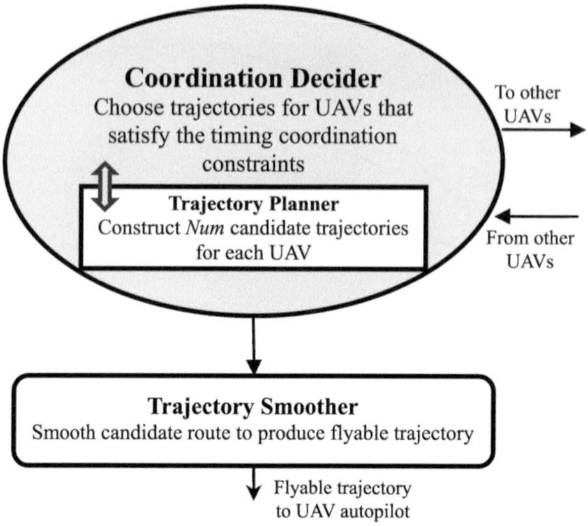

Fig. 4.20 Path-planning mechanism for a single UAV (Reprinted from Duan et al. (2009), with kind permission from Elsevier)

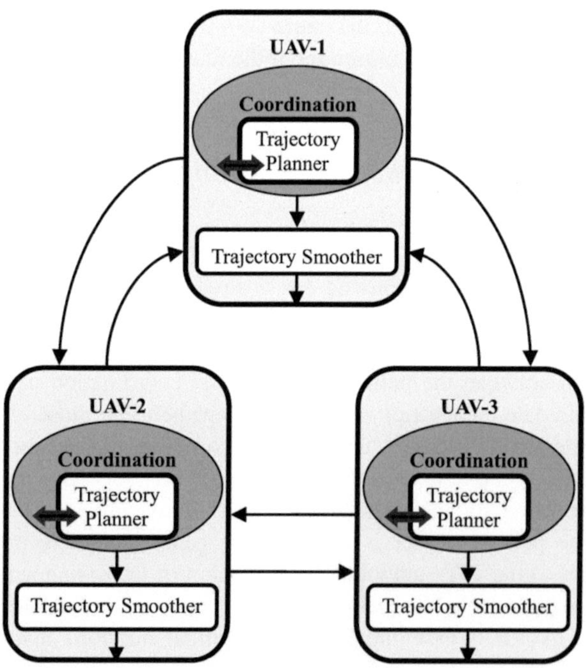

Fig. 4.21 Mechanism of multiple UAV coordination (Reprinted from Duan et al. (2009), with kind permission from Elsevier)

perform its own path-planning subsystems. The algorithm implemented on each UAV in the coordination decider is identical. Multiple UAVs communicate the coordination information in the forms of coordination function and coordination variable in the coordination decider layer. Using the coordination information, multiple UAVs calculate the coordinated arrival time for the UAVs formation, which is just the ETA. After that, the coordination decider calculates the flying velocity for each UAV according to the generated path and the ETA. Although shown for only three UAVs in Fig. 4.21, the distributed structure of the framework applies to larger numbers of multiple UAVs obviously.

Furthermore, in the modern complicated air battle, UAVs are equipped with the ability to detect their surroundings, so that they can sense the pop-up threats occurring in the mission region. The cause of path replanning lies in the following aspects: (1) The pop-up threat is detected just on the flight route ahead, and the vehicle has to change its route out of security; (2) the UAV is not threatened by the pop-up threat, but in view of coordination, it also receives the command of replanning; (3) the mission changes. For the sake of simplicity, the third cause is not involved. When the replanning is inevitable, information about the pop-up threats including the locations and threat grades will soon be shared by the multiple UAVs. Then, every UAV slows down or speed up and flies to a neighboring safe node, which will be served as the starting point of the following replanning. During the span, the new replanned path is generated for multiple UAVs, and so does the new ETA. In the replanning procedure, on the basis of original edge cost of the 2-D mesh, the first step is to update the original threat cost of those edges threatened by pop-up threats and recalculate their cost function value. After that, multiple UAVs start to implement the coordinated path replanning from the new starting points to their arranged targets. This procedure is shown in Fig. 4.22. In this procedure, a set of new trajectories ought to be reproduced, and UAVs also should recalculate the ETA and readjust respective flight velocity. The procedure of path replanning will perhaps be carried on more than once, due to the complicated combating environment that changes uncertainly.

4.5.3 Procedures of Multiple UAV Coordinated Path Replanning

4.5.3.1 Principle Model of ACO with Improved Strategies

The basic mathematical model of ACO has firstly been applied to the TSP. The aim of the TSP is to find the shortest path that traverses all cities in the problem exactly once, returning to the starting city. While the UAV path planning is to find the optimal or suboptimal safe flight path, along which UAV is able to accomplish the prearranged task and avoid the hostile threats. There are some common points between TSP and UAV path planning, and ACO is a feasible way in solving UAV path-planning problem under complicated combat field environment.

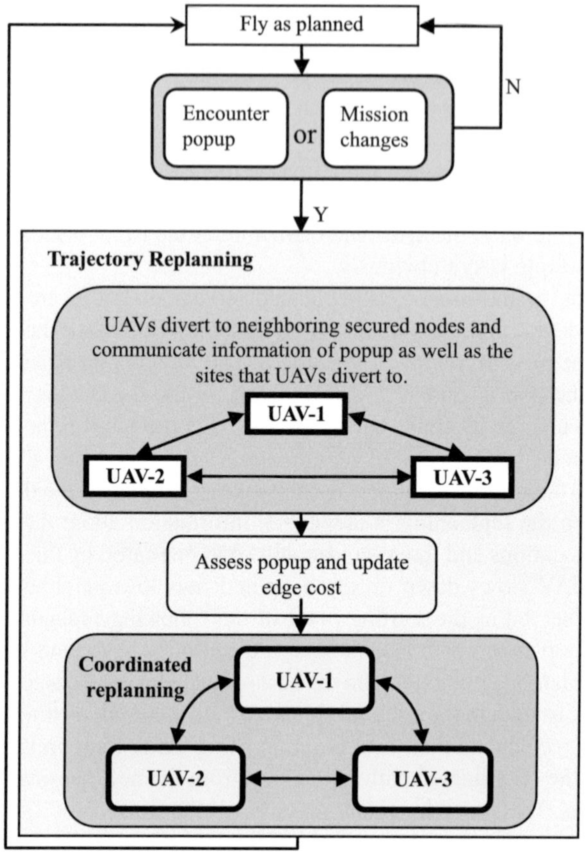

Fig. 4.22 Procedure of multiple UAV coordination path replanning (Reprinted from Duan et al. (2009), with kind permission from Elsevier)

For the *i*th vehicle in the formation of *N* UAVs, let *m* ants be in its starting point; the ants will choose the next nodes in the grid network diagram according to the transition rule. An ant-left pheromone which can be felt by the next ant as a signal to affect its action, and the pheromone which the following one left will enhance the original pheromone. Thus, the more ants a mesh edge is passed by, the bigger possibility that the edge can be selected by the other ants. This process can guarantee nearly all ants walk along the shortest UAV path in the end.

The key factors in ACO affecting the ants' behaviors are the pheromone τ and the heuristic desirability η. In our work, we described the heuristic desirability from node *s* to node *u* as follows:

$$\eta_{su} = \frac{1}{J_{su} \cdot d_{u,t\arg et}} \tag{4.27}$$

where, J_{su} is the total cost of the edge (s,u) and $d_{u,t\arg et}$ represents the distance from node u to the target, which is used to lead ants located at node s to tend to choose those nodes that are nearer to the target.

The amount of pheromone trail τ that leads ants to choose the next node consists of two parts in our work: one is the traditional edge pheromone τ^e, and the other is the point pheromone τ^p defined for the collision avoidance consideration. Ant colonies do not only leave their pheromone on the edges they passed but also deposit the pheromone on those nodes in their paths. Ants serving for the ith UAV will tend to choose those edges richer in its own edge pheromone τ_i^e and avoid the nodes with bigger point pheromone of other UAVs. Thus, total pheromone considered by the ith UAV ants from node s to u is determined by the following equation:

$$\tau_{i,su} = \tau_{i,su}^{e} \cdot \frac{N-1}{\sum\limits_{j \neq i} \tau_{j,u}^{p}} \tag{4.28}$$

We define the transition probability from node s to u for the kth ant as follows:

$$p_{i,su}(t) = \begin{cases} \dfrac{\left[\tau_{i,su}(t)\right]^{\alpha}\left[\eta_{su}\right]^{\beta}}{\sum\limits_{v \in allowed_k} [\tau_{i,sv}(t)]^{\alpha}[\eta_{sv}]^{\beta}} & if \ u \in allowed_k \\ 0 & otherwise \end{cases} \tag{4.29}$$

where $allowed_k$ denotes the feasible domain of the kth ant. α and β are parameters that control the relative importance of trail versus visibility.

After the ants in the algorithm construct their paths, the edge pheromone trail values of every edge (s,u) and the point pheromone of every point u are updated according to the following equations:

$$\tau_{i,su}^{e}(t+1) = (1-\rho) \cdot \tau_{i,su}^{e}(t) + \Delta\tau_{i,su}^{e} \tag{4.30}$$

$$\tau_{i,u}^{p}(t+1) = (1-\rho) \cdot \tau_{i,u}^{p}(t) + \Delta\tau_{i,u}^{p} \tag{4.31}$$

where ρ is the local pheromone decay parameter and $\rho \subset (0,1)$. ρ represents the evaporation rate of trail between time t and $t+1$:

$$\Delta\tau_{i,su}^{e} = \sum_{k=1}^{n} \Delta\tau_{i,su,k}^{e} \tag{4.32}$$

$$\Delta\tau_{i,u}^{p} = \sum_{k=1}^{n} \Delta\tau_{i,u,k}^{p} \tag{4.33}$$

where $\Delta\ \tau^{e}_{i,su,k}$ and $\Delta\tau^{p}_{i,su,k}$ are the quantity of pheromone trail laid on edge (s,u) and the node u by the kth ant of the ith UAV between time t and $t+1$. In the popular ant-cycle model, they can be given by

$$\Delta\tau^{e}_{i,su,k} = \begin{cases} \frac{Q}{J_{i,k}}, & if\ k-th\ ant\ use\ (s,u) \\ 0, & otherwise \end{cases} \tag{4.34}$$

$$\Delta\tau^{p}_{i,u,k} = \begin{cases} \frac{Q}{J_{i,k}}, & if\ k-th\ ant\ use\ node\ u \\ 0, & otherwise \end{cases} \tag{4.35}$$

where Q is a constant, and $J_{i,k}$ denotes the path cost of the kth ant.

4.5.3.2 Max–Min Adaptive ACO

In order to enhance performance of ant system, alleviate the problems concerning early stagnation, and expedite the rapidity of convergence, the following strategies are introduced into the ACO algorithm:

Firstly, for the m ants serving for the ith UAV, there are total m paths constructed in every iteration. The average cost of these paths is $J_{i,ave}(t) = \frac{1}{m}\sum_{k=1}^{m} J_{i,k}(t)$; when and only when the path cost of kth ant in the tth iteration satisfies $J_{i,k}(t) \le J_{i,\min}(t)$ can the kth ant update both its edge and point pheromone.

Secondly, independent of the choice between the iteration-best and the global-best ant for the pheromone trail update, search stagnation may occur. Such a stagnation situation should be avoided. One way for achieving this is to influence the probabilities for choosing the next solution component, which depends directly on the pheromone trails and the heuristic information. The heuristic information is typically problem dependent and static throughout the application of the algorithm. But by limiting the influence of the pheromone trails, one can easily avoid the relative differences between the pheromone trails during the employment of the algorithm. To achieve this goal, ACO imposes and explicitly limits $\tau_{\max}$ and $\tau_{\min}$ on the minimum and maximum pheromone trails for all pheromone trails. After updating pheromone in the end of iteration, the following operation will be applied to the pheromone on both edges and points:

$$\tau^{new}(t) = \begin{cases} \tau_{\min}, & \tau^{old}(t) < \tau_{\min} \\ \tau^{old}(t), & \tau_{\min} \le \tau^{old}(t) \le \tau_{\max} \\ \tau_{\max}, & \tau^{old}(t) > \tau_{\max} \end{cases} \tag{4.36}$$

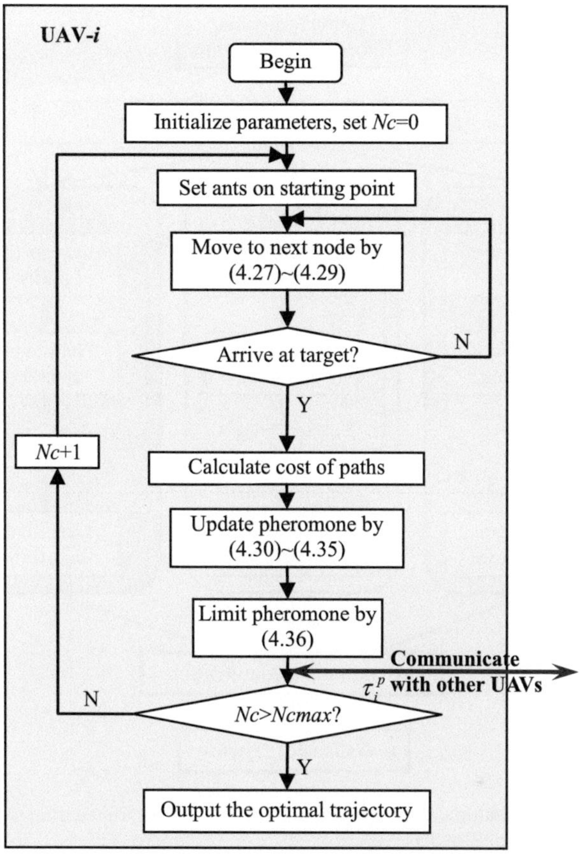

Fig. 4.23 Flowchart of ACO applied to one UAV of multiple UAVs (Reprinted from Duan et al. (2009), with kind permission from Elsevier)

4.5.3.3 Procedures of Coordinated Path Replanning Based on Max–Min Adaptive ACO

The flowchart in Fig. 4.23 describes the detailed procedure of applying the proposed Max–Min adaptive ACO to one single UAV in the practical issue of multiple UAV coordinated path replanning. Ants of the *i*th UAV have constructed their paths and finished updating both the edge and point pheromone trails. Then, the new updated pheromone is then passed to the next iteration. Meanwhile, the point pheromone is transmitted to other UAVs. The multiple UAVs' air–space coordination which is mainly to deal with the collision avoidance just depends on the point pheromone. Therefore, through the communication of each UAVs' point pheromone, the air–space coordination can be settled in the path-planner layer.

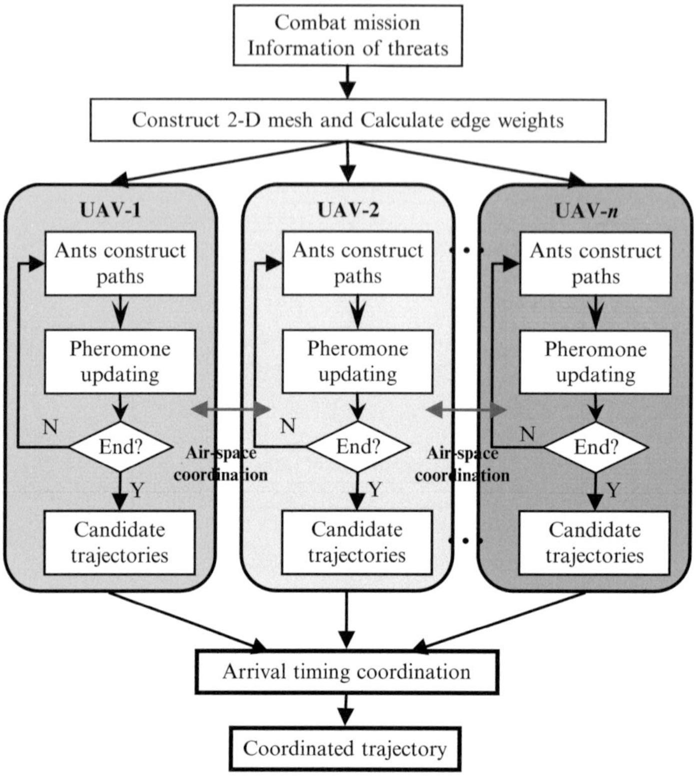

Fig. 4.24 Flowchart of multiple UAV coordination using ACO (Reprinted from Duan et al. (2009), with kind permission from Elsevier)

A series of feasible candidate routes have been determined for each UAV by the coordinated ACO; it remains to select which path each UAV will fly. We selected four optimal candidate trajectories for every UAV from each subpopulation, then determined the optimal coordination ETA by means of finding the coordination variable T_a that minimizes the total coordination cost. After gaining the team's coordinated ETA, the trajectories that multiple UAVs will fly are available, so do the flight velocities of the UAVs. The complete procedure of applying ACO to the multiple UAV coordination issue is shown in Fig. 4.24.

The programming steps of the Max–Min adaptive ACO algorithm in solving path replanning can be described as follows:

Step 1: Construct the 2-D mesh covering the mission region and calculate the cost of the edges.

Step 2: Initialize the parameters of the algorithm, including α, β, ρ, Q, $\tau_{\max}$, and $\tau_{\min}$, as well as the number of ants m and the number of iteration Nc max.

Step 3: Initialize the edge and point pheromones for every subpopulation and place ants at the respective starting point.
Step 4: For every UAV, ants choose the node until they reach the target, and then some feasible trajectories are constructed at last.
Step 5: Calculate the cost function value $J_{i,k}$ of the kth ant belonging to the ith UAV and update the point and edge pheromones.
Step 6: Pass the pheromone on to the next iteration calculation and communicate the point pheromone with other UAVs; then return to Step 4 until it satisfies the ending condition $Nc > Nc\,\max$.
Step 7: Select several feasible candidate routes for each UAV, and determine the optimal coordinated ETA for the team.
Step 8: According to the team ETA, select the path and the flight velocity for each UAV.

When the pop-up threats are detected and the original path is in danger, emergency response action of multiple UAVs will be taken according to the following steps:

Step 1: Determine the information of the pop-up threats, including the location, threatened range, and threat grade.
Step 2: Calculate the threat cost of the edges which the pop-up threats pose to, and then update the cost function value of each edge in the 2- D mesh.
Step 3: Every UAV diverts to a neighboring and safe-enough node, which will be taken as the new starting point of path replanning.
Step 4: During the time that multiple UAVs are moving as Step 3, the newly replanned path is calculated according to the procedure shown in Figs. 4.23 and 4.24.

4.5.4 Experiments

In order to investigate the feasibility and effectiveness of the proposed Max–Min adaptive ACO approach to multiple UAV coordinated path replanning, a series of simulation experiments have been conducted in dynamic and uncertain environments. These simulation experiments were all implemented in Matlab (Version 7.0) programming environment on an Intel Core 2 PC running Windows Vista; no ACO or multiple UAV tools were used in the following experiments.

In these simulation experiments, the mission region is 60 km long and 70 km wide with 5 known enemy threats. Information about these hazardous threats was set as the following (Table 4.2):

Firstly, two UAVs are assigned to reach the same target from neighboring nodes. In this scenario, only air–space coordination was considered in order to verify the collision avoidance performance of our proposed Max–Min adaptive ACO model. Trajectories optimized in these experiments are presented in Fig. 4.25, and it is

Table 4.2 Information about known threats

No.	Location (km)	Threat radius (km)	Threat grade
1	(52,32)	10	2
2	(36,26)	6	1.2
3	(22,48)	8	1.6
4	(26,56)	12	1.4
5	(30,30)	9	2

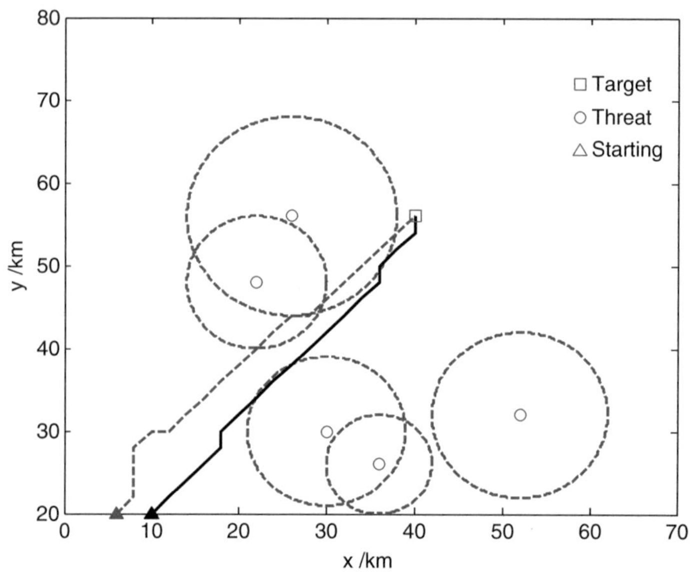

Fig. 4.25 Trajectories of 2 UAVs (Reprinted from Duan et al. (2009), with kind permission from Elsevier)

clear that there is no overlap appearing between the two neighboring trajectories. UAVs that fly along these trajectories will not hit each other, and thus the collision avoidance is achieved. Figure 4.26 shows the evolution curves of the two UAV trajectories' costs, which converge after a few iterations. The simulation results of two UAVs' air–space coordination demonstrate that the ACO, considering the point pheromone as air–space coordination factor, is feasible to produce the trajectories satisfying collision avoidance.

Assume such a mission scenario, an air combat formation, which is composed of three UAVs located on the different sites, is assigned to attack different targets simultaneously. Table 4.3 shows the mission starting points and attacking targets of multiple UAVs.

In this assumptive scenario, each UAV flies along its path with velocity between $V_{\max} = 300m/s$ and $V_{\min} = 200m/s$.

Set the initialization parameters as follows: $\alpha = 3,\ \beta = 2,\ \rho = 0.7,\ Q = 10,$ $Nc\max = 20$, $m = 20$, $\tau_{\max} = 10$, and $\tau_{\min} = 0.1$.

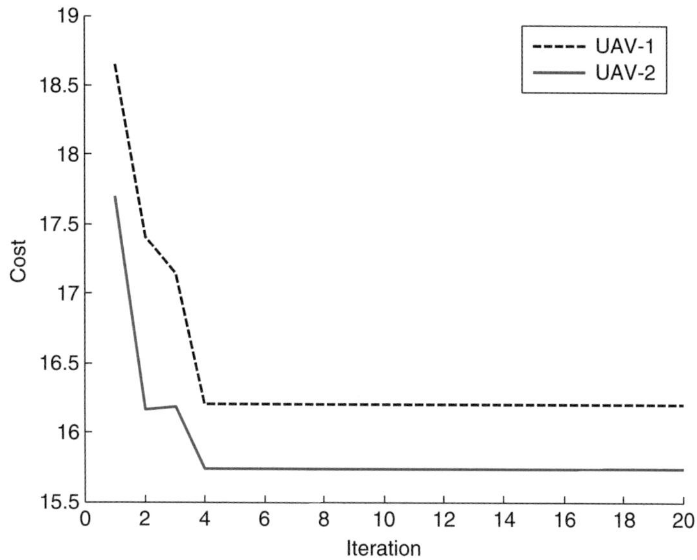

Fig. 4.26 Evolution curves of 2 UAVs using the Max–Min adaptive ACO

Table 4.3 Mission starting points and attacking targets

	UAV-1	UAV-2	UAV-3
Starting point (km)	(6,20)	(24,20)	(40,20)
Target (km)	(10,68)	(40,60)	(50,68)

After iteration calculation in the process of the coordination path preplanning, each UAV obtained four candidate trajectories. The determined arrival time ETA is shown in Fig. 4.27, in which $T_a = 173s$. According to this ETA, each UAV can select its optimized path and proper flight velocity. The selected path length and flight velocity are listed in Table 4.4.

The original trajectories planed for 3 UAVs are shown in Fig. 4.28, while the evolution curves are shown in Fig. 4.29.

As is shown in Fig. 4.30, two pop-up threats appear suddenly, which are detected by marching multiple UAVs. Information about these pop-up threats in the following table is shared immediately (Table 4.5).

Then, each UAV diverts to a neighboring secure node and treats it as the new starting point. Meanwhile, computers equipped in UAVs run the replanning program and generate new trajectories for multiple UAVs.

Comparison between the original trajectories and the new replanned trajectories is shown in Figs. 4.31, and 4.32 shows the practical trajectories flown by multiple UAVs in the air battlefield.

Figures 4.33, 4.34, and 4.35 demonstrate the other experimental results of multiple UAV coordinated path replanning under various complicated environments. Different from the mission assumed in the preceding, the target of multiple UAVs

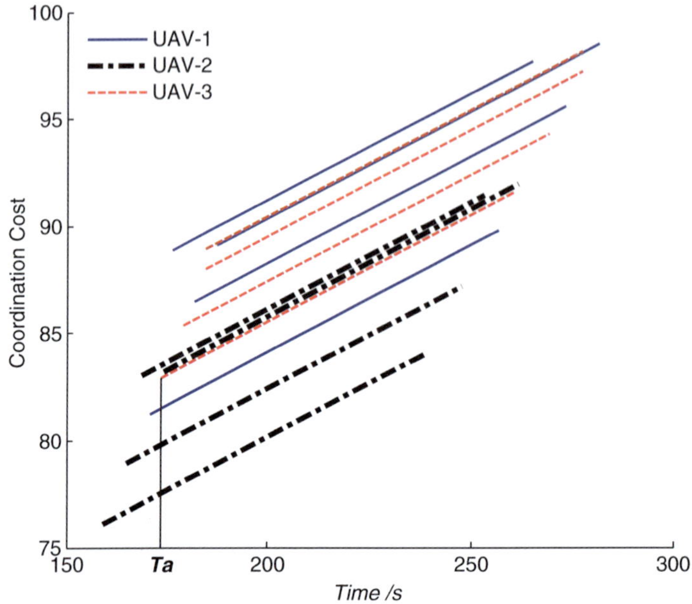

Fig. 4.27 Decision of ETA (Reprinted from Duan et al. (2009), with kind permission from Elsevier)

Table 4.4 Arrival time, path length, and flight velocity planned for multiple UAVs

	UAV-1	UAV-2	UAV-3
ETA		174s	
Path length (km)	51.31	47.80	52.14
Flight velocity (km/s)	295	275	300

is the same destination. Because the original preplanned trajectories are hardly menaced by the pop-up threats, the replanned trajectories change little compared with the original ones.

The above experimental results illustrated that the proposed Max–Min adaptive ACO algorithm can solve the multiple UAV coordinated path-replanning problems in dynamic and uncertain environments effectively, and the convergence time is also rather short.

4.6 Conclusions

This chapter presented the main properties of path planning for UAVs, involving 2-D and 3-D path planning and coordinated path replanning in dynamic and uncertain environments. Path planning is an imperative task required in the design of UAVs, which is to search out an optimal or near-optimal flight path between an initial

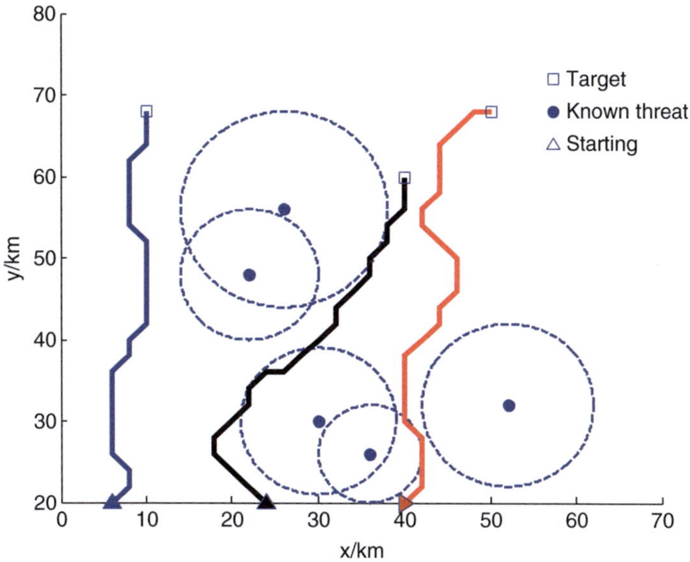

Fig. 4.28 Preplanned trajectories of 3 UAVs (Reprinted from Duan et al. (2009), with kind permission from Elsevier)

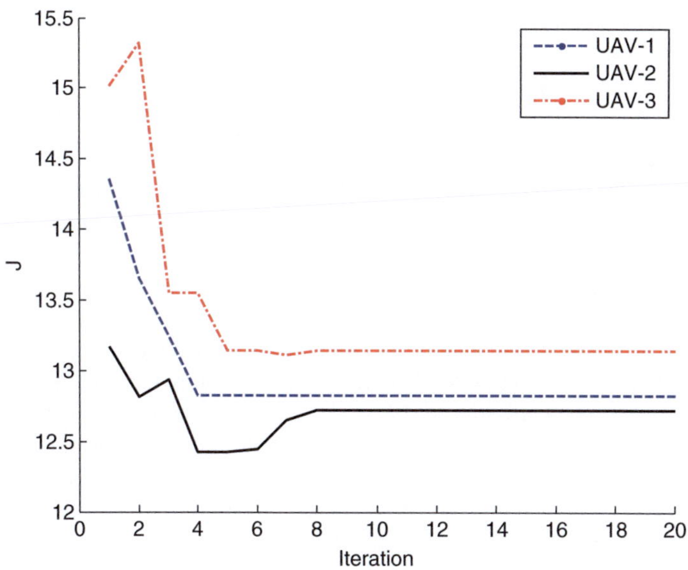

Fig. 4.29 Evolution curves of 3 UAVs using ACO (Reprinted from Duan et al. (2009), with kind permission from Elsevier)

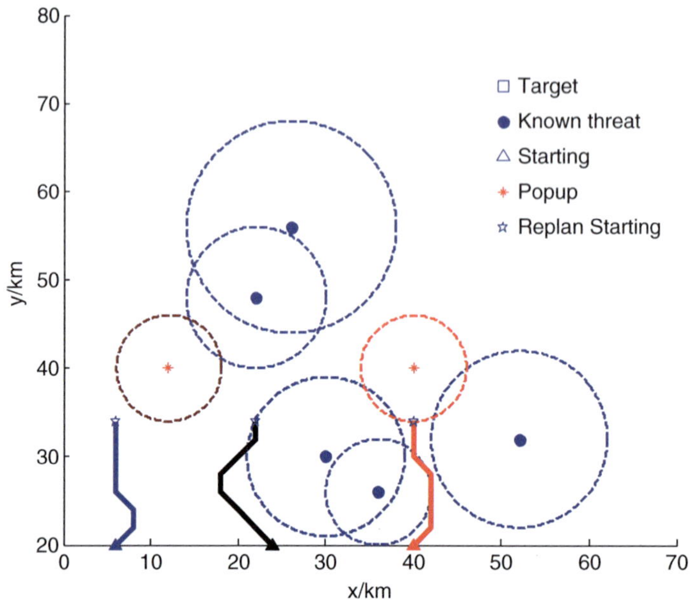

Fig. 4.30 Pop-up threats are detected by marching 3 UAVs (Reprinted from Duan et al. (2009), with kind permission from Elsevier)

Table 4.5 Information about pop-up threats

No.	Location (km)	Threat radius (km)	Threat grade
1	(12,40)	6	2.5
2	(40,40)	6	2

location and the desired destination under specific constraint conditions. There are several considerations for an ideal path planner, which includes optimality, completeness, and computational complexity, the last one of which is the most important requirement since path planning has to occur quickly due to fast vehicle dynamics.

In Sect. 4.3, 2-D path-planning problem is dealt with using a chaotic ABC approach under the assumption that the UAV maintains constant flight altitude and speed when on a mission and that the enemy's defensive areas are flat. A new hybrid meta-heuristic ACO and DE algorithm is proposed to solve the UAV three-dimensional path-planning problem in Sect. 4.4. DE is applied to optimize the pheromone trail of the improved ACO model during the process of ant pheromone updating. Then, the UAV can find the safe path by connecting the chosen nodes of the three-dimensional mesh while avoiding the threats area and costing minimum fuel. This new approach can accelerate the global convergence speed while preserving the strong robustness of the basic ACO. Based on the construction of the basic model of multiple UAV coordinated path replanning, which includes problem description, threat modeling, constraint conditions, coordinated function,

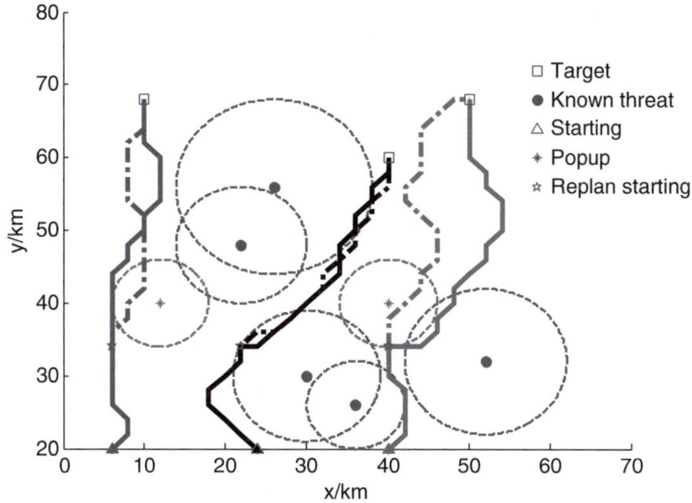

Fig. 4.31 Comparison between the original trajectories and the new replanned trajectories (Reprinted from Duan et al. (2009), with kind permission from Elsevier)

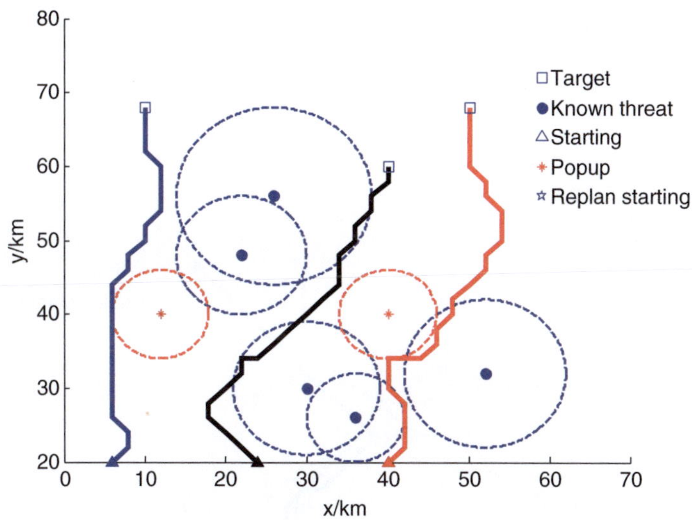

Fig. 4.32 Practical trajectories of 3 UAVs in dynamic and uncertain environments (Reprinted from Duan et al. (2009), with kind permission from Elsevier)

and coordination mechanism, a novel Max–Min adaptive ACO approach to multiple UAV coordinated path replanning is presented in detail in Sect. 4.5. In view of the characteristics of multiple UAV coordinated path replanning in dynamic and uncertain environments, the minimum and maximum pheromone trails in ACO are set to enhance the searching capability, and the point pheromone is adopted to

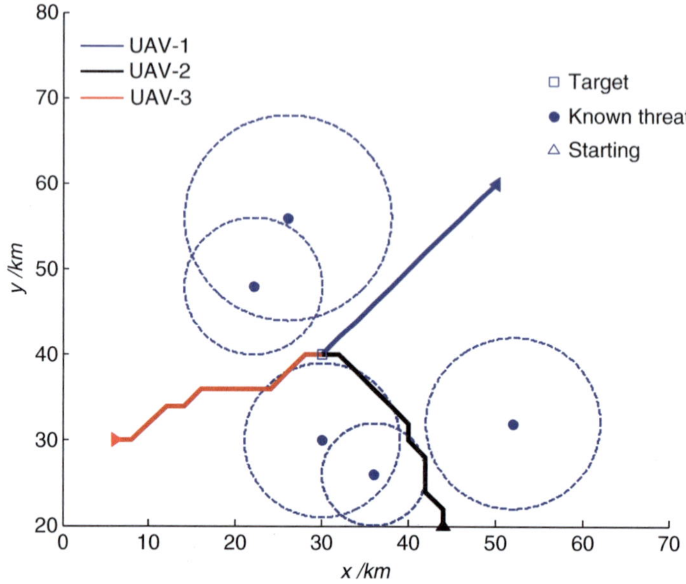

Fig. 4.33 The original trajectories (Reprinted from Duan et al. (2009), with kind permission from Elsevier)

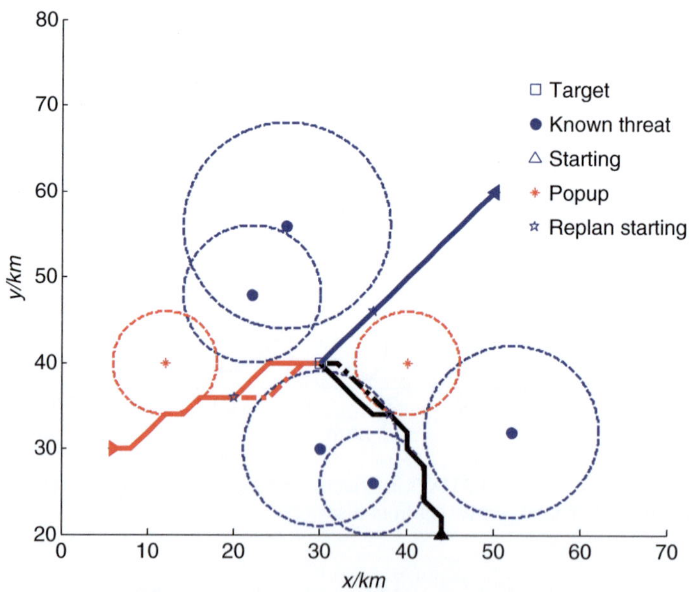

Fig. 4.34 Comparison between the original trajectories and the new replanned trajectories (Reprinted from Duan et al. (2009), with kind permission from Elsevier)

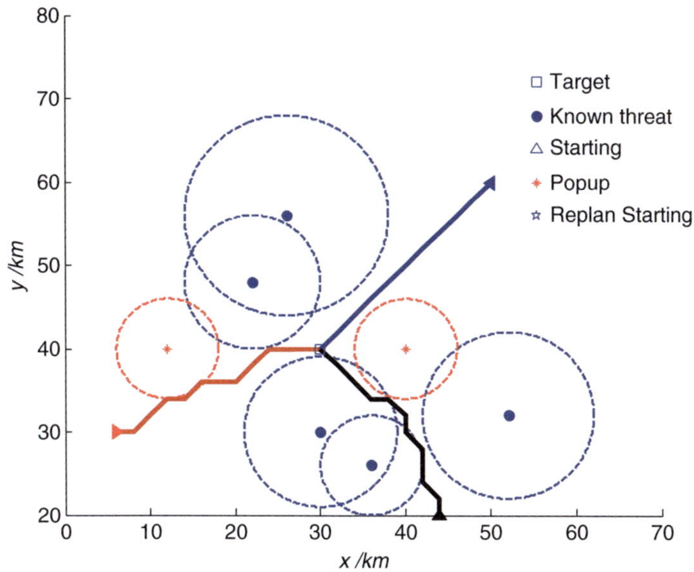

Fig. 4.35 Practical trajectories of multiple UAVs in dynamic and uncertain environments (Reprinted from Duan et al. (2009), with kind permission from Elsevier)

achieve the collision avoidance between various UAVs at the path-planner layer. Considering the simultaneous arrival and the air–space collision avoidance, an ETA is decided firstly, and then the path and flight velocity of each UAV are determined.

References

Anderson EP, Beard RW, McLain TW (2005) Real-time dynamic trajectory smoothing for unmanned air vehicles. IEEE Trans Control Syst Technol 13(3):471–477

Beard RW, McLain TW (2003) Multiple UAV cooperative search under collision avoidance and limited range communication constraints. In: Proceedings of 42nd IEEE Conference on Decision and Control, Hawaii. IEEE, pp 25–30

Beard RW, McLain TW, Goodrich MA, Anderson EP (2002) Coordinated target assignment and intercept for unmanned air vehicles. IEEE Trans Rob Autom 18(6):911–922

Dorigo M, Di Caro G (1999) Ant colony optimization: a new meta-heuristic. In: Proceedings of the 1999 Congress on Evolutionary Computation, Washington, DC. IEEE, pp 1470–1477

Dorigo M, Maniezzo V, Colorni A (1996) Ant system: optimization by a colony of cooperating agents. IEEE Trans Syst Man Cybern B Cybern 26(1):29–41

Duan H, Huang L (2013) Imperialist competitive algorithm optimized artificial neural networks for UCAV global path planning. Neurocomputing. doi:10.1016/j.neucom.2012.09.039

Duan H, Xu C, Liu S, Shao S (2010a) Template matching using chaotic imperialist competitive algorithm. Pattern Recognit Lett 31(13):1868–1875

Duan H, Xu C, Xing Z (2010b) A hybrid artificial bee colony optimization and quantum evolutionary algorithm for continuous optimization problems. Int J Neural Syst 20(1):39–50
Duan H, Yu Y, Zhang X, Shao S (2010c) Three-dimension path planning for UCAV using hybrid meta-heuristic ACO-DE algorithm. Simul Model Pract Theory 18(8):1104–1115
Duan H, Zhang X, Wu J, Ma G (2009) Max-min adaptive ant colony optimization approach to multi-UAVs coordinated trajectory replanning in dynamic and uncertain environments. J Bionic Eng 6(2):161–173
Duan H, Zhang X, Xu C (2011) Bio-inspired computing. Science Press, Beijing
Karaboga D, Basturk B (2007) A powerful and efficient algorithm for numerical function optimization: artificial bee colony (ABC) algorithm. J Glob Optim 39(3):459–471
Kito T, Ota J, Katsuki R, Mizuta T, Arai T, Ueyama T, Nishiyama T (2003) Smooth path planning by using visibility graph-like method. In: Proceedings of IEEE International Conference on Robotics and Automation (ICRA'03) Taipei. IEEE, pp 3770–3775
Li P, Duan H (2012) Path planning of unmanned aerial vehicle based on improved gravitational search algorithm. Sci China Technol Sci 55(10):2712–2719
Liu F, Duan H, Deng Y (2012) A chaotic quantum-behaved particle swarm optimization based on lateral inhibition for image matching. Optik 123(21):1955–1960
Lorenz EN (1963) Deterministic nonperiodic flow. J Atmos Sci 20(2):130–141
Price K, Storn R (1997) Differential evolution–a simple evolution strategy for fast optimization. Dr Dobb's j 22(4):18–24
Xu C, Duan H, Liu F (2010) Chaotic artificial bee colony approach to Uninhabited Combat Air Vehicle (UCAV) path planning. Aerosp Sci Technol 14(8):535–541
Yang G, Kapila V (2002) Optimal path planning for unmanned air vehicles with kinematic and tactical constraints. In: Proceedings of the 41st IEEE Conference on Decision and Control, Las Vegas. IEEE, pp 1301–1306
Yu J, Duan H (2013) Artificial bee colony approach to information granulation-based fuzzy radial basis function neural networks for image fusion. Optik 124(17):3103–3111
Zheng C-W, Ding M-Y, Zhou C-P (2002) Cooperative path planning for multiple air vehicles using a co-evolutionary algorithm. In: Proceedings of 2002 International Conference on Machine Learning and Cybernetics, Beijing. IEEE, pp 219–224
Zheng C, Li L, Xu F, Sun F, Ding M (2005) Evolutionary route planner for unmanned air vehicles. IEEE Trans Robot 21(4):609–620
Zucker M, Kuffner J, Branicky M (2007) Multipartite RRTs for rapid replanning in dynamic environments. In: Proceedings of 2007 IEEE International Conference on Robotics and Automation, Roma. IEEE, pp 1603–1609

Chapter 5
Multiple UAV Formation Control

Haibin Duan

Abstract Formation flight has long been performed by many species of birds for its social and aerodynamic benefits. As a challenging interdisciplinary research topic, autonomous formation flight for multiple unmanned aerial vehicles (UAVs) is about flying in formations with precisely defined geometries, with the benefits of fuel saving and improved efficiency in air traffic control and cooperative task allocation. This chapter mainly focuses on three important aspects associated with formation, which are respectively formation control, close formation (tight formation), and formation configuration. A chaotic particle swarm optimization (PSO)-based nonlinear dual-mode receding horizon control (RHC) method is proposed to cope with the complexity and nonlinearity of vehicle dynamics. Then a novel type of control strategy of using hybrid RHC and differential evolution (DE) algorithm is proposed based on the nonlinear model of multiple UAV close formation. Moreover, based on the Markov chain model, the convergence of DE is proved. Finally, the formation configuration, which is about diving multiple UAVs to form a new flying formation state, is explained in detail using the RHC-based DE. The global control problem of multiple UAV formation reconfiguration is transformed into several online local optimization problems at a series of receding horizons, while the DE algorithm is adopted to optimize control sequences at each receding horizon.

5.1 Introduction

Unmanned aerial vehicle (UAV), which develops in the direction of unmanned attendance and intelligence, is small in size, is light in weight, is low cost, and is able to operate autonomously. With these qualities, UAV has become one of the inevitable trends of the modern military and civilian applications. Recently there has been a considerable amount of interests in cooperative control of a group of UAVs flying in a formation. When multiple UAVs fly in formation, the formation's initial

---

The original version of this chapter was revised. A correction to this chapter is available at https://doi.org/10.1007/978-3-642-41196-0_9

H. Duan and P. Li, *Bio-inspired Computation in Unmanned Aerial Vehicles*, DOI 10.1007/978-3-642-41196-0_5, © Springer-Verlag Berlin Heidelberg 2014

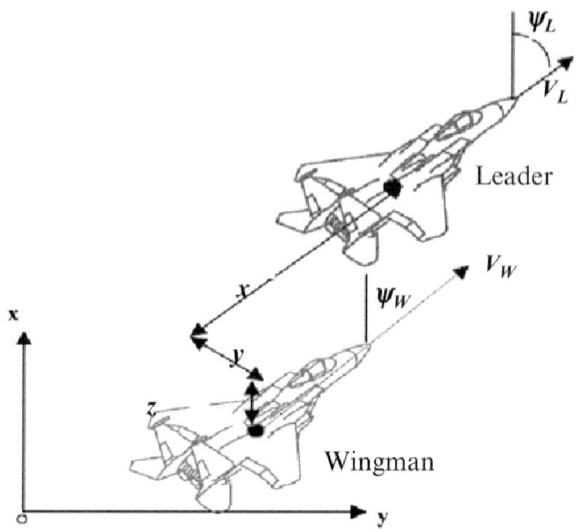

Fig. 5.1 Leader–Wingman formation (Reprinted from Zhang et al. (2010), with kind permission from Springer Science+Business Media)

geometry, including the longitudinal, lateral, and vertical separation, should be preserved during maneuvers with heading change, speed change, and altitude change.

5.1.1 Formation Control

In recent years, formation control of multiple UAVs has become a challenging interdisciplinary research topic, while autonomous formation flight is an important research area in the aerospace field (Duan et al. 2013a). The main motivation is the wide range of military and civilian applications where UAVs formations could provide a low-cost and efficient alternative to existing technology. Multiple UAV teams flying in formations with precisely defined geometries have many advantages, such as energy saving when the vortex forces are taken into account. Formation flight can also be used for airborne refueling and quick deployment of troops and vehicles. Formation flight can be regarded as a complicated control problem which computes the inputs driving the UAVs along challenging maneuvers while maintaining relative positions as well as safe distances between each UAV pair (Duan et al. 2013b). The challenge here lies in designing a formation controller that is computationally simple yet robust.

5.1.2 Close Formation

A close formation, also called "tight formation," is one in which "the lateral separation between UAV is less than a wingspan" (Pachter et al. 2001). In this case, aerodynamic coupling is introduced into the formation's dynamics. Multiple UAVs flying in a close formation can achieve a significant reduction in power demand, thereby improving cruise performances, such as range and speed, or to increase the payload (Binetti et al. 2003). The "Leader–Wingman" formation pattern can be shown with Fig. 5.1. If the Wingman flies in close formation with the leading UAV,

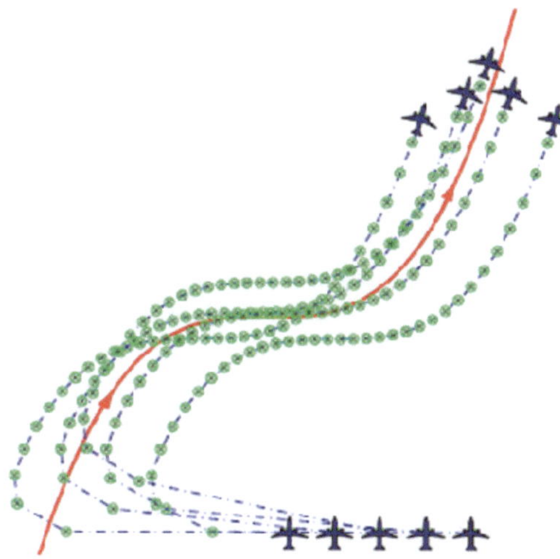

Fig. 5.2 Formation reconfiguration of five UAVs (Reprinted from Zhang and Duan (2012), with kind permission from SAGE Publications)

the Leader's vortices will produce aerodynamic coupling effects, and a reduction in the formation's drag can be achieved. According to the effects of aerodynamic interference, multiple UAV close formation flight control is a complex problem with strongly nonlinear and coupling character.

5.1.3 Formation Configuration

The formation reconfiguration problem for multiple UAVs can be described as follows: given a group of UAVs with an initial configuration, a final configuration, and a set of inter- and intra-UAV constraints, the goal is to determine a nominal control input for each vehicle such that the multiple UAV group can start from the initial configuration and reach its final configuration while satisfying the set of constraints, as is shown in Fig. 5.2. The formation reconfiguration problem can be recognized as an optimal control problem with dynamic constraints (Zelinski et al. 2003; Ueno and Kwon 2007; Duan et al. 2008). Several theoretical techniques such as graph theory (Hendrickx et al. 2008), reconfiguration maps, Dijkstra algorithm (Giulietti et al. 2000), or functional optimization have been developed to define the new/optimal positions to be occupied by the UAVs in the formation.

As a large-scale centralized control problem, formation reconfiguration aims to obtain the control input signals (such as steering angle, throttle/thrust) for each UAV through complex calculation to drive each UAV in a complicated flight maneuver. In this process, multiple UAVs must satisfy several constraints; for example, the distance between two UAVs must be greater than the safety collision distance, and also should not be too greater than the communication distance.

5.2 Dual-Mode RHC for Multiple UAV Formation Flight Based on Chaotic PSO

5.2.1 Leader-Following Formation Model

The point mass model is considered for formation flight. Each UAV is assumed to fly at a constant altitude, parallel to the 2-dimensional region to be surveyed. A commonly used nonlinear kinematics model that represents a UAV with zero or negligible velocity in the direction perpendicular to the UAV's heading is applied to our model.

$$\begin{aligned}\dot{x} &= v\cos\psi \\ \dot{y} &= v\sin\psi \\ \dot{v} &= u \\ \dot{\psi} &= \omega\end{aligned} \tag{5.1}$$

where x and y are the Cartesian coordinates of the UAV, v is the velocity, and ψ is the heading angle in the (x,y) plan (Stipanović et al. 2004). The acceleration in the longitudinal direction u and angular turn rate ω are assumed to be the control inputs to the UAV. Figure 5.3 shows the UAV position and orientation in the plane coordinate system.

In a typical multiple UAV formation flight, the Wingman follows the trajectory of the Leader UAV, taking other aircrafts as reference to keep its own position inside the formation. In a large formation, intra-aircraft distances must be kept constant (Giulietti et al. 2000). The formation model in this paper adopts Leader mode strategy (as shown in Fig. 5.4), which means each Wingman UAV takes its trajectory references from the Leader UAV, while the altitude is the same for all. The Leader UAV takes charge of formation trajectory.

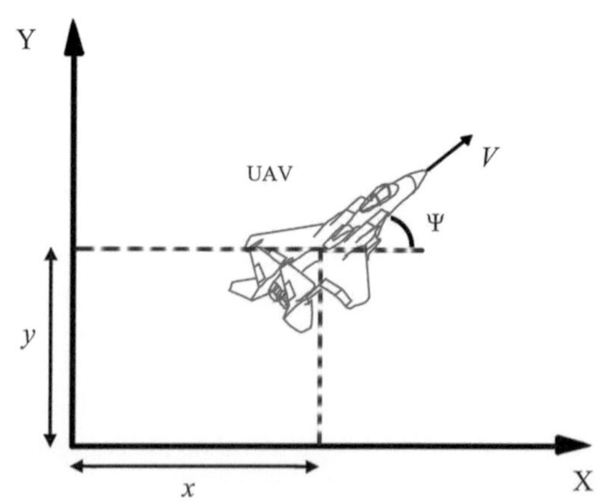

Fig. 5.3 UAV position and orientation (© [2002] IEEE. Reprinted, with permission, from Duan and Liu (2010))

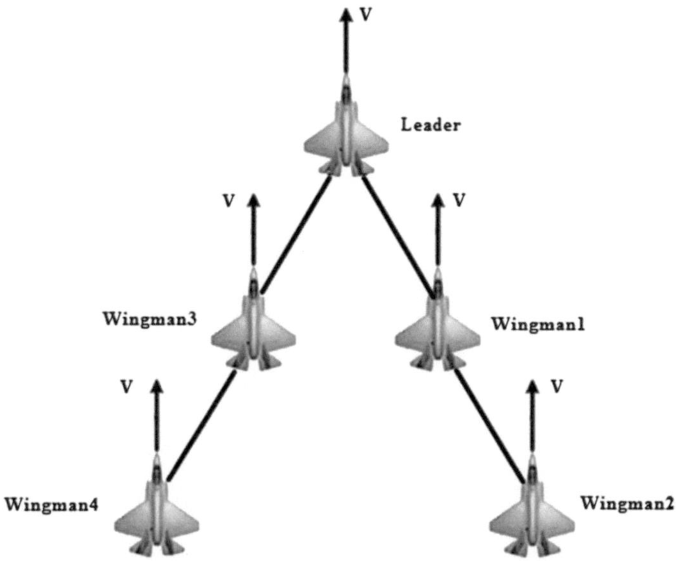

Fig. 5.4 Multiple UAV formation (© [2002] IEEE. Reprinted, with permission, from Duan and Liu (2010))

The Virtual Leader is employed in our model to replace the real UAV Leader so that UAVs adjust speed and heading angle based on the relative states of Virtual Leader (as shown in Fig. 5.5). Then a multiple UAV formation, defined with respect to all the real UAVs as well as to the Virtual Leader, should be maintained at the same time as the Virtual Leader tracks its reference trajectory. The key advantage of the Virtual Leader UAV is that a physical UAV Leader is subject to destruction, while the Virtual Leader can never be damaged. The Virtual Leader provides a stabile, robust reference for formation control.

5.2.2 Principle of RHC

Nonlinear RHC is that the finite time optimal control law is computed by solving an online optimization problem. And linear RHC theory is quite mature so far (Kwon and Han 2005). Generally, many systems are inherently nonlinear, and they are often required to operate over a wide range of operating conditions. Linear models are often inadequate to describe the process dynamics, and nonlinear models have to be used. This motivates the use of nonlinear RHC. The optimization problems over the finite horizon, on which the RHC is based, can be applied to a broad class of

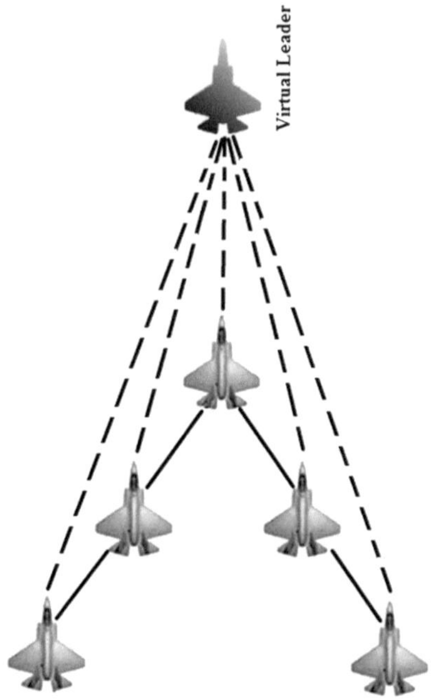

Fig. 5.5 Multiple UAV formation with the Virtual Leader (© [2002] IEEE. Reprinted, with permission, from Duan and Liu (2010))

systems, including nonlinear systems and time-delayed systems. Thus, the RHC has the same broad applications even for nonlinear systems:

$$\begin{aligned} & \min_{u} J = f\left(x, u; t_c, T_p\right) \\ & J = \int_{t_c}^{t_c+T_p} F\left(x, u\right) dt + \Phi\left(x_{t_c+T_p}\right) \\ & subject\ to\ \dot{x} = f\left(x, u\right) \\ & L \leq \begin{bmatrix} x \\ u \end{bmatrix} \leq U \end{aligned} \tag{5.2}$$

where x and u are respectively state vector and control sequence; t_c and T_p represent the control and the prediction horizon with $t_c \leq T_p$; L and U are lower and upper bounds; and δ is the predicted time step.

Receding optimization is the most important idea of RHC, which is also the typical difference between RHC and optimum control, as shown in Fig. 5.6 (Duan and Liu 2010; Zhang et al. 2011; Duan et al. 2011). The whole control process can be divided into a series of optimizing intervals called rolling window or receding horizon. RHC method forms the closed-loop rolling mechanism, including observation, planning, implementation, and reobservation. RHC is a p -step-ahead online optimization strategy. At each time interval, RHC optimizes the specific problem for the following p intervals based on current available information.

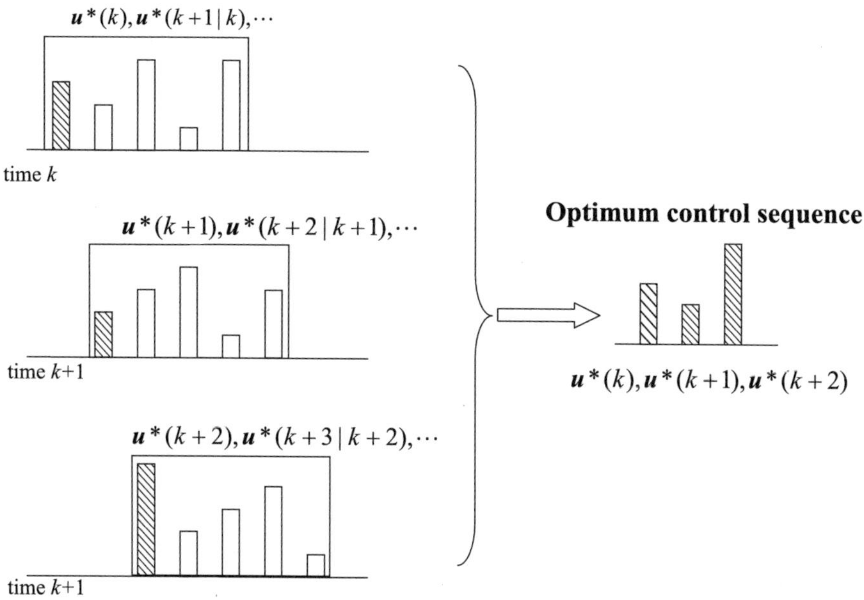

Fig. 5.6 Receding optimization (Reprinted from Zhang et al. (2010), with kind permission from Springer Science+Business Media)

5.2.3 *Chaotic PSO-Based Dual-Mode RHC Formation Controller Design*

5.2.3.1 A Dual-Mode Formation Controller Design

In this subsection, we will introduce the framework of the multiple UAV formation flight controller (Duan and Liu 2010). In our proposed formation flight control strategy, each UAV follows the Virtual UAV Leader.

The ith UAV state vector and control input sequence in (5.1) are

$$\boldsymbol{x}_i = (v_i, \Psi_i, x_i, y_i), \boldsymbol{u}_i = (u_i, \omega_i) \tag{5.3}$$

The Virtual Leader state is $\boldsymbol{x}_{VL}$ and control inputs are $\boldsymbol{u}_{VL}$. According to the UAV Leader, the ith UAV relative state is $\boldsymbol{x}_{ri} = \boldsymbol{x}_i - \boldsymbol{x}_{VL}$. We define formation state and input sequence:

$$X = (\boldsymbol{x}_{VL}, \boldsymbol{x}_1, \ldots, \boldsymbol{x}_N), X_r = (\boldsymbol{x}_{r1}, \ldots, \boldsymbol{x}_{rN}), U = (\boldsymbol{u}_{VL}, \boldsymbol{u}_1, \ldots, \boldsymbol{u}_N) \tag{5.4}$$

In nonlinear RHC, the input applied to the system is usually given by the solution of the following finite horizon optimal control problem according to (5.2), which is solved at every sampling instant:

$$
\begin{aligned}
J &= \min_{U} \int_{t}^{t+T_p} \mathrm{F}\left(X_r(\tau), U(\tau)\right) d\tau + \mathrm{V}\left(X_r\left(t+T_p\right)\right) \\
s.t.\ \dot{X}(\tau) &= \mathrm{f}\left(X(\tau), U(\tau)\right) \\
& U(\tau) \in U \forall \tau \in [t, t+t_c] \\
& X(\tau) \in X \forall \tau \in \left[t, t+T_p\right] \\
& X\left(t+T_p\right) \in \Omega
\end{aligned}
\tag{5.5}
$$

where

$$F(X_r, U) = X_r{}^T Q X_r + U^T R U \tag{5.6}$$

$$V\left(X_r\left(t+T_p\right)\right) = X\left(t+T_p\right)_r{}^T P X\left(t+T_p\right)_r \tag{5.7}$$

The deviation from the desired values is weighted by the positive-definite matrices Q, R, and P; the time step is δ. V is the terminal penalty.

$$\boldsymbol{\Omega} = \left\{X \left| X^T P X \le \alpha \right.\right\} \tag{5.8}$$

The terminal region $\boldsymbol{\Omega}$ is chosen such that it is invariant for the nonlinear system control by using a linear state feedback. As control systems become more complex and performance requirements more demanding, the invariant sets are widely employed to design stabilizing controllers and, in particular, for applying RHC strategies. In order to enlarge the solution range and make search process of PSO easier, the dual-mode control strategy is chosen in this paper, for this strategy provides an efficient way to guarantee the stability of RHC with input constraints. The basic idea is to use a finite horizon of allowable control inputs to steer the state into an invariant set. The terminal region $\boldsymbol{\Omega}$ and terminal penalty matrix P can be determined off-line.

A local linear control law which stabilizes the nonlinear system in $\boldsymbol{\Omega}$ is obtained as follows:

$$\dot{X} = \frac{\partial f}{\partial x}(0,0)\, X + \frac{\partial f}{\partial u}(0,0)\, u \tag{5.9}$$

Substitute the linear state feedback $u = KX$ into (5.9), we can get

$$\dot{X} = \boldsymbol{f} X \tag{5.10}$$

$$\boldsymbol{f} = \frac{\partial f}{\partial x}(0,0) + \frac{\partial f}{\partial u}(0,0)\, K \tag{5.11}$$

Define the following Lyapunov equation:

$$\boldsymbol{f} P + P \boldsymbol{f}^T + Q^* = 0 \tag{5.12}$$

where $Q^* = Q + K^T RK$ and the solution P is a positive-definite symmetric matrix. For any vector $X \in R^n$, $\|X\|$ denotes Euclidean norm. There exists a constant $\alpha \in (0,\infty)$ to fix the terminal region $\boldsymbol{\Omega}$ at the origin as (5.8). The constant α satisfies $KX \in \boldsymbol{U}$ for all $x \in \boldsymbol{\Omega}$ and the following condition, according to (5.8) and $u = KX$:

$$\begin{aligned} \|KX\| &= \left\| K\left(\alpha^{\frac{1}{2}} P^{-\frac{1}{2}}\right)\left(\alpha^{-\frac{1}{2}} P^{\frac{1}{2}} X\right) \right\| \leq \left\| K\left(\alpha^{\frac{1}{2}} P^{-\frac{1}{2}}\right) \right\| \cdot \left\| \alpha^{-\frac{1}{2}} P^{\frac{1}{2}} X \right\| \\ &\leq \left\| K\alpha^{\frac{1}{2}} P^{-\frac{1}{2}} \right\| = \alpha K^T PK \end{aligned} \tag{5.13}$$

It follows the input constraints that

$$\alpha K^T PK \leq u_{\max}^2 \ \|u\| \leq u_{\max} \tag{5.14}$$

As multiple UAV formation state X enter the terminal region $\boldsymbol{\Omega}$, X will be kept in this region all the while and tend to the origin gradually.

5.2.3.2 Collision Avoidance

In multiple UAV formation flight system, each UAV moves in an environment in which there are obstacles and other UAVs. Thus the multiple UAVs, at the same time, have to consider the problem of formation control and collisions avoidance. Collision avoidance is assumed to be the most important task: only when UAV is at safe distance from the other UAVs and the obstacles can it take care of maintaining the formation.

To achieve collision avoidance with other UAVs, a priority indexing scheme is used (Wang et al. 2007): all UAVs are tagged, and the UAV with a lower index creates an imaginary obstacle around the UAV with a higher index (as seen in Fig. 5.7) and tries to avoid it. Thus, collision avoidance is achieved.

The UAVs with a lower index must react rapidly when neighboring UAVs with a higher index approach within unsafe range or when obstacles are detected as they appear within the sensor range, to avoid any collision. Consequently, a multiple UAV formation control strategy that ensures avoidance of collisions is achieved by adding a constraint to (5.5).

$$J_1 = J + Penalty_c \cdot \sum_{j=1}^{Co_{num}} Dis_{\text{constraint}}\left(d_{ij}\right) \tag{5.15}$$

$$Dis_{\text{constraint}}\left(d_{ij}\right) = \begin{cases} 1 \ d_{ij} \leq d_{safe} \\ 0 \ d_{ij} > d_{safe} \end{cases} \tag{5.16}$$

where $Penalty_c$ is the collision penalty coefficient, Co_{num} is the total account of collisions that UAV_i has to avoid, and d_{ij} is the distance between UAV_i and the jth collision center. (The obstacle shape is specified as a circle.) As long as UAV_i spatial

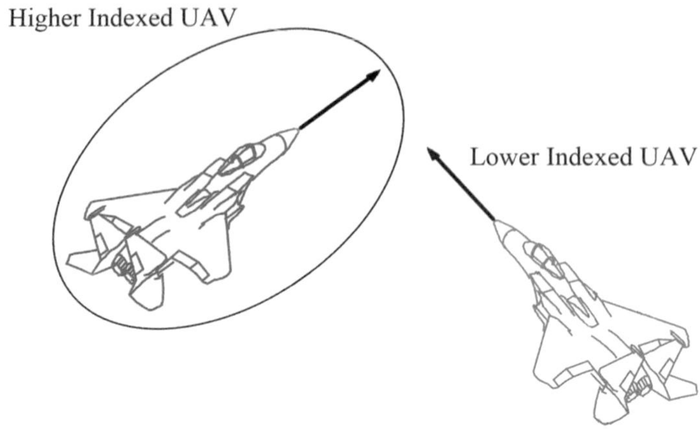

Fig. 5.7 Collision avoidance (© [2002] IEEE. Reprinted, with permission, from Duan and Liu (2010))

horizon overlaps the *j*th obstacle, $Dis_{constraint}(d_{ij}) = 1$ and the value of (5.15) will be very large. Therefore, this constraint is adopted for effectively avoiding collisions.

5.2.3.3 Chaotic Particle Swarm Optimization

In RHC, the cost function (5.15) plays a role of an evaluation function in PSO. The future control input sequence U is obtained by minimizing (5.15) via a particle swarm optimization.

In PSO design, the optimization concepts based on chaotic sequences can be a good alternative to provide diversity in PSO populations. The application of chaotic sequences instead of random sequences in PSO is a powerful strategy to diversify the population of particles and improve the PSO's performance in preventing premature convergence to local minimum. Chaos optimization is realized through chaos variables which can be obtained by many ways. One of the simplest maps which is brought to the attention of scientists by May (1976), which appears in nonlinear dynamics of biological population evidencing chaotic behavior, is logistic map:

$$Z_{n+1} = \mu Z_n (1 - Z_n) \tag{5.17}$$

where Z_n is the *n*th chaotic number where n denotes the iteration number. Obviously, $Z_n \in (0,1)$ under the conditions that the initial $Z_0 \in (0,1)$, and $Z_n \notin \{0.0, 0.25, 0.5, 0.75, 1.0\}$. $\mu = 4$ has been used in our algorithm.

The population diversity measured the average particle distance, which describes population diversity with dispersion degree between particles. Assume L is the maximum length of search space, ps is the population of particles, `ln` is the dimensions of solution space, p_{id} is the dth coordinate of particle p_i, P_d is the average of the dth coordinate, and the average particles distance $D(iter_k)$ at the kth iteration is defined as follows:

$$D\left(iter_k\right) = \frac{1}{ps \cdot L}\sum_{i}^{ps}\sqrt{\sum_{d=1}^{ln}\left(p_{id} - P_d\right)^2} \tag{5.18}$$

In our chaotic PSO approach, the PSO algorithm is first run to find the global-best position as a candidate solution, and once particles collide, $D(iter_k) < \varepsilon$, where ε is a positive constant. Then, the better solution generated from chaotic systems substitute random numbers for the PSO particles, where it is necessary to make a random-based choice. In this way, the global convergence can be improved, and falling into local-best solution can be prevented.

As we have mentioned, PSO can also be improved by a modification of the inertia weight w in (2.11) (Shi and Eberhart 1998). The inertia weight can be used to balance the local and global search during the optimization process. If the inertia weight is big, it is possible to enhance global search. Otherwise, smaller inertia weight will enhance the local search. While the value of w is made to decrease gradually with the increase in the number of iterations by the following equation at the kth iteration $iter_k$:

$$w = w_{\max} - iter_k\frac{w_{\max} - w_{\min}}{iter_{\max}} \tag{5.19}$$

where $iter_{\max}$ is maximum iteration and $w_{\max}$ and $w_{\min}$ are separately maximum and minimum of w.

In order to guarantee the stability and enhance the efficiency of the control algorithm, the initial value of each particle chooses the last control input sequence after the achievement of primary sequence, such as the lth particle at a time t is initialed by $U(t-\delta)$.

The process of our proposed nonlinear dual-mode RHC method based on chaotic PSO for solving multiple UAV formation flight problem can be described as follows:

Step 1. Initialize UAVs states X_0 and the nonlinear dual-mode RHC parameters used in formation system.

Step 2. Evaluate the terminal region $\boldsymbol{\Omega}$ and terminal penalty matrix P by (5.9), (5.10), (5.11), (5.12), (5.13), and (5.14).

Step 3. Detect if formation state $X(t)$ enters the terminal region $\boldsymbol{\Omega}$ or not by (5.8). If it is true, then go to Step 8; else go to Step 4.

Step 4. Initialize particle swarm adopted with the last predictive control sequence $U(t-\delta)$ optimized by PSO, while particle swarm is initialed randomly at the first time.

Step 5. Evaluate the value of each particle by computing the cost function (5.15), and update the particle swarm and the global-best particle gbest according to (2.11).

Step 6. Detect if PSO precociously converge to local minima with (5.18). If it is true, then go to the next step; else go to Step 8.

Step 7. Use chaotic systems to generate a better solution to substitute random numbers for the PSO particles in next iteration and then go to Step 4.

Step 8. Detect the PSO terminate conditions (reaching the maximal generation or finding the idea optimum). If the terminate conditions are met, end the PSO algorithm and return the global-best particle *gbest* as the control sequence $U(t)$, or continue the computation.

Step 9. Apply the first part of the optimal control sequence to update formation state $X(t)$, and then go to Step 1.

Step 10. Use a linear state feedback $u = KX$ to control the nonlinear formation system to guarantee the stability of multiple UAV formation.

Step 11. Detect the formation stability conditions. If certain conditions are achieved, end the formation control method, or go to Step 2.

Figure 5.8 displays the flowchart of the chaotic PSO-based nonlinear dual-mode RHC formation control scheme for multiple UAV formation flight

5.2.4 Experiments

In this section, series experiments have been performed to investigate the performance of the proposed chaotic PSO-based nonlinear dual-mode RHC formation control scheme for multiple UAV formation flight. We use (5.1) to represent states of UAV model respectively. The multiple UAV group consists of 5 agents, with input constraints $[-5, 5]\text{m/s}^2$ for the acceleration u and $[-\pi/18, \pi/18]$ rad/s for the angular turn rate ω. To improve performance and avoid collisions, a safe distance between UAVs is defined, $d_{safe} = 3$. Collisions between UAVs are solved with the priority index strategy. Each UAV is tagged with a serial number.

The initial conditions of the nonlinear dual-mode formation controller are prediction horizon $T_p = 8\text{s}$, time step $\delta = 1\text{s}$ (the simulation time), weighting matrices $Q = \text{diag}(1,1,1,1)$, $R = \text{diag}(1,1)$, $P = (0.0596\ 0\ -0.0063\ 0;0\ 0.0596\ 0\ -0.0063;-0.0063\ 0\ 0.0129\ 0;0\ -0.0063\ 0\ 0.0129)$. The improved PSO parameter setting is the size of the particle swarm $p_s = 20$, inertia weight $w_{\max} = 1.2$, $w_{\min} = 0.1$, particle maximum velocity $vp_{\max} = 4$, $c_1 = 0.5$, $c_2 = 0.5$, and the maximum iteration $iter_{\max} = 200$.

In order to fully illustrate the efficiency of the proposed algorithm, we compare its performance with the standard GA under the same conditions. In both CPSO and GA, the population size is 20. The GA is real valued with random initialization and updates the population and search for the optimum with random techniques. Crossover and mutation probabilities are set as 0.8 and $1/n$, respectively, where n is the dimension of the problem.

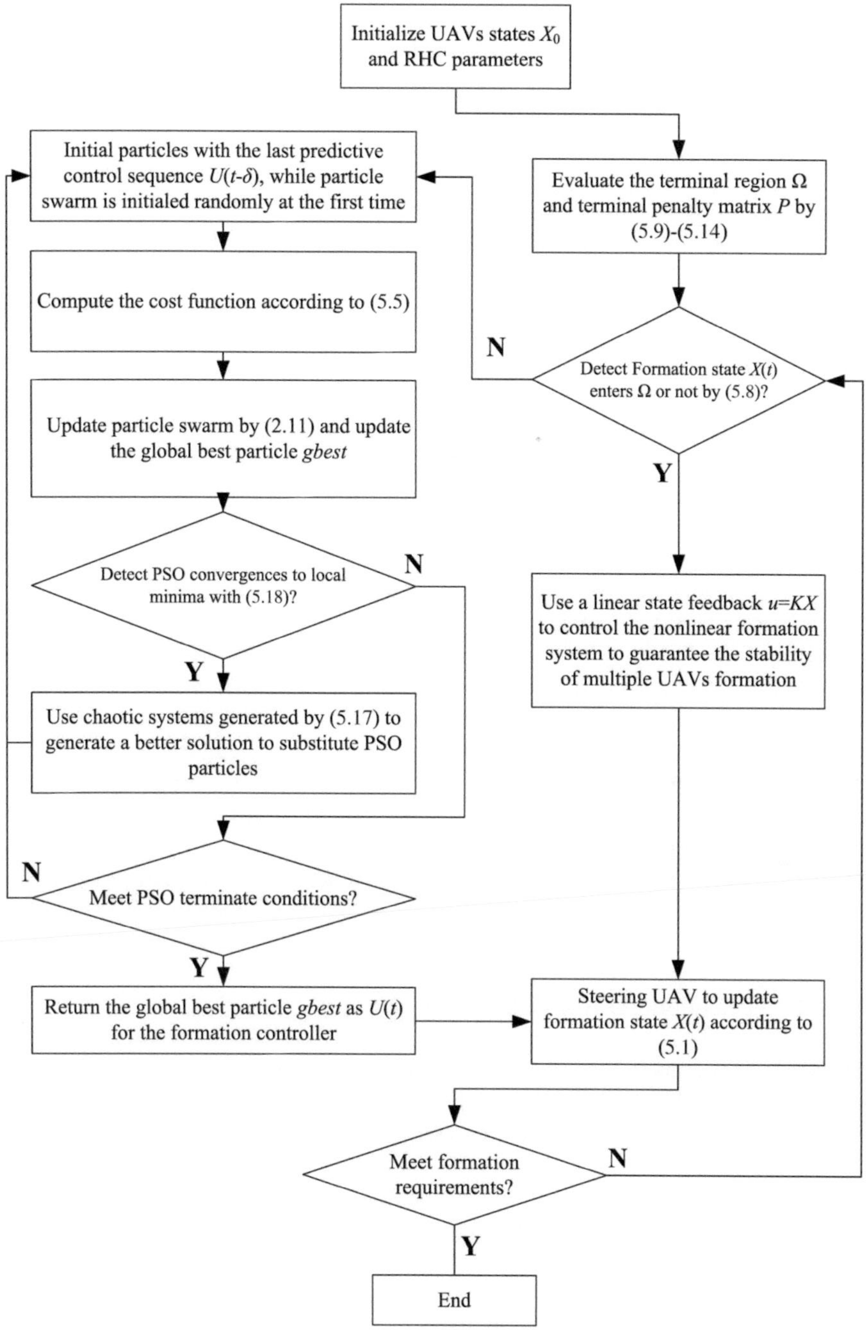

Fig. 5.8 Chaotic PSO-based RHC formation control scheme for multiple UAV formation flight (© [2002] IEEE. Reprinted, with permission, from Duan and Liu (2010))

5.2.4.1 Formation Control

Multiple UAV states are initialed as $\boldsymbol{x}_{VL}$ =(15,0,30,60), $\boldsymbol{x}_1$ =(15,0,5,65), $\boldsymbol{x}_2$=(15,0,5,75), $\boldsymbol{x}_3$ =(15,0,5,85), $\boldsymbol{x}_4$ =(15,0,5,45), $\boldsymbol{x}_5$ =(15,0,5,55), and the relative distances are $\boldsymbol{x}_{r1}$ =(0,0,0,0), $\boldsymbol{x}_{r2}$ =(0,0,–15,15), $\boldsymbol{x}_{r3}$ =(0,0,–30,30), $\boldsymbol{x}_{r4}$ =(0,0,–15,–15), $\boldsymbol{x}_{r5}$ =(0,0,–30,–30). The Virtual Leader is marked with "o," while UAV is "Δ". The five UAVs reconfigure from a "|"initial shape to forming a "V" formation. Assume that the Virtual Leader speed is 15 m/s in the x-direction and its heading angle is 0 rad during the simulations. Figures 5.9 and 5.10 show the detailed results generated by the control sequence optimized by GA and CPSO, respectively.

The UAV group has to follow the Virtual Leader as seen in Fig. 5.9a, which shows the (x,y) positions of UAVs generated by using GA to optimize the control sequence. Note that the UAV group is traveling from left to right in the figure. The results are shown in Fig. 5.9, which illustrates that both the path tracking and formation maintenance tasks are not achieved. Figure 5.9b–e shows the multiple UAV convergence time. It is obvious that the 5th UAV cannot move to the initial relative position to follow the Virtual Leader, while other UAVs can converge to designated position. The same experiment is tested 5 times with similar results that the satisfactory tasks are not achieved.

In Fig. 5.10, UAVs follow the Virtual Leader with a constant velocity in x -direction and the desired separation between UAVs is 15 m in both the x - and y -directions. Under the control inputs optimized by CPSO, the multiple UAVs converge to the desired formation from the same initial configuration. Formation results are presented in Fig. 5.10a. The formations converge in 50s as seen in Fig. 5.10b. Compared with the results generated by the control sequence optimized by GA, CPSO performs better for its comprehensive ability to search and high precision, which demonstrates that CPSO is suitable for the dual-mode RHC controller.

5.2.4.2 Formation Control with an Obstacle

The second experiment illustrates the effectiveness of the proposed method in Sect. 4.2.3 for five UAVs performing obstacle avoidance. The parameters in the dual-mode RHC controller are chosen to be exactly the same as in the first experiment. The second experiment is initialed: $\boldsymbol{x}_{VL}$ =(15,0.7854,20,40), $\boldsymbol{x}_1$ =(5,0,–20,20), $\boldsymbol{x}_2$ =(5,0,20,20), $\boldsymbol{x}_3$ =(5,0,–40,20), $\boldsymbol{x}_4$ =(5,0,0,20), $\boldsymbol{x}_5$ =(5,0,–60,20), and the relative distances are $\boldsymbol{x}_{r1}$ =(0,0,0,0), $\boldsymbol{x}_{r2}$ =(0,0,–15,15), $\boldsymbol{x}_{r3}$ =(0,0,–30,30), $\boldsymbol{x}_{r4}$ =(0,0,–15,–15), $\boldsymbol{x}_{r5}$ =(0,0,–30,–30). And we change the Virtual Leader's heading angle from 0 to 0.7854 (π/4) and add a circular obstacle in the simulation environment. The obstacle center is located at (20, 40) and the radius is 12 m. Figure 5.11 shows the results generated by the control sequence optimized by GA, while Fig. 5.12 shows the results generated by the proposed algorithm.

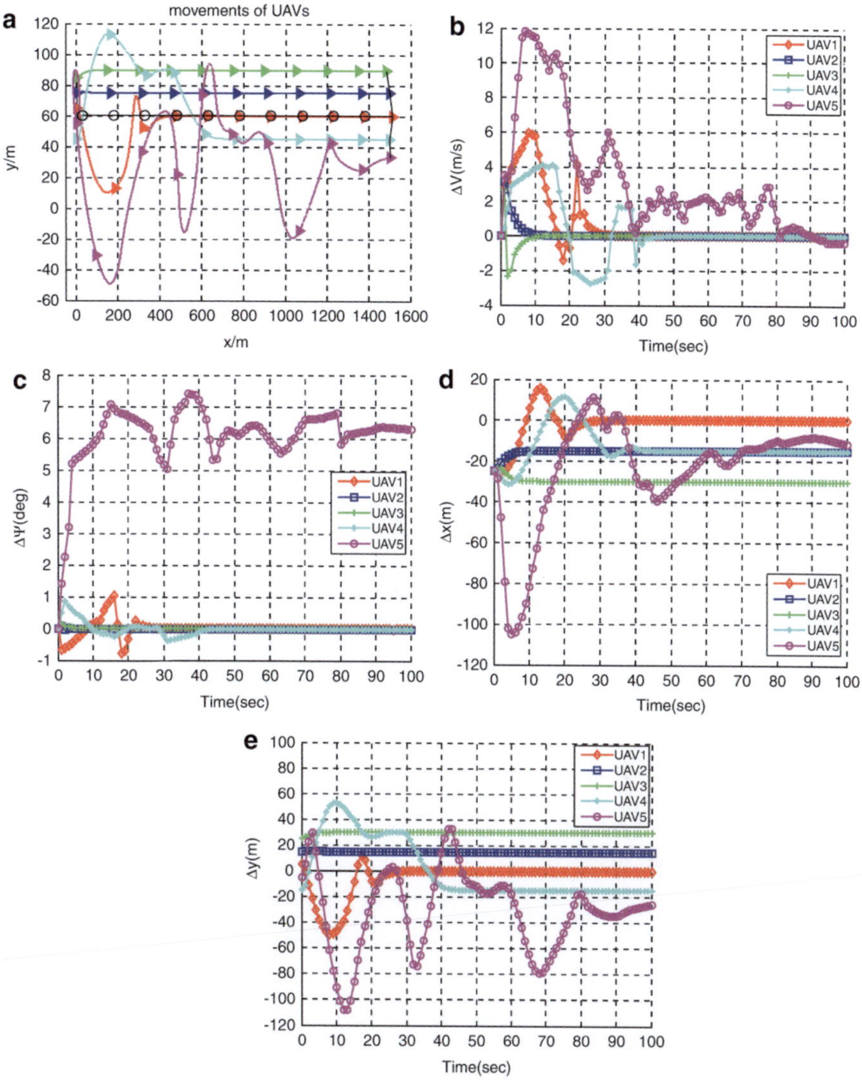

Fig. 5.9 The detailed results generated by the control sequence optimized by GA. (**a**) Five UAVs merge to a V-formation while following a Virtual Leader. (**b**) Relative velocities of 5 UAVs. (**c**) Relative heading angles. (**d**) Relative distances in the x-direction. (**e**) Relative distances in the y-direction (© [2002] IEEE. Reprinted, with permission, from Duan and Liu (2010))

The optimal obstacle avoidance trajectory with the dual-mode controller is generated assuming that UAVs can sense the circle obstacle. The trajectory of five UAVs after 20 time steps is presented in Figs. 5.11a and 5.12a. The results generated by control sequence optimized by GA as seen in Fig. 5.11b–e are far from satisfactory. Figure 5.12a shows the formation trajectory after 20 time steps.

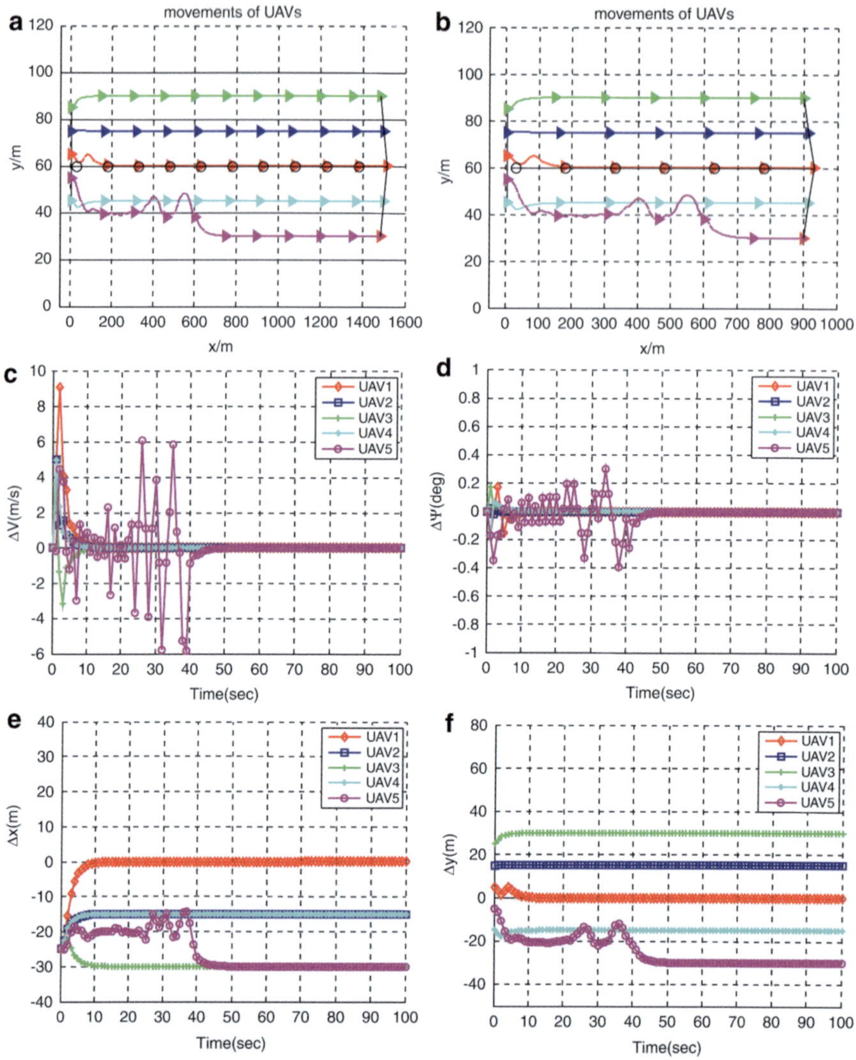

Fig. 5.10 The detailed results generated by the control sequence optimized by CPSO. (**a**) Formation trajectory under the same initial conditions. (**b**) Formation trajectory during the first sixty time steps. (**c**) Relative velocities of UAVs. (**d**) Relative heading angles. (**e**) Relative distances in the x -direction. (**f**) Relative distances in the y -direction (© [2002] IEEE. Reprinted, with permission, from Duan and Liu (2010))

The results are shown in Fig. 5.12c–d, which demonstrates that both the path tracking and formation maintenance tasks are successfully achieved. Figure 5.12c–e represents the multiple UAV convergence time. It is obvious that converge in 20s. It clearly shows the superiority of the proposed algorithm over GA.

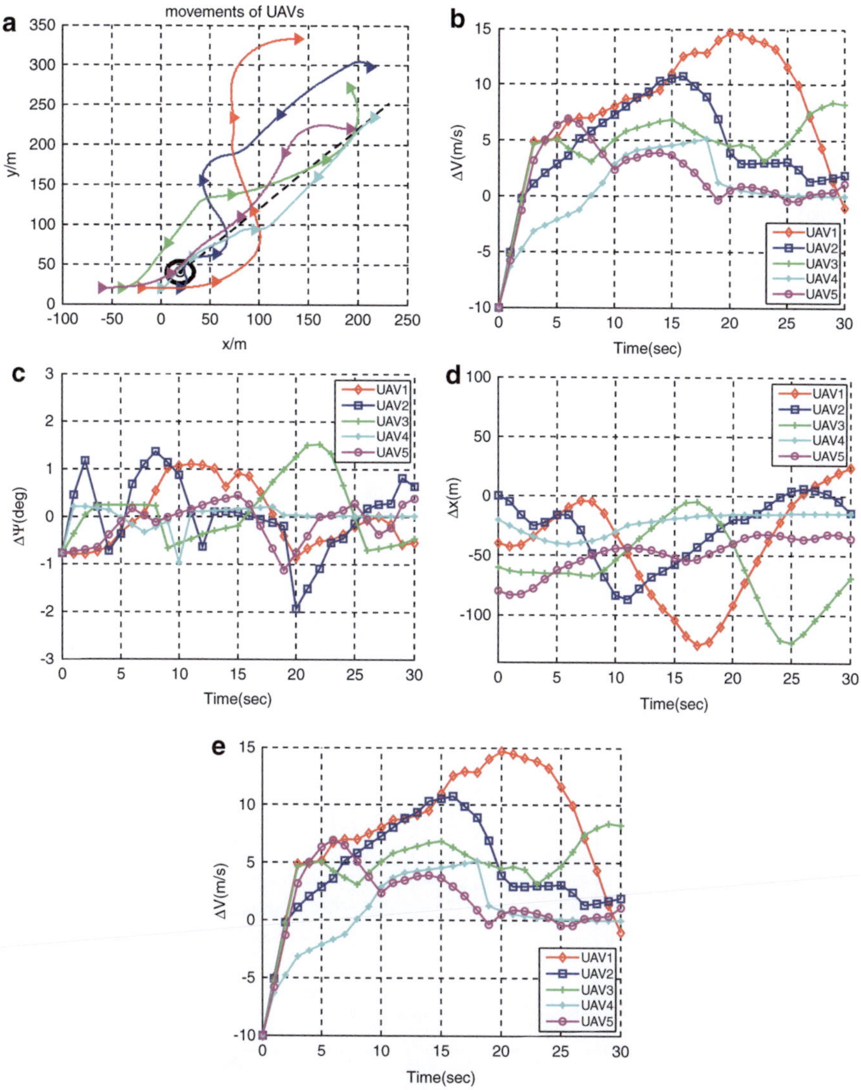

Fig. 5.11 The detailed results of GA for the second experiment. (**a**) Formation trajectory in a complicated environment with obstacle. (**b**) Relative velocities of UAVs. (**c**) Relative heading angles. (**d**) Relative distances in the x -direction. (**e**) Relative distances in the y -direction (© [2002] IEEE. Reprinted, with permission, from Duan and Liu (2010))

The simulation results obtained by applying the proposed dual-mode RHC algorithm show the UAVs are capable of flying in the desired formation. Simulations with different conditions are conducted to verify the feasibility and effectiveness of the proposed controller.

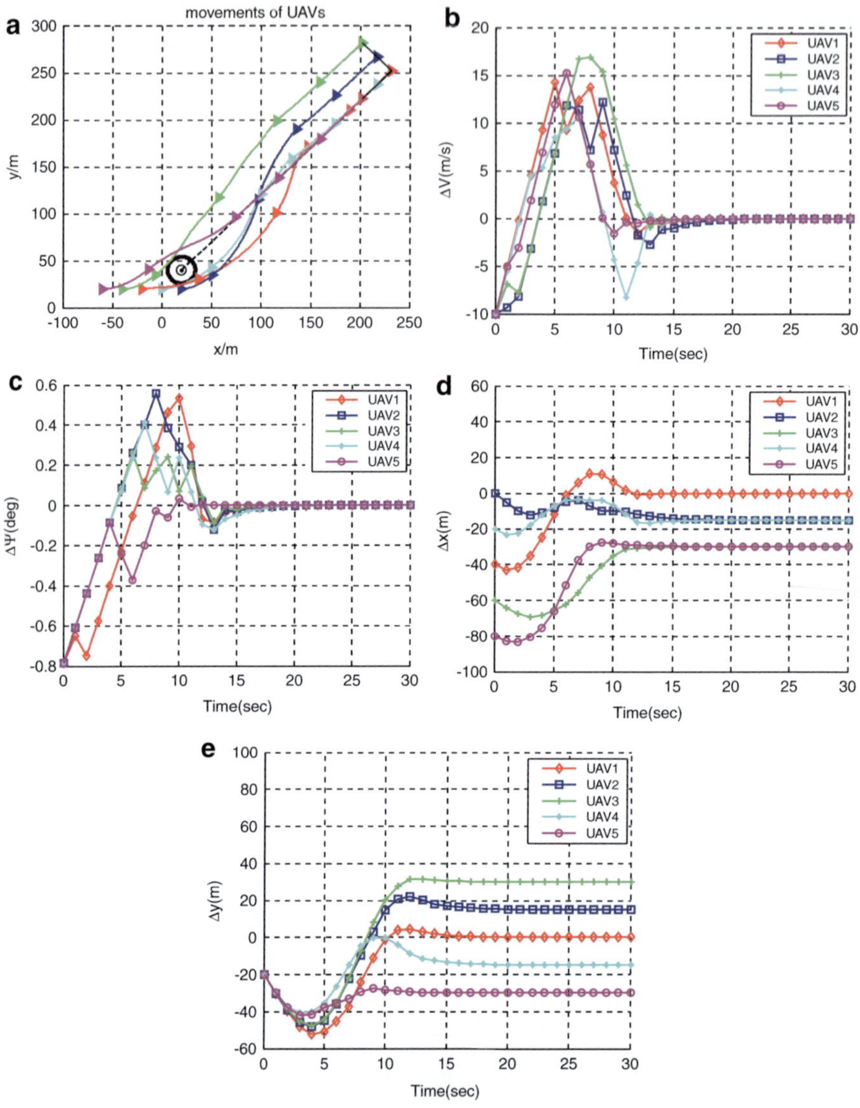

Fig. 5.12 The detailed results of CPSO for the second experiment. (**a**) Formation trajectory in a complicated environment with obstacle. (**b**) Relative velocities of UAVs. (**c**) Relative heading angles. (**d**) Relative distances in the x -direction. (**e**) Relative distances in the y -direction (© [2002] IEEE. Reprinted, with permission, from Duan and Liu (2010))

5.3 DE-Based RHC Controller for Multiple UAV Close Formation

5.3.1 *Model of Multiple UAVs for Close Formation*

In this section, a typical multiple UAV close formation model established by Proud et al. (1999) and Pachter et al. (2001) are adopted.

$$\begin{aligned}
\dot{x} &= -\frac{\overline{y}}{\tau_{\psi_W}} \cdot \psi_W - V_W + V_L \cos\psi_E + \frac{\overline{y}}{\tau_{\psi_W}} \cdot \psi_{W_C} + \overline{y}\frac{\overline{q}S}{mV}\left[\Delta C_{Y_{W_y}} \cdot y + \Delta C_{Y_{W_z}} \cdot z\right] \\
\dot{y} &= \frac{\overline{x}}{\tau_{\psi_W}} \cdot \psi_W + V_L \cdot \sin\psi_E - \frac{\overline{x}}{\tau_{\psi_W}} \cdot \psi_{W_C} - \overline{x}\frac{\overline{q}S}{mV}\left[\Delta C_{Y_{W_y}} \cdot y + \Delta C_{Y_{W_z}} \cdot z\right] \\
\dot{\psi}_W &= -\frac{1}{\tau_{\psi_W}} \cdot \psi_W + \frac{1}{\tau_{\psi_W}} \cdot \psi_{W_C} + \frac{\overline{q}S}{mV}\left[\Delta C_{Y_{W_y}} \cdot y + \Delta C_{Y_{W_z}} \cdot z\right] \\
\dot{V}_W &= -\frac{1}{\tau_{V_W}} \cdot V_W + \frac{1}{\tau_{V_W}} \cdot V_{W_C} + \frac{\overline{q}S}{m} \cdot \Delta C_{D_{W_z}} \cdot z \\
\dot{z} &= \zeta \\
\dot{\zeta} &= -\left(\frac{1}{\tau_a} + \frac{1}{\tau_b}\right) \cdot \zeta - \frac{1}{\tau_a \tau_b} z + \frac{1}{\tau_a \tau_b} h_{W_C} - \frac{1}{\tau_a \tau_b} h_{L_C} + \frac{\overline{q}S}{m} \cdot \Delta C_{L_{W_y}} \cdot y
\end{aligned} \tag{5.20}$$

The optimal separation between the Wingman and Leader UAV can be described with $\overline{x} = 2b, \overline{y} = \pi b/4, \overline{z} = 0$, where b is the wingspan of the Leader. The close formation model is established based on the aerodynamic forces on the Wing UAV near the optimal relative position, with a rotating reference frame affixed to the Wingman's instantaneous position and aligned with the Wingman's velocity vector used.

In the close formation model shown in (5.20), $(x, y, \psi_W, V_W, z, \zeta)$ are the state vectors, where x, y, and z denote the longitudinal, lateral, and vertical separation between the Leader and Wingman, respectively. ψ_W and V_W denote the heading angle and velocity of the Wingman, respectively. $(\psi_{W_C}, V_{W_C}, h_{W_C})$ are the control inputs to Wingman's heading hold, mach hold, and altitude hold autopilot channels, respectively. The Leader's maneuvers are regarded as a disturbance, which can be expressed with (ψ_L, V_L, h_{L_C}).

5.3.2 Description of RHC-Based Multiple UAV Close Formation

When multiple UAVs fly in a close formation, the Wingman must maintain itself at the optimal separation which is measured with respect to the Leader's position. Thus, the cost function (or objective function) of multiple UAV close formation flight can be described with a quadratic form (Zhang and Duan 2012):

$$\begin{aligned} \min J &= \int_0^T \left((\boldsymbol{X}_{ref} - \boldsymbol{x}(t))^T \cdot \boldsymbol{Q} \cdot (\boldsymbol{X}_{ref} - \boldsymbol{x}(t)) dt \right. \\ s.t.\ \boldsymbol{x}(t) &= \int_0^t \boldsymbol{f}(\boldsymbol{x}(0), \boldsymbol{u}(\tau), \boldsymbol{d}(\tau)) d\tau \\ \boldsymbol{U}_{\min} &\le \boldsymbol{u}(t) \le \boldsymbol{U}_{\max} \end{aligned} \tag{5.21}$$

where $\boldsymbol{x}(t) = [x, y, z, V_W, \psi_W]^T$ denotes the formation state, and the control inputs of Wingman's autopilot is represented by $\boldsymbol{u}(t) = [V_{W_C}, \psi_{W_C}, h_{W_C}]^T$. $\boldsymbol{X}_{ref} = [x_C, y_C, z_C, V_L, \psi_L]^T$ represents the reference state of the close formation system, and x_C, y_C, z_C determines the formation geometry. In the close formation model adopted in our work, Leader's flight states' change can break the formation's stability. In this way, Leader's maneuver can be regarded as the disturbance to the flight formation, described as $\boldsymbol{d}(t)$. $\boldsymbol{Q} = diag\{q_1, \cdots, q_5\}$ is a positive-definite matrix.

RHC divides the global control problem into some local optimization problems at receding time horizons. These local optimization problems have the same optimization objectives with the global control problem. In the kth sampling instant, the dynamic of the close formation can be written as:

$$\begin{aligned} \boldsymbol{x}(k+1) &= \boldsymbol{x}(k) + \int_{kT}^{(k+1)T} \boldsymbol{f}(\boldsymbol{x}(k), \boldsymbol{u}(k)) dt = \boldsymbol{f}_d(\boldsymbol{x}(k), \boldsymbol{u}(k), \boldsymbol{d}(k)) \\ \boldsymbol{x}(0) &= \boldsymbol{x}_0 \end{aligned} \tag{5.22}$$

The control inputs of the Wingman are subject to the following constraints:

$$U = \{\boldsymbol{u}(k) \,||\, \boldsymbol{U}_{\min} \le \boldsymbol{u}(k) \le \boldsymbol{U}_{\max}\} \tag{5.23}$$

where $\boldsymbol{x}(k) \in \mathbf{R}^5$ represents the formation state at the kth sampling time, and the control action, keeping constant until next predictive horizon, is represented by $\boldsymbol{u}(k) \in \boldsymbol{R}^3$. $\boldsymbol{d}(k) \in \boldsymbol{R}^3$ describes the Leader's state. T denotes the span of one time horizon, or sampling interval.

Assumption 1 *The multiple UAV close formation system given in* (5.20) *is controllable and stabilizable.*

Taking into account various practical constraints, such as the UAV's physical performance and the flight mission requirements, Wingman's control action $\boldsymbol{u}$ is always available with the Leader's state $\boldsymbol{d}$, which can stably maintain the formation state as the reference $\boldsymbol{X}_{ref}$, i.e., $\forall \boldsymbol{d}, \exists \boldsymbol{u} \in U$, subject to $\boldsymbol{X}_{ref} = \boldsymbol{f}_d(\boldsymbol{X}_{ref}, \boldsymbol{u}, \boldsymbol{d})$.

At time k, RHC controller computes predictive control sequences of current and future p predictive time horizons according to the formation's current state, which can be represented as $\boldsymbol{u}(k|k)$, $\boldsymbol{u}(k+1|k)$, $\cdots$, $\boldsymbol{u}(k+p-1|k)$. Suppose that Leader's state will not change in the following p time horizons, namely, $\boldsymbol{d}(k+i) = \boldsymbol{d}(k)$. Then the states of the formation in these time horizons $\boldsymbol{x}(k+1|k)$, $\boldsymbol{x}(k+2|k)$, $\cdots$, $\boldsymbol{x}(k+p|k)$ can be obtained. The next p time horizons are named as predictive time horizon.

Denote the quadratic cost by the following fitness function at the kth time:

$$
\begin{aligned}
&\min J(k) = \sum_{i=1}^{p} \left[\left(\boldsymbol{X}_{ref} - \boldsymbol{x}\left(k+i \middle| k\right) \right)^T \cdot \boldsymbol{Q} \cdot \left(\boldsymbol{X}_{ref} - \boldsymbol{x}\left(k+i \middle| k\right) \right) \right. \\
&s.t.\ \boldsymbol{x}\left(k+1 \middle| k\right) = \boldsymbol{f}_d\left(\boldsymbol{x}(k|k), \boldsymbol{u}(k|k), \boldsymbol{d}(k)\right) \\
&\boldsymbol{x}\left(k+j+1 \middle| k\right) = \boldsymbol{f}_d\left(\boldsymbol{x}(k+j|k), \boldsymbol{u}(k+j|k), \boldsymbol{d}(k)\right) \\
&\boldsymbol{U}_{\min} \le \boldsymbol{u}\left(k+j \middle| k\right) \le \boldsymbol{U}_{\max}
\end{aligned}
\tag{5.24}
$$

Minimize fitness function (5.24); the optimal solution to the local optimization problem at time k can be obtained, which is represented by $\boldsymbol{u}^*(k+j-1|k)$, $j = 1, \cdots, p$. Apply the preceding m control actions $\boldsymbol{u}^*(k|k)$, $\boldsymbol{u}^*(k+1|k)$, $\cdots$, $\boldsymbol{u}^*(k+m-1|k)$), $(0 \le m \le p)$ to the formation-hold control system residing on Wingman successively in current and following $m-1$ time horizons. Subsequently, at time $k+m$, repeat sampling, predicting, optimization, and implementing. By using this receding optimization technique, multiple UAV close formation state can approximate to the reference value finally. Dunbar and Murray (2006) theoretically demonstrated the stability of distributed MPC with a sufficiently fast update period. This process can be described as Fig. 5.13.

RHC treats the global control problem as a series of online local optimization problems. However, multiple UAV formation reconfiguration problem is actually constrained nonlinear optimization problems and is very difficult to be solved by using the traditional approaches. However, numerous population-based optimization approaches can provide good solutions to these complicated problems. DE algorithm is utilized to optimize the fitness function, and the predictive control law can be optimized directly.

5.3.3 DE-Based RHC Controller Design for Close Formation

5.3.3.1 Controller Design

The formation flight controller is equipped on the Wing UAV. It is an outer-loop controller that receives measurements of separation between the Leader and

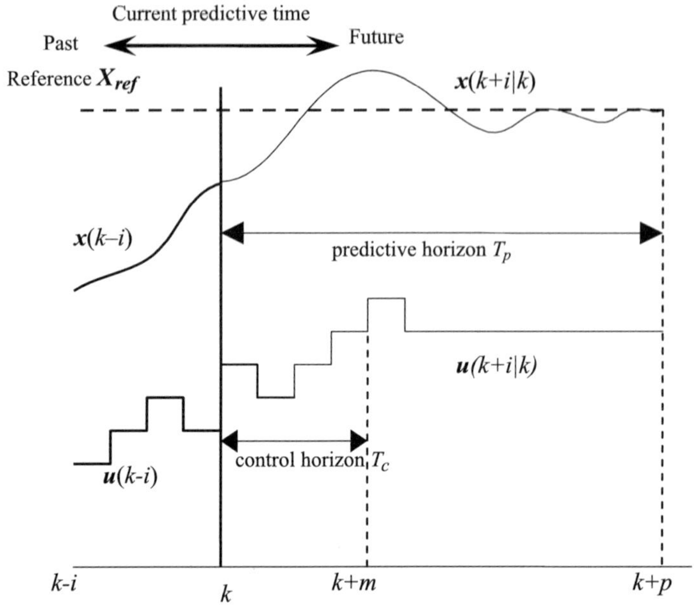

Fig. 5.13 Basic ideas of RHC (Reprinted from Zhang et al. (2010), with kind permission from Springer Science+Business Media)

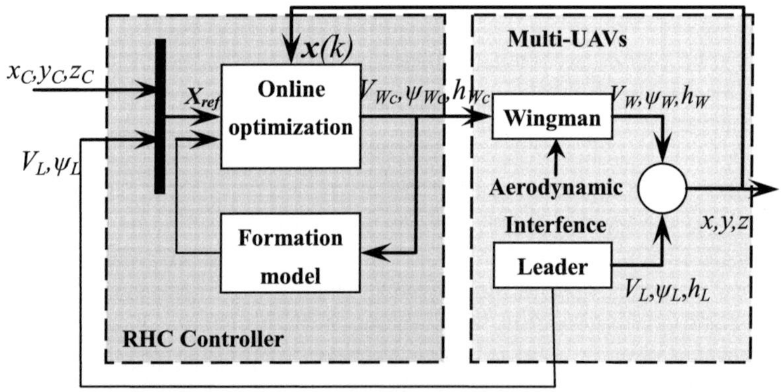

Fig. 5.14 Frame of multiple UAV close formation controller-based RHC (Reprinted from Zhang et al. (2010), with kind permission from Springer Science+Business Media)

Wingman and drives the control signals of the Wingman's three channels: mach hold, heading hold, and altitude hold autopilot. The block diagram of DE-based RHC controller for multiple UAV close formation is shown in Fig. 5.14 (Zhang and Duan 2012):

Models of each single UAV are low-order models: the heading hold and the mach hold autopilot are first order, and the altitude hold autopilot is second order.

$$\begin{aligned}
\dot{\psi}_W &= -\frac{1}{\tau_{\psi_W}} \cdot \psi_W + \frac{1}{\tau_{\psi_W}} \cdot \psi_{W_C} \\
\dot{V}_W &= -\frac{1}{\tau_{V_W}} \cdot V_W + \frac{1}{\tau_{V_W}} \cdot V_{W_C} \\
\ddot{h}_W &= -\left(\frac{1}{\tau_a} + \frac{1}{\tau_b}\right) \cdot \dot{h}_W - \frac{1}{\tau_a \cdot \tau_b} \cdot h_W + \frac{1}{\tau_a \cdot \tau_b} \cdot h_{W_C}
\end{aligned} \tag{5.25}$$

In the online optimization process, set (5.24) as the fitness function of DE (Price and Storn 1997), and set predictive control sequence $\boldsymbol{u}(k+i-1|k)$, $i=1,\cdots,p$ as the individual vector, which is just the objective of DE optimization operations. Here, the length of the predictive horizon is p, and thus DE's search region is a $D=3p$ -dimensional space. At time k, the jth individual of DE algorithm is represented by $\boldsymbol{x}_j$ $=[V_{Wc}{}^j(k|k),\ \psi_{Wc}{}^j(k|k),\ h_{Wc}{}^j(k|k),\ \ldots,\ V_{Wc}{}^j(k+p-1|k),$ $\psi_{Wc}{}^j(k+p-1|k), h_{Wc}{}^j(k+p-1|k)]$. Apply DE's evolutionary operators to $\boldsymbol{x}_j$ until the terminal criterion is satisfied, and then choose the individual with the smallest fitness function value as the optimal control sequence. After that, implement the preceding $3 \cdot m$ control actions to the Wing UAV's autopilot at each time horizon respectively.

With the purpose of improving the online searching efficiency and making full use of all aspects of information, for the solution $\boldsymbol{x}_j{}^{\kappa}$ $=[V_{Wc}{}^j(k+\kappa|k),$ $\psi_{Wc}{}^j(k+\kappa|k),\ h_{Wc}{}^j(k+\kappa|k)]$, $\kappa=0,\cdots,p-1$ at time $k+\kappa$ with respect to the individual $\boldsymbol{x}_j$ of DE population, one feasible method used is to assign their initial value in the following three steps: (i) Set them as Leader's current state $[V_L(k+\kappa), \psi_L(k+\kappa), h_L(k+\kappa)]$. (ii) Set them as former one time horizon's control action $[V_{Wc}{}^j(k+\kappa-1|k),\ \psi_{Wc}{}^j(k+\kappa-1|k),\ h_{Wc}{}^j(k+\kappa-1|k)]$. (iii) Assign their values randomly.

In order to reduce the computing complexity, we adopted the one step predictive control, i.e., $m=p=1$.

When multiple UAVs fly in a close formation, at the kth time horizon, the DE-based RHC controller implements the process online in the following steps:

Step 1: Input the formation's current state $\boldsymbol{x}(k)=[x,y,z,V_W,\psi_W]^T$ as well as Leader's state $[V_L(k), \psi_L(k), h_L(k)]$, compare them with the reference state $\boldsymbol{X}_{ref}(k)$, and then optimize the predictive control law.

Step 2: Initialize the DE population (each individual of the DE's population is a reference solution to $\boldsymbol{u}(k|k)$), and set their initial values as follows:

$$\boldsymbol{x}_j = \begin{cases} [V_L(k), \psi_L(k), h_L(k)], & j=1 \\ [V_{Wc}(k-1), \psi_{Wc}(k-1), h_{Wc}(k-1)], & j=2 \\ rand \cdot (\boldsymbol{U}_{\max} - \boldsymbol{U}_{\min}) + \boldsymbol{U}_{\min}, & else \end{cases}$$

where $j=1,\cdots D$.

Step 3: Compute the fitness function value $f(\boldsymbol{x}_j)$.

Step 4: Apply DE's mutation and recombination operators to $\boldsymbol{x}_j$, and generate the trial vector $\boldsymbol{u}_j$, and then compute its fitness function value $f(\boldsymbol{u}_j)$.

Step 5: Compare $f(\boldsymbol{u}_j)$ and $f(\boldsymbol{x}_j)$, implement DE's selection operator, and then move the individual with the lowest fitness value to the next generation. Go to Step 3 until stopping criterion is satisfied.

Step 5: The best individual of DE population is just the optimal control input $\boldsymbol{u} * (k|k)$; output and apply it to each autopilot.

Step 7: Go to **Step 1** and move forward into the $(k+1)$th time horizon.

5.3.3.2 Stability Analysis

Let $\boldsymbol{u} * (k|k)$ be the optimal predictive control sequence at time k. If this control sequence continues to work until time $k+2$, the multiple UAV close formation system state at time $k+2$ can be obtained by

$$\boldsymbol{x}\,(k+2) = \int_{kT}^{(k+2)T} \boldsymbol{f}\left(\boldsymbol{x}\,(\tau)\,,\boldsymbol{u}*\left(k\middle|k\right),\boldsymbol{d}\,(k)\right)d\tau \tag{5.26}$$

According to the further analysis, the fitness function $J(k)$ of $\boldsymbol{u}(k|k)$ depends on the state $\boldsymbol{x}(k+1)$ at time $k+1$. Using $\boldsymbol{u} * (k|k)$ yielded by the DE optimization, the multiple UAV formation state at time k+1 will be the optimal:

$$\boldsymbol{x}*(k+1) = \int_{kT}^{(k+1)T} \boldsymbol{f}\left(\boldsymbol{x}\,(\tau)\,,\boldsymbol{u}*\left(k\middle|k\right),\boldsymbol{d}\,(k)\right)d\tau \tag{5.27}$$

Continue using $\boldsymbol{u} * (k|k)$ at the next time horizon, i.e., let $\boldsymbol{u}(k+1|k+1)=\boldsymbol{u} * (k|k)$. However, this cannot guarantee that the fitness function value $J(k+1)$ is the optimal. Therefore, the system state at time $k+2$:

$$\boldsymbol{x}\,(k+2) = \int_{(k+1)T}^{(k+2)T} \boldsymbol{f}\left(\boldsymbol{x}\,(\tau)\,,\boldsymbol{u}*\left(k\middle|k\right),\boldsymbol{d}\,(k)\right)d\tau \tag{5.28}$$

is not optimal.

At time $k+1$, optimize $\boldsymbol{u}(k+1|k+1)$ by using DE algorithm. Since the initial population has contained last time's control input $\boldsymbol{u} * (k|k)$, together with Lemma 2, the system state at time $k+2$

$$\boldsymbol{x}^*\,(k+2) = \int_{(k+1)T}^{(k+2)T} \boldsymbol{f}\left(\boldsymbol{x}\,(\tau)\,,\boldsymbol{u}^*\left(k+1\middle|k+1\right),\boldsymbol{d}\,(k+1)\right)d\tau \tag{5.29}$$

is superior to $\boldsymbol{x}(k+2)$ determined by (5.28) and $J^*(k+1) \leq J(k+1)$.

In which, the control sequence $\boldsymbol{u} * (k|k), \boldsymbol{u} * (k+1|k+1), \ldots, \boldsymbol{u} * (k+N|k+N)$ driving the system's fitness function value is always superior to that at the last time horizon. Given an appropriate control step, the system's fitness function will converge to the stable value. Thus, the multiple UAV close formation system can be stabilized at the reference state.

5.3.4 Experiments

In order to investigate the feasibility and effectiveness of our proposed DE-based RHC approach to multiple UAV close formation control, two experiments were conducted. The proposed approach was coded in MATLAB language and implemented on PC-compatible with 2 GB of RAM under the Microsoft Windows Vista.

In all experiments, the initial separations between Leader and Wingman were set with $x_C = 60$ft, $y_C = 23.6$ft, $z_C = 0$ft. The model parameters in (5.20) were from Proud et al. (1999). Given Leader's heading angle and velocity were $\psi_L = 0^o$ and $V_L = 200$ft/s at the beginning of the experiments, while at time $t = 5$s, the Leader's flight states turned to $\psi_L = 20^o$ and $V_L = 250$ft/s. The experiments were performed with 30 s, and the following results show the response curves of the longitudinal and lateral separation, Wing UAV's heading and velocity, as well as the final fitness value at each prediction horizon.

The parameters of DE-based RHC controller are set as follows: $F = 1.4$, $CR = 0.5$, $NP = 10$, $NC = 20$, $m = p = 1$, $Q = \text{diag}[100,100,1,1]$.

1. Consider the aerodynamic interference introduced by Leader, and set the sampling interval at $Ts = 0.1$s. The time response is shown as Fig. 5.15.
2. Change the sampling time interval as $Ts = 0.01$s.The simulation results are shown in Fig. 5.16.

The comparative results in Figs. 5.15 and 5.16 show that the shorter the sampling period, the more stable the time response for the multiple UAV close formation system. It is obvious that better performance can be obtained by shortening the prediction horizon.

5.4 DE-Based RHC Controller for Multiple UAV Formation Reconfiguration

5.4.1 Model of Multiple UAVs for Formation Configuration

We consider a group of N UAVs are flying at the same altitude without sideslip, and they turn through the coordinated turn. For each UAV in the flight formation, its state variable is set as $\boldsymbol{x}_i = (v_i, \psi_i, x_i, y_i)^T$, $i = 1, \cdots, N$, and the dynamic of the single UAV can be written as

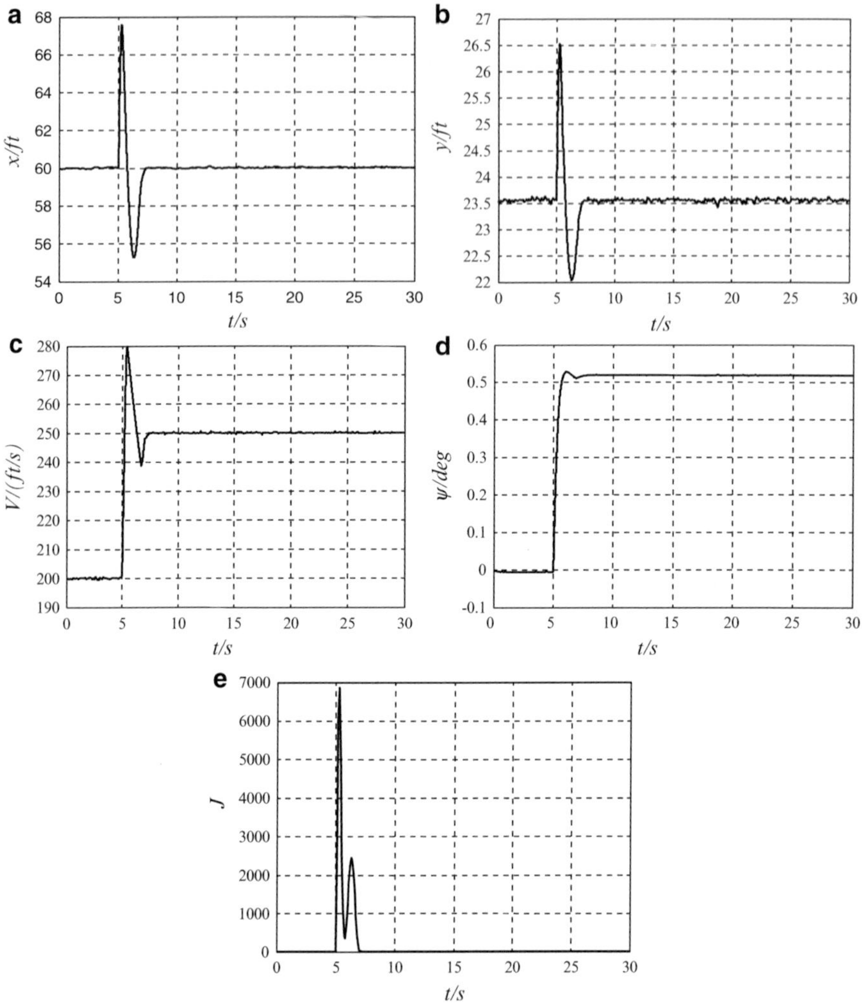

Fig. 5.15 Time response of multiple UAV close formation system ($T_s = 0.1$s). (**a**) Longitudinal separation x. (**b**) Lateral separation y. (**c**) Wingman's velocity V_W. (**d**) Wingman's heading ψ_W. (**e**) Optimal fitness value $J^*(k)$ (Reprinted from Zhang et al. (2010), with kind permission from Springer Science+Business Media)

$$
\begin{aligned}
\dot{v} &= (T - D) / W \\
\dot{\psi} &= g \cdot \sin \varphi / v \\
\dot{x} &= v \cdot \cos \psi \\
\dot{y} &= v \cdot \sin \psi
\end{aligned}
\tag{5.30}
$$

where v is the horizontal flying velocity of each UAV in the flight formation, ψ denotes the heading angle, and the UAV's horizontal location is represented by

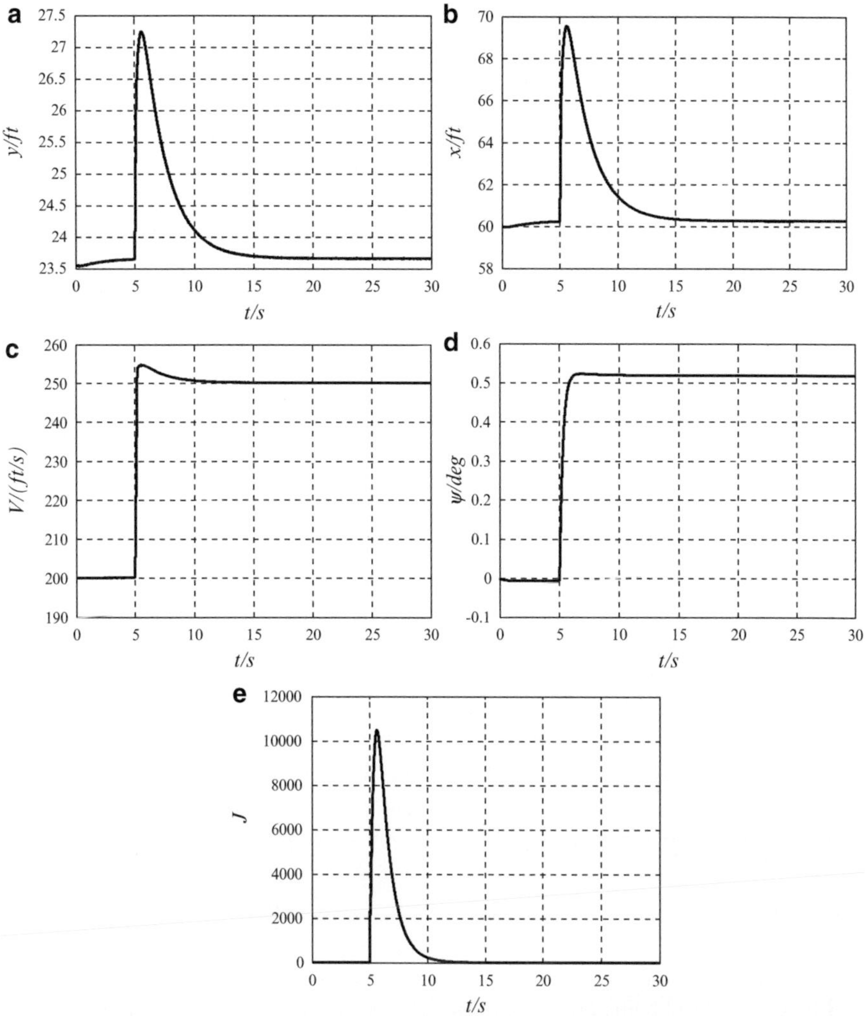

Fig. 5.16 Time response of multiple UAV close formation system ($Ts = 0.01$s). (**a**) Longitudinal separation x. (**b**) Lateral separation y. (**c**) Wingman's velocity V_W. (**d**) Wingman's heading ψ_W. (**e**) Optimal fitness value $J^*(k)$ (Reprinted from Zhang et al. (2010), with kind permission from Springer Science+Business Media)

(x,y) in the earth-surface inertial reference frame. The control inputs of each UAV's autopilot are represented by $\boldsymbol{u} = (T,\varphi)^T$, which contain the thrust T and the roll angle φ. D represents the aerodynamic drag, which is simply regarded as a constant. The weight of each UAV is W, and gravity acceleration $g = 9.8m/s^2$. Thus, dynamics of the ith UAV can be described as

$$\dot{x}_i(t) = f\left(x_i(t), u_i(t)\right), i = 1, \cdots, N \tag{5.31}$$

Set one UAV in the formation as the Leader, which is treated as the reference point, and its state vector is represented by $\boldsymbol{x}_L$. Relative to the Leader UAV, $(\boldsymbol{x}_i - \boldsymbol{x}_L)$ denotes the relative state of the ith UAV. Assume the initial time of the reconfiguration process is $t_0 = 0$, and the terminal time is $t = T$. The process of the formation reconfiguration is regarded as a control optimization problem, and so, the goal is to find the continuous control action $\boldsymbol{U} = (\boldsymbol{u}_1, \cdots, \boldsymbol{u}_N)$ that minimize a cost function that enables the terminal positions of every UAV to reach their desired value. Set the cost function as the quadratic form, which is described in the following equation:

$$\begin{aligned}
\min J(U) = \sum_{i=1}^{N} & \left(x_{ref,i} - \left(x_i \left(T \middle| u_i \right) - \boldsymbol{x}_L \left(T \middle| u_L \right) \right) \right)^T \\
& Q \left(x_{ref,i} - \left(x_i \left(T \middle| u_i \right) - x_L \left(T \middle| u_L \right) \right) \right) \\
s.t.\ \boldsymbol{x}_i \left(t \middle| \boldsymbol{u}_i \right) &= \boldsymbol{x}_i(0) + \int_0^t f\left(\boldsymbol{x}_i(\tau), \boldsymbol{u}_i(\tau) \right) d\tau \\
\boldsymbol{U}_{\min} \leq \boldsymbol{U} & \leq \boldsymbol{U}_{\max}
\end{aligned} \tag{5.32}$$

where $\boldsymbol{X}_{ref} = [\boldsymbol{x}_{ref,1}, \cdots, \boldsymbol{x}_{ref,N}]$ represents the terminal reference state of the multiple UAV system, which defines the terminal shape of the multiple UAV formation. $\boldsymbol{x}_i(T|\boldsymbol{u}_i)$ denotes the ith UAV's terminal state driving by the control input $\boldsymbol{u}_i$. For each UAV, given the initial state $\boldsymbol{x}_i(0)$, its state $\boldsymbol{x}_i(t)$ at any time t can be uniquely determined by $\boldsymbol{u}_i$. The control inputs are constrained by the performance of UAVs. $\boldsymbol{Q} = \text{diag}\{q_1, q_2, q_3, q_4\}$ is a positive-definite matrix.

Denote the distance between any two UAV as $d_{i,j}(t)$, $i,j = 1, \cdots, N$, which is computed as

$$d_{i,j}(t) = \sqrt{\left(x_i(t) - x_j(t) \right)^2 - \left(y_i(t) - y_j(t) \right)^2}$$

In order to avoid collision between two UAVs, $d_{i,j}(t)$ must be greater than the safe anti-collision distance D_{safe}, that is,

$$d_{i,j}(t) \geq D_{safe}, \forall t \in [0, T], \forall_{i \neq j} i, j \in \{1, \cdots, N\} \tag{5.33}$$

In order to ensure the real-time communication to achieve the information sharing, $d_{i,j}(t)$ must be less than the communication distance D_{com}, that is,

$$d_{i,j}(t) \leq D_{com}, \forall t \in [0, T], \forall_{i \neq j} i, j \in \{1, \cdots, N\} \tag{5.34}$$

Comprehensively consider the distance-restrictive conditions as shown in (5.33) and (5.34); the extended cost function can be rewritten as

$$\min J_{extend}(\boldsymbol{U}) = J(\boldsymbol{U}) + \omega \cdot \int_0^T \sum_{i \neq j} \Big[\max\big(0, D_{safe} - d_{i,j}(t)\big) + \max\big(0, d_{i,j}(t) - D_{com}\big)\Big] dt \tag{5.35}$$

where ω denotes the distance punishment constant coefficient, and it should be great enough so that the distance restricts of multiple UAV formation can be satisfied and $\omega = 10^{10}$.

5.4.2 Description of RHC-Based Multiple UAV Formation Reconfiguration

RHC divides the global control problem into some local optimization problems at receding time horizons. These local optimization problems have the same optimization objectives as the global control problem. In the kth sampling instant, the dynamic of the ith UAV of multiple UAV formation can be written as

$$\boldsymbol{x}_i(k+1) = \boldsymbol{x}_i(k) + \int_{kT}^{(k+1)T} f(\boldsymbol{x}_i(k), \boldsymbol{u}_i(k))dt \tag{5.36}$$

The control inputs subject to the following constraints:

$$U = \{\boldsymbol{u}_i(k) \,||\, \boldsymbol{u}_{\min} \le \boldsymbol{u}_i(k) \le \boldsymbol{u}_{\max}\} \tag{5.37}$$

where $\boldsymbol{x}_i(k) = [v_i(k), \psi_i(k), x_i(k), y_i(k)] \in \boldsymbol{R}^4$ represents the ith UAV's state at the kth sampling time, and the control input of the ith UAV, keeping constant until the next predictive horizon, is represented by $\boldsymbol{u}_i(k) = [T_i(k), \varphi_i(k)] \in \boldsymbol{R}^3$. T denotes the span of one time horizon or sampling interval.

At time k, RHC controller computes predicted control sequences of the current and future p predicted time horizons for each UAV according to multiple UAVs' current state and the constraint, and these control inputs can be represented by $\boldsymbol{u}_i(k|k)$, $\boldsymbol{u}_i(k+1|k)$, $\cdots$, $\boldsymbol{u}_i(k+p-1|k)$. Then, the predictive states of each UAV in the next p time horizons can be obtained, which are represented by $\boldsymbol{x}_i(k+1|k)$, $\boldsymbol{x}_i(k+2|k)$, $\cdots$, $\boldsymbol{x}_i(k+p|k)$. The next p time horizons are called predictive time horizon.

Denote the quadratic cost by the following fitness function at the kth time:

$$\min J(k) = \sum_{j=1}^{p}\sum_{i=1}^{N}\left(x_{ref,i} - \left(\boldsymbol{x}_i\left(k+j\middle|k\right) - x_L\left(k+j\middle|k\right)\right)\right)^T Q$$
$$\left(x_{ref,i} - \left(x_i\left(k+j\middle|k\right) - x_L\left(k+j\,|\,k\right)\right)\right)$$
$$s.t.\ \boldsymbol{x}_i\left(k+1\middle|k\right) = \boldsymbol{x}_i(k) + \int_0^T f\left(\boldsymbol{x}_i(k), \boldsymbol{u}_i\left(k\middle|k\right)\right)dt$$
$$\boldsymbol{x}_i\left(k+j+1\middle|k\right) = \boldsymbol{x}_i\left(k+j\,|\,k\right) + \int_0^T f\left(\boldsymbol{x}_i\left(k+j\middle|k\right), \boldsymbol{u}_i\left(k+j\middle|k\right)\right)dt$$
$$\boldsymbol{u}_{\min} \le \boldsymbol{u}_i\left(k+j\middle|k\right) \le \boldsymbol{u}_{\max} \tag{5.38}$$

Minimize fitness function (5.38); the optimal control solution to the local optimization problem at time k can be obtained, which is represented by $\boldsymbol{u}_i*(k+j-1|k)$, $j=1,\cdots,p$. Apply the preceding m control actions $\boldsymbol{u}*(k|k)$, $\boldsymbol{u}*(k+1|k)$, $\cdots$, $\boldsymbol{u}*(k+m-1|k))$, $(0\le m\le p)$ to each UAV's autopilot successively in current and following $m-1$ time horizons. Subsequently, at time $k+m$, repeat sampling, predicting, optimization, and implementing. By using this receding technique, multiple UAV formation's state can approximate to the reference value finally.

RHC treats the global control problem as a series of online local optimization problems. However, multiple UAV formation reconfiguration problem is actually constrained nonlinear problems and is difficult to be solved by using the traditional approaches. The method to compute the control law is a key technique to RHC. Numerous population-based optimization approaches provide good solutions to these complicated problems. DE algorithm is utilized to optimize the fitness function, and the RHC control law can be worked out directly.

5.4.3 DE-Based RHC Controller Design for Formation Reconfiguration

DE algorithm is utilized to solve the predictive control law directly. The block diagram of DE-based RHC controller for multiple UAV formation reconfiguration process is shown in Fig. 5.17 (Zhang et al. 2010).

In the online reconfiguration process, set (5.37) as the fitness function of DE, and set predicted control sequence $\boldsymbol{u}(k+i-1|k)$, $i=1,\cdots,p$ as the individual vector, which is just the objective of DE operators. For the flight formation of N UAVs, the length of the predicted horizon is p and the control input has two actions: thrust and roll angle, so the DE's search region is a $D=2\cdot N\cdot p$ -dimensional space. At time k, the ath individual of DE is represented by $x_a=[T_1(k|k),\varphi_1(k|k),\ldots,\ T_i(k|k),\varphi_i(k|k),\ldots,\ T_N(k|k),\varphi_N(k|k),\ldots,$

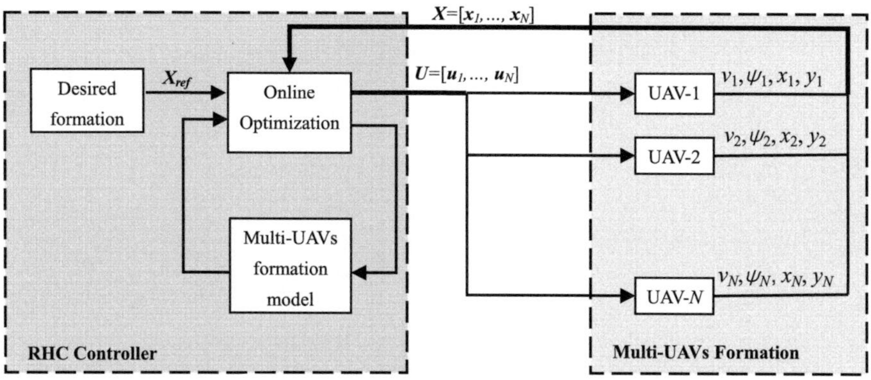

Fig. 5.17 Block diagram of formation reconfiguration (Reprinted from Zhang and Duan (2012), with kind permission from SAGE Publications)

$T_i(k+j-1|k), \varphi_i(k+j-1|k), \ldots]$, where $a = 1, \cdots NP$, $i = 1, \cdots N$, $j = 1, \cdots p$. Apply DE's evolutionary operators to x_a until the terminal criterion is satisfied, and then choose the individual with the lowest fitness function value as the optimal control sequence. After that, implement the preceding $2 \cdot N \cdot m$ control actions to corresponding UAV at each time horizon respectively.

In order to reduce the computing complexity, we adopt the one step predicted control, i.e., $m = p = 1$.

When multiple UAV is flying in a formation, at the kth time horizon, the flight formation receives the command that a new formation is inevitable, and thus, the DE-based RHC controller implements the process online in the following steps:

Step 1: Set the parameters of RHC and DE.

Step 2: Input each UAV's current state $\boldsymbol{X}(k) = [\boldsymbol{x}_1, \cdots, \boldsymbol{x}_N]$ as well as the desired formation shape and then carry on the optimization process.

Step 3: Initialize the DE population (each individual of the population is a candidate solution to $\boldsymbol{U}(k|k)$). In order to improve the online searching efficiency and make full use of all aspects of information, half individuals of the population are chosen randomly, and others are set as the control actions $\boldsymbol{U}(k-1)$ at the former one time horizon.

Step 4: Compute the fitness function value $f(x_a)$.

Step 5: Apply DE's mutation and recombination operators to x_a and generate trial vector u_a, and then compute its fitness function value $f(u_a)$.

Step 6: Compare $f(u_a)$ and $f(x_a)$, implement DE's selection operator, and then preserve the individual with the lower fitness value in the next generation. Go to Step 4 until the stopping criterion is satisfied.

Step 7: The best individual of DE population is just the optimal control sequence $\boldsymbol{U}*(k|k)$; output and apply them to each UAV respectively.

Step 8: Go to Step 2 and move forward into next $(k+1)$th time horizon.

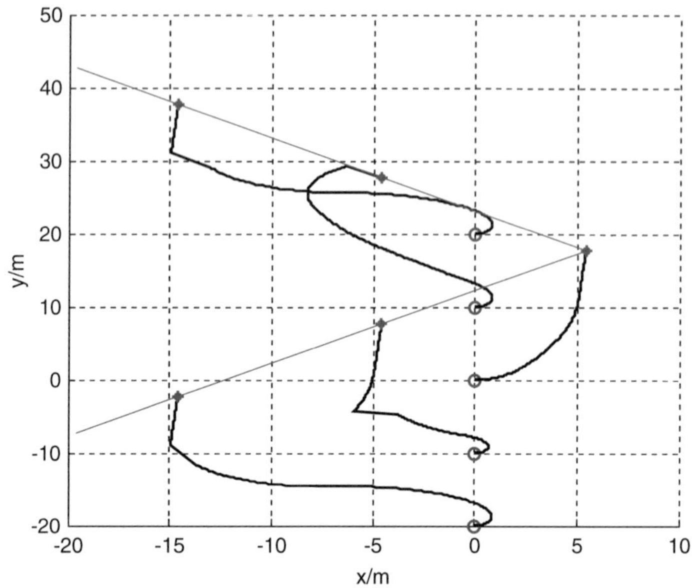

Fig. 5.18 Reconfiguration trajectory of multiple UAVs; "o" is the initial position, and "•" is the terminal position (Reprinted from Zhang and Duan (2012), with kind permission from SAGE Publications)

5.4.4 Experiments

In order to investigate the feasibility and effectiveness of our proposed DE-based RHC approach to the problem of the multiple UAV formation reconfiguration control, a series of experiments were conducted under the complicated combating environment. The proposed approach was coded in MATLAB language and implemented on PC-compatible with 2 GB of RAM under the Microsoft Windows Vista.

In all experiments, parameters of DE and RHC are set as follows: $F=0.9$, $CR=0.5$, $NP=100$, the number of iteration $Nc=100$, $m=p=1$, $\boldsymbol{Q}=diag\{1,1,1,1\}$, $T=1s$. In all experiments, for each UAV, aerodynamic drag $D=2000$, the thrust $T\in[0,6000]$, the roll angle $\varphi\in\left[-\frac{\pi}{3},\frac{\pi}{3}\right]$, and UAV's weight $W=10000$; $D_{safe}=5$, $D_{com}=50$.

Given the flight formation has 5 UAVs, they fly as an initial formation shape "|", and the terminal formation is a V-shape formation. Initial states of each UAV are [2,0,0,20], [2,0,0,10], [2,0,0,0], $[2,0,0,-10]$ and $[2,0,0,-20]$. Desired formation can be described as $\boldsymbol{X}_{ref}=[0,0,-20,20;0,0,-10,10;0,0,0,0;0,0,-10,-10;0,0,-20,-20]$; choose the UAV-3 as the Leader of the flight formation. In the experiment, after four time horizons, multiple UAVs maneuver to the desired V-shape successfully, and Fig. 5.18 shows the reconfiguration trajectory of multiple UAVs by DE-based RHC controller. Figure 5.19a–d shows the evolution curve of DE algorithm within each time horizon. Figure 5.20 describes the optimal fitness

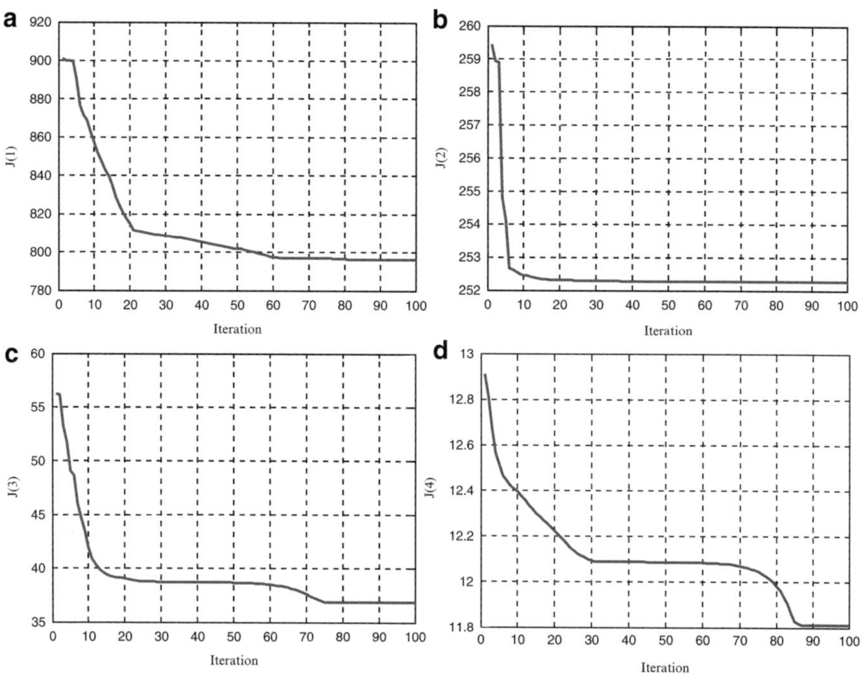

Fig. 5.19 (**a**) Evolution curve at the first time horizon. (**b**) Evolution curve at the second time horizon. (**c**) Evolution curve at the third time horizon. (**d**) Evolution curve at the fourth time horizon (Reprinted from Zhang and Duan (2012), with kind permission from SAGE Publications)

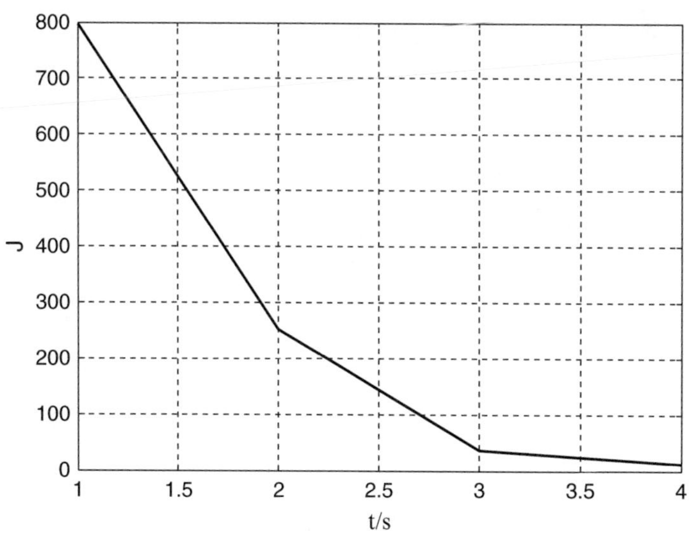

Fig. 5.20 Fitness function value (Reprinted from Zhang and Duan (2012), with kind permission from SAGE Publications)

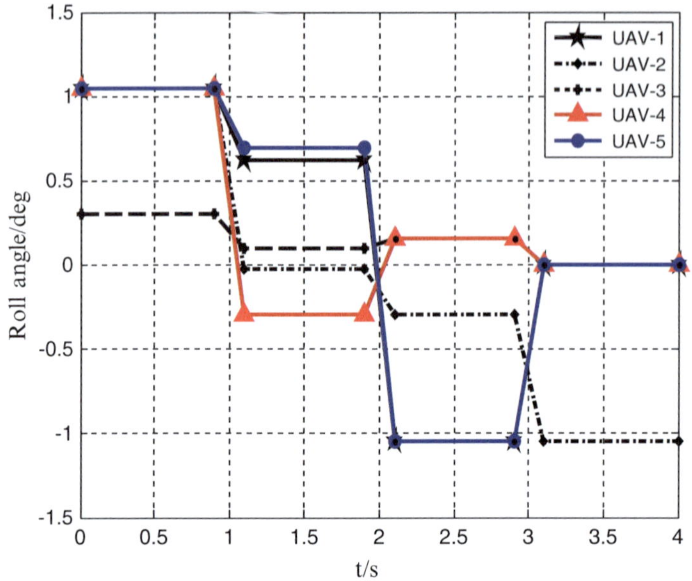

Fig. 5.21 Roll angle at every time horizon (Reprinted from Zhang and Duan (2012), with kind permission from SAGE Publications)

value at each time horizon. Figures 5.21 and 5.22 show the optimal control actions implemented to each UAV, including the thrust and the roll angle. Figure 5.23 shows the distance between two UAVs, which is less than the communication distance and greater than safe anti-collision distance.

Another simulation is as follows. Initial states of each UAV are set as [2,π,0,20], [2,π,0,10], [2,π,0,0], [2, π, 0, − 10], and [2, π, 0, − 20]. Still choose the UAV-3 as the Leader, and the desired formation is $\boldsymbol{X}_{ref} = [0, 0, -20, 20; 0, 0, -10, 10; 0, 0, 0, 0; 0, 0, -10, -10; 0, 0, -20, -20]$, which is also a V-shape formation. Figures 5.24, 5.25, 5.26, 5.27, and 5.28 show the reconfiguration results after five time horizons.

From these experiment results, it is obvious that our proposed DE-based RHC controller can solve the multiple UAV formation reconfiguration problem efficiently.

5.5 Conclusions

This chapter deals with three significant problems in the multiple UAV formation flight problem, which are respectively formation control, close formation (tight formation), and formation configuration.

In Sect. 5.2, a chaotic PSO-based nonlinear dual-mode RHC method is proposed for solving the constrained nonlinear systems. The presented chaotic PSO derives

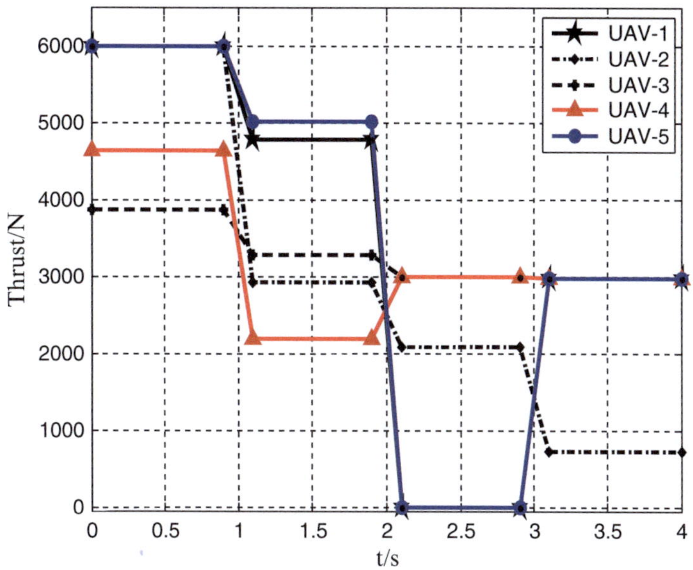

Fig. 5.22 Thrust at every time horizon (Reprinted from Zhang and Duan (2012), with kind permission from SAGE Publications)

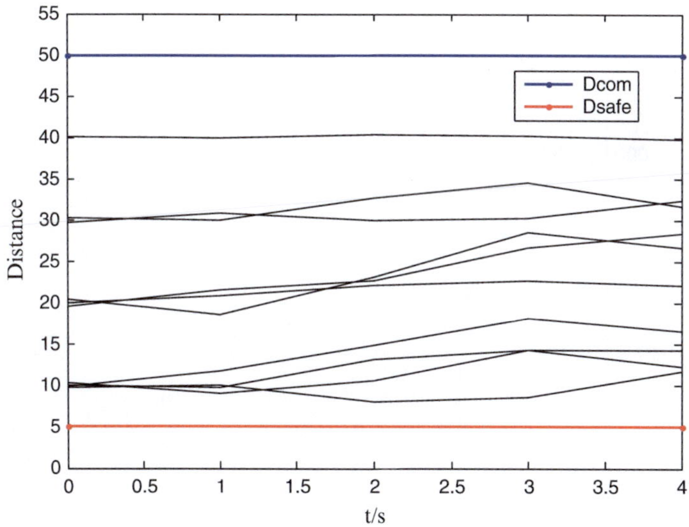

Fig. 5.23 Distance between UAVs (Reprinted from Zhang and Duan (2012), with kind permission from SAGE Publications)

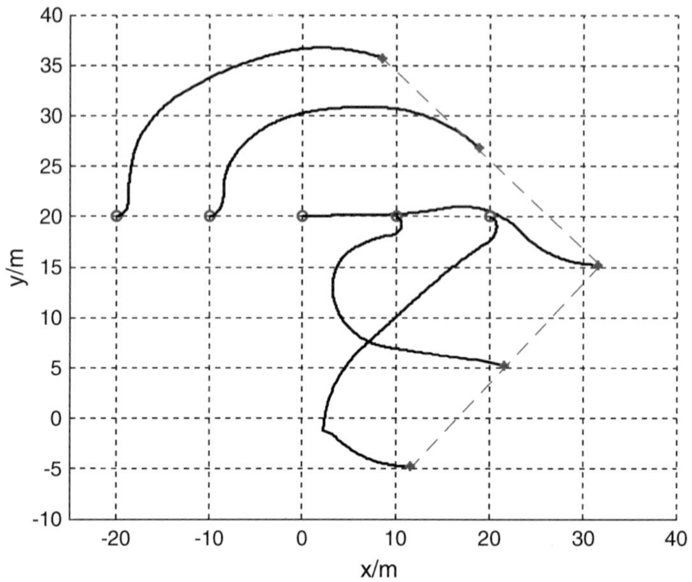

Fig. 5.24 Reconfiguration trajectory of multiple UAVs; "o" is the initial position, and "•" is the terminal position (Reprinted from Zhang and Duan (2012), with kind permission from SAGE Publications)

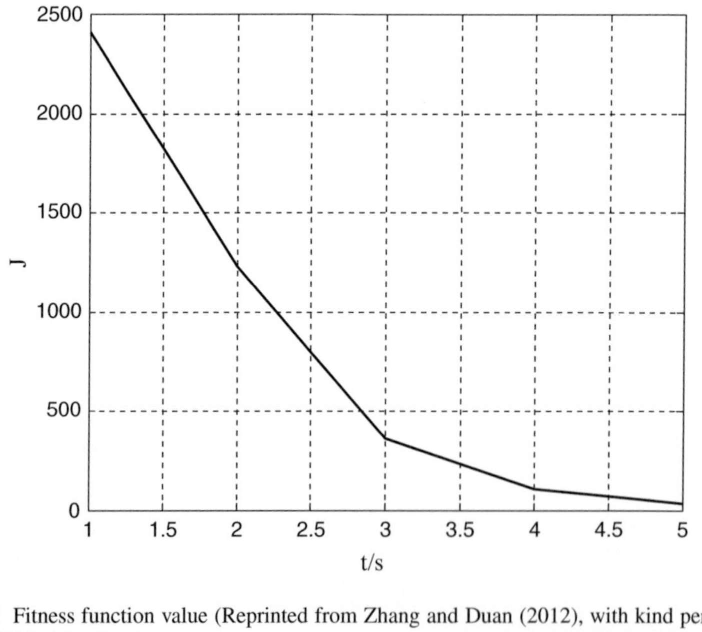

Fig. 5.25 Fitness function value (Reprinted from Zhang and Duan (2012), with kind permission from SAGE Publications)

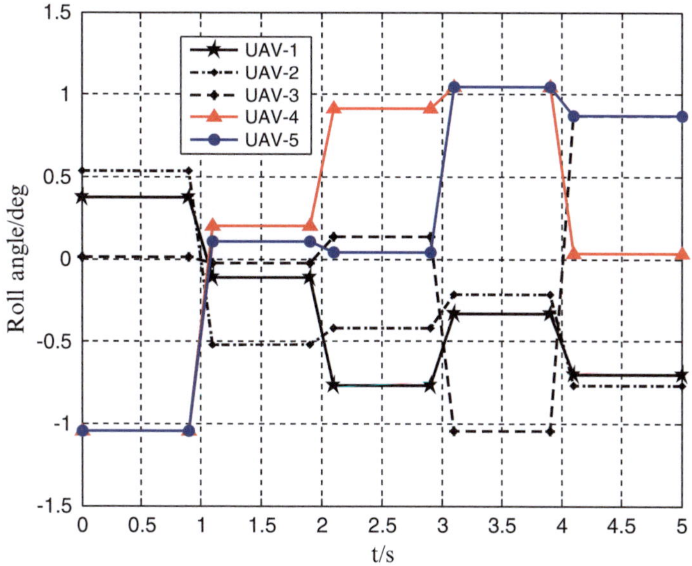

Fig. 5.26 Roll angle at every time horizon (Reprinted from Zhang and Duan (2012), with kind permission from SAGE Publications)

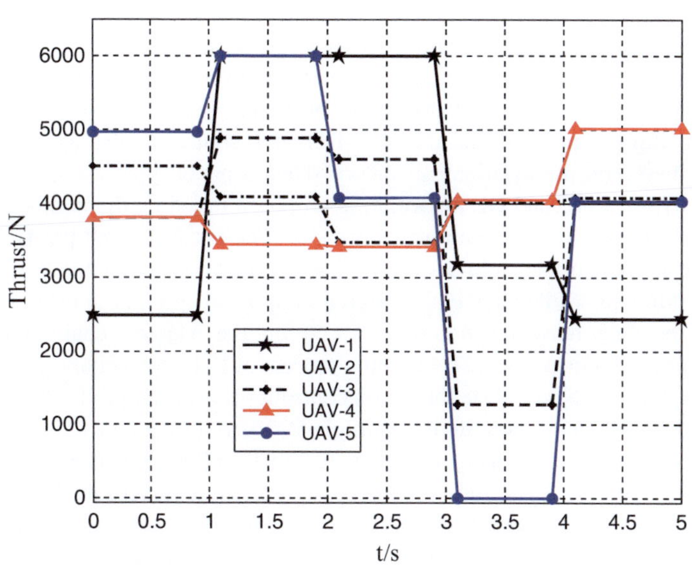

Fig. 5.27 Thrust at every time horizon (Reprinted from Zhang and Duan (2012), with kind permission from SAGE Publications)

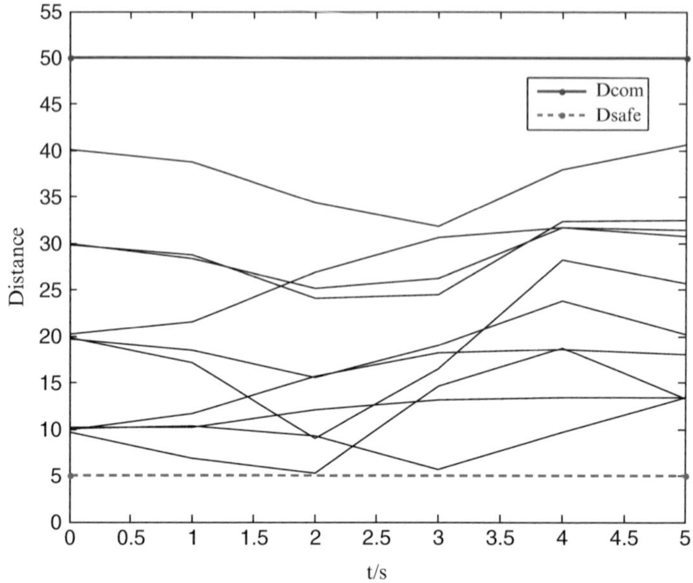

Fig. 5.28 Distance between UAVs (Reprinted from Zhang and Duan (2012), with kind permission from SAGE Publications)

both formation model and its parameter values, and the control sequence is predicted in this way, which can also guarantee the global convergence speed. A dual-model control strategy is used to improve the stability and feasibility for multiple UAV formation flight controller, and the state-feedback control is also adopted, which model is based on the invariant set theory. Then a novel type of control strategy of using hybrid RHC and DE algorithm is proposed based on the nonlinear model of multiple UAV close formation in Sect. 5.3. The issue of multiple UAV close formation is transformed into several online optimization problems at a series of receding horizons, while the DE algorithm is adopted to optimize control sequences at each receding horizon. Moreover, based on the Markov chain model, the convergence of DE is proved. The working process of RHC controller is presented in detail, and the stability of close formation controller is also analyzed. The formation configuration, which is about diving multiple UAVs to form a new flying formation state, is explained in detail using the DE-based RHC in Sect. 5.4. The global control problem of multiple UAV formation reconfiguration is transformed into several online local optimization problems at a series of receding horizons, while the DE algorithm is adopted to optimize control sequences at each receding horizon. Both the feasibility and effectiveness of these three proposed methods are verified by series of experiments.

References

Binetti P, Ariyur KB, Krstic M, Bernelli F (2003) Formation flight optimization using extremum seeking feedback. J Guid Control Dyn 26(1):132–142

Duan H, Liu S (2010) Non-linear dual-mode receding horizon control for multiple unmanned air vehicles formation flight based on chaotic particle swarm optimisation. IET Control Theory Appl 4(11):2565–2578

Duan H, Ma G, Luo D (2008) Optimal formation reconfiguration control of multiple UCAVs using improved particle swarm optimization. J Bionic Eng 5(4):340–347

Duan H, Zhang Y, Liu S (2011) Multiple UAVs/UGVs heterogeneous coordinated technique based on Receding Horizon Control (RHC) and velocity vector control. Sci China Technol Sci 54(4):869–876

Duan H, Luo Q, Ma G, Shi Y (2013a) Hybrid particle swarm optimization and genetic algorithm for multi-UAVs formation reconfiguration. IEEE Comput Intell Mag 8(3):16–27

Duan H, Luo Q, Yu Y (2013b) Trophallaxis network control approach to formation flight of multiple unmanned aerial vehicles. Sci China Technol Sci 56(5):1066–1074

Dunbar WB, Murray RM (2006) Distributed receding horizon control for multi-vehicle formation stabilization. Automatica 42(4):549–558

Giulietti F, Pollini L, Innocenti M (2000) Autonomous formation flight. IEEE Control Syst 20(6):34–44

Hendrickx JM, Fidan B, Yu C, Anderson BD, Blondel VD (2008) Formation reorganization by primitive operations on directed graphs. IEEE Trans Autom Control 53(4):968–979

Kwon WH, Han SH (2005) Receding horizon predictive control: model predictive control for state models. Springer, New York

May RM (1976) Simple mathematical models with very complicated dynamics. Nature 261(5560):459–467

Pachter M, D'Azzo JJ, Proud AW (2001) Tight formation flight control. J Guid Control Dyn 24(2):246–254

Price K, Storn R (1997) Differential evolution–a simple evolution strategy for fast optimization. Dr Dobb's j 22(4):18–24

Proud AW, Pachter M, D'Azzo JJ (1999) Close formation flight control. Air Force Inst of Tech Wright-Patterson AFB OH School of Engineering, Ft. Belvoir Defense Technical Information Center

Shi Y, Eberhart R (1998) A modified particle swarm optimizer. In: Proceedings of the 1998 IEEE International Conference on Evolutionary Computation (IEEE World Congress on Computational Intelligence), Anchorage, AK, 1998. IEEE, pp 69–73

Stipanović DM, Inalhan G, Teo R, Tomlin CJ (2004) Decentralized overlapping control of a formation of unmanned aerial vehicles. Automatica 40(8):1285–1296

Ueno S, Kwon SJ (2007) Optimal reconfiguration of UAVs in formation flight. In: Proceedings of 2007 International Conference on Instrumentation, Control, Information Technology and System Integration (SICE, 2007 annual conference), Takamatsu. IEEE, pp 2611–2614

Wang X, Yadav V, Balakrishnan S (2007) Cooperative UAV formation flying with obstacle/collision avoidance. IEEE Trans Control Syst Technol 15(4):672–679

Zelinski S, Koo TJ, Sastry S (2003) Optimization-based formation reconfiguration planning for autonomous vehicles. In: Proceedings of IEEE International Conference on Robotics and Automation (ICRA'03), Taipei. IEEE, pp 3758–3763

Zhang X, Duan H (2012) Differential evolution-based receding horizon control design for multi-UAVs formation reconfiguration. Trans Inst Meas Control 34(2–3):165–183

Zhang X, Duan H, Yu Y (2010) Receding horizon control for multi-UAVs close formation control based on differential evolution. Sci China Inf Sci 53(2):223–235

Zhang Y, Duan H, Zhang X (2011) Stable flocking of multiple agents based on molecular potential field and distributed receding horizon control. Chin Phys Lett 28(4):040503

Chapter 6
Multiple UAV/UGV Heterogeneous Control

Haibin Duan

Abstract Multiple unmanned aerial vehicles (UAVs) can be used to cover large areas searching for targets. However, sensors on UAVs are typically limited in operating airspeed and altitude, combined with attitude uncertainty, placing a lower limit on their ability to resolve and localize ground features. Unmanned ground vehicles (UGVs) can be deployed to accurately locate ground targets, but they have the disadvantage of not being able to move rapidly or see through such obstacles as buildings or fences. This chapter mainly focuses on heterogeneous coordinated control for multiple UAVs/UGVs and cooperative search problem for multiple UAVs. On the basis of introduction of UAV/UGV mathematical model, the characteristics of heterogeneous flocking is analyzed in detail. Two key issues are considered in multiple UGV subgroups, which are Reynolds rule and Virtual Leader (VL). Receding horizon control (RHC) with particle swarm optimization (PSO) is proposed for multiple UGV flocking, and velocity vector control approach is adopted for multiple UAV flocking. Thus, multiple UAV and UGV heterogeneous tracking can be achieved by these two approaches. Then a time-delay compensation approach of heterogeneous network control for multiple UAVs and UGVs is described to handle the time delay in network control system. What's more, a differential evolution (DE)-based RHC design for cooperative area search using multiple UAVs is presented. In this approach, an extended search map is used to represent the environment information on the search region.

6.1 Introduction

Unmanned aerial vehicle (UAV) has the advantages of zero casualties, high-speed overload, good stealth performance, short operational preparation time, and relatively low life-cycle cost. These advantages increase the capability of high-risk targets penetration, suppressing enemy air defense, deep target attacking, and dominating the battlespace (Duan et al. 2013). Unmanned ground vehicle (UGV) is generally capable of operating outdoors and over a wide variety of terrain, functioning

H. Duan and P. Li, *Bio-inspired Computation in Unmanned Aerial Vehicles*,
DOI 10.1007/978-3-642-41196-0_6,

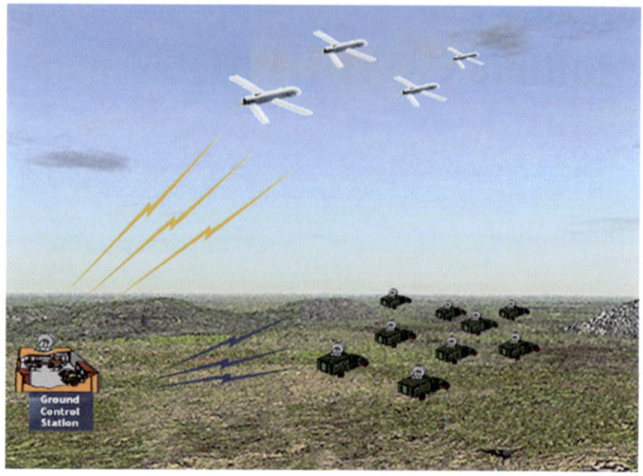

Fig. 6.1 Multiple UAV and UGV heterogeneous cooperation scenario (Reprinted from Duan et al. (2011), with kind permission from Springer Science + Business Media)

in place of humans. Multiple UAVs can be used to cover large areas searching for targets. However, sensors on UAVs are typically limited in operating airspeed and altitude, combined with attitude uncertainty, placing a lower limit on their ability to resolve and localize ground features. UGVs on the other hand can be deployed to accurately locate ground targets, but they have the disadvantage of not being able to move rapidly or see through such obstacles as buildings or fences. Therefore, multiple UAV/UGV heterogeneous cooperation provides a new breakthrough for the effective application of UAVs and UGVs (Tanner and Christodoulakis 2006; Hsieh et al. 2007; Duan and Liu 2010; Duan et al. 2010a, b). Multiple UAV and UGV heterogeneous cooperation scenario can be shown with Fig. 6.1.

For multiple UAV/UGV heterogeneous system, dynamic tracking is a typical problem. The ground moving target should be positioned dynamically and automatically, which make the target within the perception of the airborne camera (Ariyur and Fregene 2008). Trentini and Beckman (2010) summarize the two advanced research projects of about UAV rotorcraft and UGV cooperating in the battlespace currently approved for funding by Defence R&D Canada. Phan and Liu (2008) proposed a cooperative control framework for a hierarchical UAV/UGV platform for wildfire detection and fighting.

In order to realize the multiple UAV and UGV heterogeneous coordinated movement, the following control objectives should be satisfied:

1. The UGV subgroups should be in flocking motion.
2. The control center receives information from the UGV subgroups and sends the central position information to the UAV subgroups at the same time.
3. The subgroups of multiple UAVs should follow and hover over the subgroups of multiple UGVs stably.

The constraints of the heterogeneous movements are as follows:

1. The subgroups of multiple UGVs could satisfy the Reynolds rule: cohesion, separation, and alignment (Duan et al. 2010a). In this way, the flocking movement can be realized.

2. The subgroups of multiple UAVs could receive the changing position center information of the UGV subgroups and follow the movements of multiple UGVs.
3. The subgroups of multiple UAVs should avoid collisions during flight.

6.2 Multiple UAV/UGV Heterogeneous Coordinated Control

6.2.1 *Mathematical Model for UAVs and UGVs*

In multiple UAV/UGV heterogeneous coordinated motion, UAVs and UGVs are all considered as particles. UGVs are moving on a plane, and the status variable of UGV_i is $x_{\mathrm{ugv}i} = (x_i, y_i, \dot{x}_i, \dot{y})^T$, $i = 1, 2, \ldots, Numugv$. The emotion equation can be expressed according to the following dynamics (Duan et al. 2011):

$$\begin{cases} \dot{r}_i = v_i \\ \dot{v}_i = u_i \end{cases} \tag{6.1}$$

For $i = 1, 2, \ldots, Numugv$, $r_i = (x_i, v_i)$ is the position vector of the ith UGV, $v_i = (\dot{x}_i, \dot{y}_i)$ is the velocity vector, and $ug_i = (u_{xi}, u_{yi})$ is the control input. Thus, the state vector of the ith UGV can be defined as $x_{\mathrm{ugv}i} = (x_i, y_i, \dot{x}_i, \dot{y}_i)$.

In the multiple UAV flocking motion without a leader, all the UGVs have the same status; thus, the velocity of the whole group is random when a stable and coordinated motion is achieved (Jadbabaie et al. 2003; Palejiya and Tanner 2006). In this section, a virtual UGV acting as a Virtual Leader (VL) is adopted to lead the whole UGV group to move in the right direction. The VL is a simulation of the instructions sent by the control center, and then received by each UGV. Since the VL will not be broken down, the instructions of the control center can be ensured to be executed by the UGV group. The motion equations of the VL are the same as those of the other UGVs, as shown in (6.1), and its state vector is $x_{vl} = (x_{vl}, y_{vl}, \dot{x}_{vl}, \dot{y}_{vl})^T$, where $r_{vl} = (x_{vl}, y_{vl})$ and $v_{vl} = (\dot{x}_{vl}, \dot{y}_{vl})$.

In the heterogeneous coordinated motion, the UAV group will follow the UGV group. The state vector of the jth UAV is $x_{\mathrm{uav}j} = (v_j, \gamma_j, \chi_j, x_j, y_j, z_j)$ for $j = 1, 2, \ldots, Numuav$, and its control inputs are thrust T_j, load factor n_j, and bank angle μ_j. The following equations are adopted as the dynamic model of a UAV:

$$\begin{aligned} \dot{v} &= g\left[(T - D)/W - \sin\gamma\right] \\ \dot{\gamma} &= (g/v)(n\cos\mu - \cos\gamma) \\ \dot{\chi} &= (gn\sin\mu)/(v\cos\gamma) \\ \dot{x} &= v\cos\gamma\cos\chi \\ \dot{y} &= v\cos\gamma\sin\chi \\ \dot{z} &= -v\sin\gamma \end{aligned} \tag{6.2}$$

where v is the airspeed of the UAV; γ is the flight path angle; χ is the flight path heading; x, y, and z denote the position; g denotes the acceleration of gravity; D denotes the drag; and W denotes the weight.

6.2.2 Multiple UGV Coordinated Control Based on RHC

The initial speed and direction of each UGV in the group are different from each other, the designed controller should make the UGVs gradually meet Reynolds rule in the motion of following the VL, and the effect of multiple UGV cooperative movement must be achieved. For the ith UGV, the control input is $ug_i = (u_{xi}, u_{yi}), i = 1, 2, \ldots, Numugv$. The control input of the whole UGV group can be defined as $U_{\text{ugv}} = (ug_1, ug_2, \ldots, ug_{Numugv}) = \{U_{ugv}(t) | \forall t \in [0, T]\}$, and the state of the whole UGV group is $X_{ugv} = (x_{ugv1}, x_{ugv2}, \ldots, x_{ugvNumugv}) \in \Re^{4 \times Numugv}$. Thus, the multiple UGV motion equations can be described as

$$\dot{X}_{ugv}(t) = f\left(t, X_{ugv}(t), U_{ugv}(t)\right) \tag{6.3}$$

Let $X_{ugv}(0) = X_{ougv}$ represent the initial state of the UGV group, then the state at any time $t \in (0, T]$ can be determined by the following equation:

$$X_{ugv}(t) = X_{ugv}(0) + \int_0^t f\left(\tau, X_{ugv}(\tau), U_{ugv}(\tau)\right) d\tau \tag{6.4}$$

If the initial states are definite, $X_{ugv}(t)$ can only be obtained by U_{ugv}, which can be also expressed by $X_{ugv}(t | U_{ugv})$.

In the research on flocking conducted by Tanner and Olfati-Saber (Tanner et al. 2003; Olfati-Saber 2006), the cohesion and separation rules of Reynolds are satisfied by designing a proper artificial potential field. Each UGV matches its velocity to its neighbors, and the alignment rule can be fully satisfied. For the flocking motion following a VL, the control input of each UGV ($u_i, i = 1, 2, \ldots, Numugv$) will include an additional part to coordinate its position and velocity with the VL.

There are complexities and diversities of control objectives in the movement of multiple UGV heterogeneous movement, which orient the optimal control strategy from the unconstrained quadratic optimization problem to multi-objective optimization problem. RHC has been proved to be more successfully optimized online in a dynamic environment, which is based on the simple idea of repetitive solution of an optimal control problem and state updating after the first input of the optimal command sequence (Zhang et al. 2010; Zhang and Duan 2012). The main idea of RHC is the online receding/moving optimization. It breaks the global control problem into several local optimization problems of smaller sizes, which can significantly decrease the computing complexity and computational expense. Particle swarm optimization (PSO) is a population-based stochastic optimization technique, which is inspired by the social behavior of bird flocking or fish schooling.

It is demonstrated that PSO can find better results in a faster, cheaper way compared with other methods. Hybrid RHC and PSO approach is developed for multiple UGV movement in this section.

Consider searching space with n dimensions, the particle population is m, and the position of the ith particle can be expressed with $X_i = (x_{i1}, x_{i2}, \cdots, x_{in})$, and the velocity is $V_i = (v_{i1}, v_{i2}, \cdots, v_{in})$. The current best solution is $P_i = (p_{i1}, p_{i2}, \cdots, p_{in})$, and the global-best solution is $P_g = (p_{g1}, p_{g2}, \cdots, p_{gn})$. For the tth generation, the position and velocity updating rule can be expressed as follows:

$$\begin{aligned} &V_i(t+1) = \chi \cdot (V_i(t) + c_1 \cdot r_1 \cdot (P_i(t) - X_i(t)) + c_2 \cdot r_2 \cdot (P_g(t) - X_i(t))) \\ &X_i(t+1) = X_i(t) + V_i(t+1) \end{aligned} \tag{6.5}$$

where r_1 and r_2 are two independent random numbers between (0,1), c_1 and c_2 are two learning factors, and χ is a constant number between (0,1). The object function for RHC can be expressed as follows:

$$\begin{aligned} \min_u J &= f\left(X_{ugv}, U_{ugv}; t_c, T_p\right) \\ J &= \int_{t_c}^{t_c+T_p} F\left(X_{ugv}, U_{ugv}\right) dt \\ \textit{subject to } \dot{X}_{ugv} &= f\left(t, X_{ugv}, U_{ugv}\right) \\ LL_{ugv} &\le \begin{bmatrix} X_{ugv} \\ U_{ugv} \end{bmatrix} \le UL_{ugv} \end{aligned} \tag{6.6}$$

where t_c is control horizon, T_p is predictive horizon, $t_c \le T_p$, and LL_{ugv} and UL_{ugv} denote the upper and lower bound, respectively.

The topology of the wireless network connecting the UGVs is an adjacency graph $G = \{V,E\}$. The set of vertices $V = \{n_1, n_2, \ldots, n_{Numugv}\}$ represent the UGVs, and the set of edges $E = \{(n_i, n_j) \in V \times V | n_i \sim n_j\}$ represent the adjacency relation between the UGVs. Let $A = (a_{ij})$ denote the adjacency matrix of G, then $a_{ij} \neq 0 \iff (i,j) \in E$, and A is symmetric, $A^T = A$. Let N_i denote the set of the UGVs that are adjacent to the ith UGV:

$$N_i = \left\{j \in V : a_{ij} \neq 0\right\} = \{j \in V : (i, j) \in E\} \tag{6.7}$$

Let R_{ugv} represent the maximum detecting range of a UGV, and $R_{ugv} > 0$; thus N_i can be described as

$$N_i = \left\{j \in V : \|r_i - r_j\| \le R_{ugv}\right\} \tag{6.8}$$

where $||\cdot||$ is the Euclidean norm. Then, for $R_{ugv} > 0$, the set of edges E can be described as

$$E = \left\{(i, j) \in V \times V : \|r_i - r_j\| < R_{ugv}, i \neq j\right\} \tag{6.9}$$

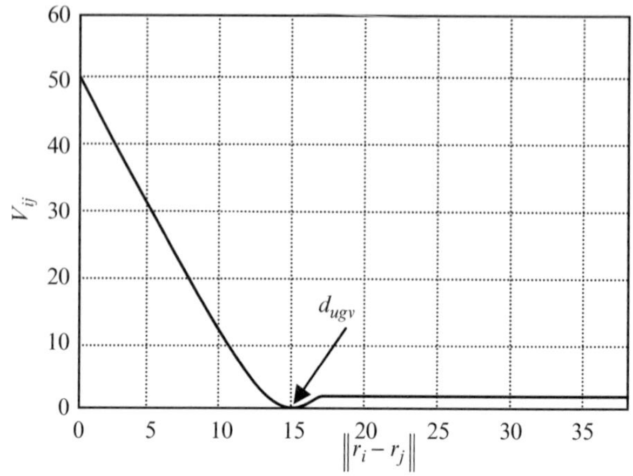

Fig. 6.2 The curve of the potential function V_{ij} (Reprinted from Duan et al. (2011), with kind permission from Springer Science + Business Media)

(1) *Potential Cost*

A potential field is established between UGVs in order that the cohesion and separation rules can be satisfied. The potential function between UGV i and UGV j is V_{ij} (see Fig. 6.2). V_{ij} is a nonnegative, differentiable, and unbounded function of the distance $\overline{v}_{ci}^{tk}$. V_{ij} obtains its unique minimum when the distance $\|r_i - r_j\|$ is the desired distance d_{ugv}.

$$V_{ij} = \frac{1}{2}\sum_{i}\sum_{j \neq i} \psi\left(\|r_i - r_j\|\right) \tag{6.10}$$

In this section, we adopt the potential function introduced by Reza and define the σ-norm as follows:

$$\|x\|_{\sigma} = \frac{1}{\xi}\left[\sqrt{1 + \xi\|x\|^2} - 1\right] \tag{6.11}$$

where the constant $\xi > 0$. The gradient of σ-norm can be expressed by

$$\sigma_{\xi}(x) = \frac{x}{\sqrt{1 + \xi\|x\|^2}} = \frac{x}{1 + \xi\|x\|_{\sigma}} \tag{6.12}$$

According to the definition of σ-norm, V_{ij} can also be rewritten as $V_{\sigma ij}$:

$$V_{\sigma ij} = \frac{1}{2}\sum_{i}\sum_{j \neq i} \psi_{\alpha}\left(\|r_i - r_j\|_{\sigma}\right) \tag{6.13}$$

When $\|r_i - r_j\| \geq R_{ugv}$, the following function of $\varphi_a(\|r_i - r_j\|)$ is introduced:

$$\varphi_\alpha\left(\|r_i - r_j\|\right) = \rho_h\Big(\|r_i - r_j\|/\|R_{ugv}\|_\sigma\Big)\varphi\left(\|r_i - r_j\| - \|d_{ugv}\|\right)$$
$$\varphi(x) = \frac{1}{2}\big[(a+b)\sigma_1(x+c) + (a-b)\big] \tag{6.14}$$

where $\sigma_1(x) = x/\sqrt{1+x^2}$. a, b, c satisfy $b \geq a > 0$, $c = |a-b|/\sqrt{4ab}$, and $\varphi(0) = 0$.

$$\rho_h(x) = \begin{cases} 1 & x \in [0, h) \\ \frac{1}{2}\left[1 + \cos\left(\pi\frac{(x-h)}{(1-h)}\right)\right] & x \in [h, 1] \\ 0 & otherwise \end{cases} \tag{6.15}$$

where $h \in (0,1)$.

$\psi_a(\|r_i - r_j\|)$ and $\varphi_a(\|r_i - r_j\|)$ satisfy the following equation:

$$\psi_\alpha\left(\|r_i - r_j\|\right) = \int_{\|d_{ugv}\|}^{\|r_i - r_j\|} \varphi_\alpha(s)ds \tag{6.16}$$

Then, the collective potential cost of the whole UGV group can be described as

$$F_{potential} = \sum_{i=1}^{Numugv} \sum_{j \in N_i} \varphi_\alpha\left(\|r_i - r_j\|_\sigma\right) n_{ij} \tag{6.17}$$

where $n_{ij} = \sigma_\xi(r_j - r_i)$.

The UGVs will maneuver to lower the collective potential, until the group converges to a stable and coordinated flocking motion, which has the lowest collective potential.

(2) *Consensus Cost*

Each UGV will match its velocity with its neighboring flockmate to satisfy the alignment rule. The consensus cost is defined as

$$F_{consensus} = \sum_{i=1}^{Numugv} \sum_{j \in N_i} \left|a_{ij}(r)\left(v_j - v_i\right)\right| \tag{6.18}$$

where $a_{ij}(r)$ in the adjacent matrix A can be obtained by

$$a_{ij}(r) = \rho_h\left(\|r_j - r_i\|_\sigma / \|R_{ugv}\|_\sigma\right) \in [0, 1] \quad j \neq i \tag{6.19}$$

When the multiple UGVs match their velocities with neighbors, the consensus cost of the whole group will be lowered. When a stable flocking motion is achieved, the consensus cost $F_{\text{consensus}}$ is close to zero.

(3) *Following Cost*

All the UGVs should regulate their motions to follow the VL; thus, the following cost is defined as

$$F_{follow} = \sum_{i=1}^{Numugv} |c_1 (r_i - r_{vl})| + |c_2 (v_i - v_{vl})| \quad c_1, c_2 > 0 \tag{6.20}$$

In the multiple UGV flocking motion, each UGV follows a VL, and the UGVs will regulate their velocity according to the position and velocity of the VL to lower the following cost.

Finally, the cost function of RHC can be described as

$$F\left(X_{ugv}, U_{ugv}\right) = F_{potential} + F_{consensus} + F_{follow} \tag{6.21}$$

This cost function will be used as the total objective function, which can be optimized by PSO algorithm. The solution is the optimal control input of each UGV, which will lower the cost value of the whole group gradually and lead to a coordinated flocking motion.

6.2.3 Multiple UAV Coordinated Control Based on Velocity Vector Control

According to the heterogeneous mission requirements, the multiple UAV subgroups need stability in the movement to follow and hover over the multiple UGVs, and each UAV should avoid collision during flight. In this way, the multiple UAV and UGV heterogeneous coordinated movement is formed.

According to the control input $ua_i = (T_i, n_i, \mu_i)$, where $i = 1, 2, \ldots, Numuav$, T_i denotes the thrust of UAV_i, n_i denotes the overload, and μ_i denotes the banking angle. The input control vector can be expressed with $U_{uav} = (ua_1, ua_2, \ldots, ua_{Numuav}) = \{U_{uav}(t) | \forall t \in [0,T]\}$, and the state vector can be defined as $X_{uav} = (x_{uav1}, x_{uav2}, \ldots x_{uavNumvua}) \in R^{6 \times Numuav}$. Then, the dynamics for UAVs can be written as follows:

$$\dot{X}_{uav}(t) = f\left(t, X_{uav}(t), U_{uav}(t)\right) \tag{6.22}$$

The control policy of changing the control input of U_{uav} into velocity vector U_{uav} can ensure that all agents eventually align with each other and have a common heading direction while at the same time avoid collisions and group into a tight formation (Gowtham and Kumar 2005), and $\overline{v}_c = (\overline{v}_{c1}, \overline{v}_{c2}, \ldots, \overline{v}_{cNumuav})$. The velocity vector $\overline{v}_{ci}$ of UAV_i includes velocity v_{ci}, banking angle γ_{ci}, and yaw angle

χ_{ci}, i.e., $\overline{\boldsymbol{v}}_{ci} = (v_{ci}, \gamma_{ci}, \chi_{ci})$. Suppose the velocity vector satisfies the following equations:

$$\begin{aligned} \dot{v}_{ci} &= \omega_v (v_{ci} - v_i) \\ \dot{\gamma}_{ci} &= \omega_r (\gamma_{ci} - \gamma_i) \\ \dot{\chi}_{ci} &= \omega_\chi (\chi_{ci} - \chi_i) \end{aligned} \tag{6.23}$$

where ω_v, ω_γ, and ω_χ are gain constants corresponding to velocity, banking angle, and yaw angle, respectively.

According to (6.2) and (6.23), the thrust T_{ci} can be obtained as the following:

$$T_{ci} = D_i + \omega_v W_i (v_{ci} - v_i)/g + W_i \sin \gamma_i \tag{6.24}$$

The overload n_{ci} can be expressed with

$$n_{ci} = \sqrt{(\omega_\gamma v_i (\gamma_{ci} - \gamma_i) + \cos \gamma_i)^2 + (\omega_\chi v_i (\chi_{ci} - \chi_i) \cos \gamma_i / g)^2} \tag{6.25}$$

The pitch angle μ_{ci} can be expressed with

$$\mu_{ci} = \arctan \left(\frac{\omega_\chi v_i (\chi_{ci} - \chi_i) \cos \gamma_i / g}{\omega_\gamma v_i (\gamma_{ci} - \gamma_i) + \cos \gamma_i} \right) \tag{6.26}$$

The resistance D_i can be obtained according to the following equation:

$$D_i = 0.5 v_i{}^2 S C_{D0} + 2kn^2 W_i{}^2 / (\rho v_i{}^2 S) \tag{6.27}$$

where S denotes the reference square of the UAV, C_{D0} denotes zero lift drag coefficient, k denotes induced drag coefficient, and ρ denotes the density of atmosphere.

The corresponding velocity vector can be expressed as the following:

$$\overline{\boldsymbol{v}}_{ci} = c_a \overline{v}_{ai} + c_{tc} \overline{v}_{tci} \tag{6.28}$$

where $\overline{v}_{ai}$ and $\overline{v}_{tci}$ denote collision avoidance vector and hovering velocity vector, respectively, $i = 1, 2 \ldots, Numuav$, c_a and c_{tc} are the corresponding weight coefficients, and $0 < c_a < 1$, $0 < c_{tc} < 1$, $c_a + c_{tc} = 1$.

(1) *Collision Avoidance Velocity Vector*

Multiple UAVs hover over multiple UGVs, and collision should be avoided to ensure safe flight of UAVs. The collision avoidance strategy of priority mechanism is adopted in this section. The priority number level is assigned to each UAV in the multiple UAV groups, and the small number with high priority and the UAVi with low priority can avoid the $UAV_j (j < i)$ with high priority. The collision avoidance velocity vector $\overline{v}_{ai}$ of the UAV_i with low priority can be calculated by the average velocity of UAVi and UAVj, and the direction of $\overline{v}_{ai}$ is pointing to UAV_i along the UAV_j (see Fig. 6.3).

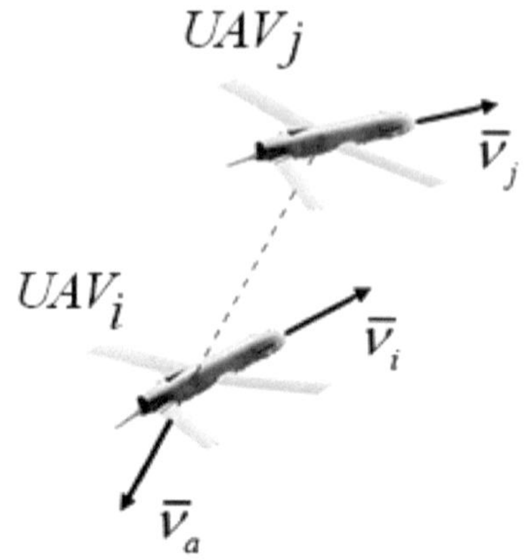

Fig. 6.3 The collision avoidance velocity vector $\overline{v}_{ai}$ of UAV$_i$ (Reprinted from Duan et al. (2011), with kind permission from Springer Science + Business Media)

The weight coefficient c_a of collision avoidance is decided by the distance da_{ij} between two UAVs and the security collision distance d_{avoid}, which can be expressed by

$$c_a = \exp\left(\left|da_{ij} - d_{avoid}\right| / \sigma_a\right) \tag{6.29}$$

where $\sigma_a > 0$ and $0 < c_a < 1$.

(2) *Track Hovering Velocity Vector*

Multiple UAV subgroup can receive messages from the control center and position the center of UGV subgroups. Suppose the minimum hovering velocity of the UAV is $v_{\min}$, the hovering velocity is ω_{circle}, and the minimum hovering radius can be expressed by $r_{circle} = v_{\min}/\omega_{circle}$. The track hovering velocity vector depends on the distance dtc_i between UAV_i and the center C in the horizontal direction.

Case 1: When $dtc_i > 3r_{circle}$, UAV_i is far away from multiple UGV subgroups, the vector $\overline{v}_{tci}$ may maintain the current velocity value or increase a little. The direction points to the center C of multiple UGV subgroups (see Fig. 6.4, marked with ★).

Case 2: When $dtc_i \le 3r_{circle}$, UAV_i hovers in the vicinity of multiple UAVs with $\overline{v}_{tci}$. When the direction and speed of UAV_i are the same with the multiple UGV subgroups, we can determine whether UAV_i follows the center C. If yes, then UAV_i continues to hover in the vicinity of multiple UAVs. Otherwise, UAV_i will follow the center C. In the following process, UAV_i maintains the same value or increases a little, while the direction keeps unanimous with the multiple UGVs (see Fig. 6.4). With $\overline{v}_{ai}$ and $\overline{v}_{tci}$, the velocity vector $\overline{v}_{ci}$ of UAVi can be obtained by (6.28), and the vector command group $\overline{v}_c$ of multiple UAV subgroup can also be obtained.

Multiple UAV and UGV heterogeneous cooperation process can be illustrated by Fig. 6.5.

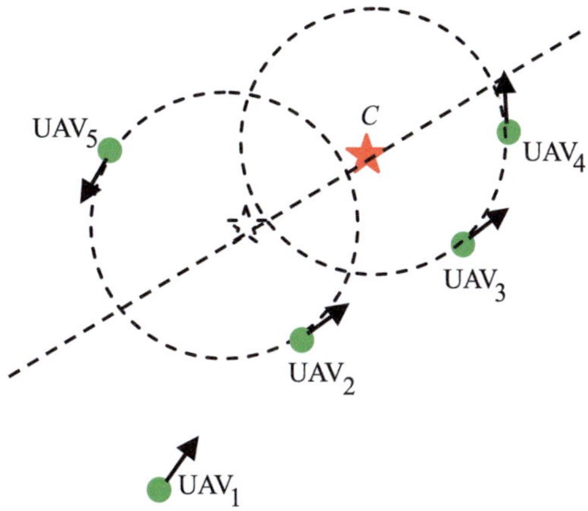

Fig. 6.4 Multiple UAV subgroups hovering over the center C of multiple UGV subgroups (Reprinted from Duan et al. (2011), with kind permission from Springer Science + Business Media)

6.2.4 Multiple UAV/UGV Heterogeneous Cooperation

The feasibility and effectiveness of our proposed method are verified by series of comparative experiments with artificial potential field method. In the experiments, there are 6 UAVs, 10 UGVs, and 1 control center. The initialized parameters of multiple UGV subgroup are set as follows: $d_{ugv} = 15$, $R_{ugv} = 1.2d_{ugv}$, $\xi = 0$, $a = 5$, $b = 5$, $h = 0.9$, $c_1 = 1$, $c_2 = 1$, $T_p = 3s$, $\delta = 1s$, $t_c = 1s$, $ps = 20$, $w_{\max = 1.2}$, $w_{\min} = 0.1$, $vp_{\max} = 4$, $pc_1 = 0.5$, $pc_2 = 0.5$, $Nc_{\max} = 80$.

The initialized parameters of multiple UAV subgroup are set as follows: $\rho = 1.25\text{kg/m}^3$, $W = 14{,}400$ kg, the reference square $= 30\ \text{m}^2$, $T_{\max} = 15{,}000$ kg, $n_{\max} = 7$, $k = 0.1$, $C_{D0} = 0.02$, $\omega_v = 1$, $\omega_\gamma = 0.2$, $\omega_\chi = 1$, $g = 9.8\ \text{m/s}^2$. $v_{\min} = 100$ m/s, $v_{\max} = 200$ m/s, $\omega_{circle} = (\pi/12)$rad/s, $\omega_{\chi\max} = (\pi/9)$rad/s, $\omega_{\gamma\max} = (\pi/9)$rad/s. $d_{avoid} = 30$ m, $\sigma_a = 10$.

The initialized position of VL in multiple UGV subgroup is (1,020,1,020)m, and the initialized velocity is 25 m/s. The initialized status of multiple UGV subgroup and VL are shown with Fig. 6.6 ("■" denotes VL).

The initialized status of multiple UAV subgroup is listed by Table 6.1, and the initialized status of multiple UAV subgroup is shown with Fig. 6.7 ("•" denotes UAV).

Figure 6.8 gives the multiple UAV and UGV heterogeneous cooperation results by using artificial potential field method.

Figure 6.9 gives the multiple UAV and UGV heterogeneous cooperation results by the hybrid method proposed in this paper.

The results in Figs. 6.8 and 6.9 demonstrate that the proposed approach in this paper can guarantee stable convergence, robust tracking, and high efficiency. It clearly shows the superiority of the proposed algorithm over the traditional artificial potential field method. Simulations with different conditions are also conducted to verify the feasibility and effectiveness of the proposed controller.

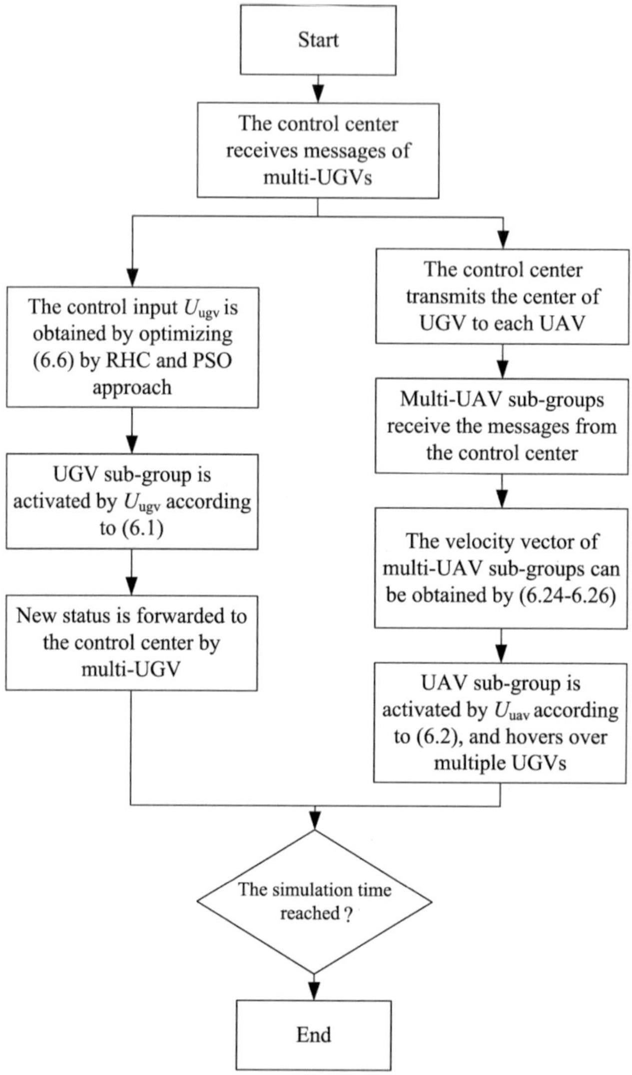

Fig. 6.5 Flowchart of multiple UAV and UGV heterogeneous cooperation (Reprinted from Duan et al. (2011), with kind permission from Springer Science + Business Media)

Besides, experiments about heterogeneous coordinated control for multiple UAVs/UGVs have been conducted by applying a low-cost quadrotor and three ground vehicles. The red vehicle acts as the target. The quadrotor and the other two vehicles aim at pursuing the red vehicle by complementing each other's advantages. As we have explained, UAV can be used to cover large areas searching for target

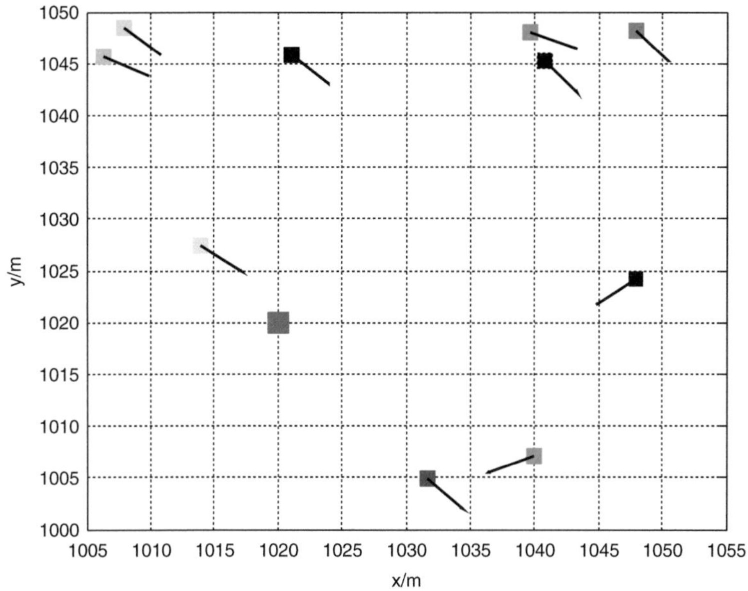

Fig. 6.6 The initialized status of multiple UGV subgroup and VL (Reprinted from Duan et al. (2011), with kind permission from Springer Science + Business Media)

Table 6.1 The initialized status of multiple UAV subgroup

UAV no.	v (m/s)	γ (rad)	χ (rad)	x, y, z (m)
1	120	$\pi/18$	0	550, 0, 500
2	120	$\pi/18$	0	500, 0, 500
3	120	0	0	0, 500, 500
4	120	0	0	0, 0, 500
5	120	0	0	0, 550, 500
6	120	0	0	0, 50, 500

while sensors on UAV are typically limited in operating airspeed and altitude. UGV can be deployed to accurately locate ground targets. Screenshots of the experiment video are illustrated in Fig. 6.10.

6.2.5 *Time-Delay Compensation of Heterogeneous Network Control*

In recent years, network-based control has emerged as a topic of significant interest in the control community. It is well known that in many practical systems, the physical plant, controller, sensor, and actuator are difficult to be located at the same

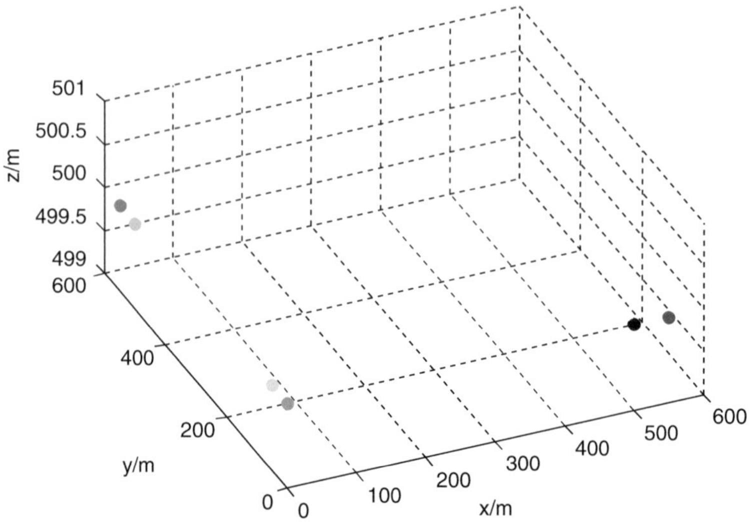

Fig. 6.7 The initialized status of multiple UAV subgroup (Reprinted from Duan et al. (2011), with kind permission from Springer Science + Business Media)

place, and thus signals are required to be transmitted from one place to another. The network- induced time delay in network control system (NCS) occurs when sensors, actuators, and controllers exchange data across the networks. This delay can degrade the performance of control systems designed without considering it and even destabilize the system.

The use of multiple UAVs in concert with UGVs affords a number of synergies. First, UAVs with cameras and other sensors can obtain views of the environment that are complementary to views that can be obtained by cameras on UGVs. Second, UAVs carry over obstacles while keeping UGVs in their field of view, providing a global perspective, and monitoring the positions of UGVs while keeping track of the goal target. This is especially advantageous in three dimensions where UAVs can obtain global maps and the coordination of UAVs and UGVs can enable efficient solutions to the mapping problem. Third, if UAVs can see the UGVs and the UGVs can see UAVs, the resulting three-dimensional sensor network can be used to solve the simultaneous localization and mapping problem, while being robust to failures in sensors like GPS and to errors in dead reckoning. We describe our work in time-delay compensation of heterogeneous network control for multiple UAVs and UGVs.

Suppose the sampling period for the multiple UAVs and UGVs is T, and the maximum time delay is $n_\tau T$, where n_τ is an integer and $n_\tau > 1$. The control sequences for UGV_i can be denoted with

$$Ug_i^{t_k} = \left(ug_i^{t_k}, ug_i^{t_{k+1}}, \ldots, ug_i^{t_{k+Tp-1}} \right) \tag{6.30}$$

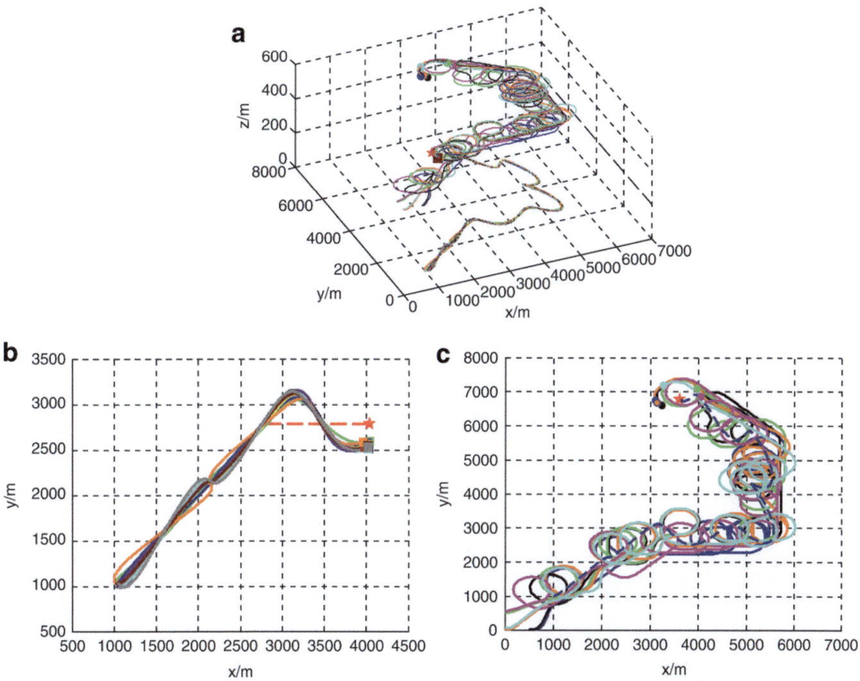

Fig. 6.8 Multiple UAV and UGV heterogeneous cooperation results by using artificial potential field method. (**a**) Traces of multiple UAVs and UGVs heterogeneous cooperation. (**b**) Traces of multiple UGV subgroup before 150 s. (**c**) Traces of multiple UAV subgroup before 150 s (Reprinted from Duan et al. (2011), with kind permission from Springer Science + Business Media)

where $ug_i^{t_k}, ug_i^{t_k+1}, \ldots, ug_i^{t_k+Tp-1}$ is the input sequence of UGV_i. The predictive control sequences for all the UGVs at time t_k is

$$U_{ugv}^{t_k} = \left(Ug_1^{t_k}, Ug_2^{t_k}, \ldots, Ug_{Numugv}^{t_k} \right) \tag{6.31}$$

Due to the time delay, we can obtain the motion equation for UGV:

$$\dot{X}_{ugv}(t) = f\left(t, X_{ugv}(t), U_{ugv}^{t_{k-n}}(t) \right) \tag{6.32}$$

where $0 \le n \le n_\tau$ and $n \in \Re$.

In the absence of the presence of network delay, UGV_i only uses $ug_i^{t_k}$, and $\left(ug_i^{t_k+1}, \ldots, ug_i^{t_k+Tp-1} \right)$ is abandoned. The new control input is obtained in the next iteration. While in the case of random long-period time delay, UGV_i may not receive the control input at the moment t_{cur}, and multiple UGVs can hardly meet the requirements of cooperative motion. In this case, the predictive control sequences can be all sent to the UGVs and are stored in various UGVs.

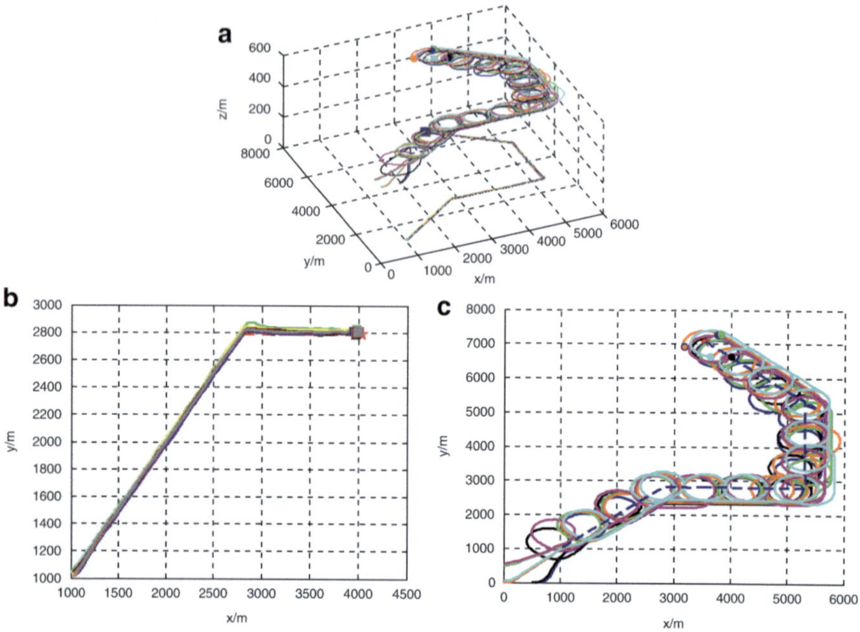

Fig. 6.9 Multiple UAV and UGV heterogeneous cooperation results by the hybrid method proposed in this paper. (**a**) Traces of multiple UAVs and UGVs heterogeneous cooperation. (**b**) Traces of multiple UGV subgroup following VL before 150 s. (**c**) Traces of multiple UAV subgroup before 150 s (Reprinted from Duan et al. (2011), with kind permission from Springer Science + Business Media)

Receding horizon control (RHC) and PSO are adopted in this approach. The objective function can be defined as

$$
\begin{aligned}
&\min_u J = f\left(X_{ugv}^{t_{k-n_1}}, U_{ugv}; t_c, T_p\right)\\
&J = \int_{t_c}^{t_c+T_p} F\left(X_{ugv}^{t_{k-n_1}}, U_{ugv}\right) dt\\
&subject\ to\ \dot{X}_{ugv}^{t_{k-n_1}} = f\left(t, X_{ugv}^{t_{k-n_1}}, U_{ugv}\right)\\
&LL_{ugv} \le \begin{bmatrix} X_{ugv}^{t_{k-n_1}} \\ U_{ugv} \end{bmatrix} \le UL_{ugv}
\end{aligned}
\tag{6.33}
$$

where $0 \le n_1 \le n_\tau$ and $n_1 \in \Re$. For multiple UAV subgroup, whose speed is much larger than UGV, there is a tracking delay between multiple UAV subgroups of multiple tracking UGV subgroups. However, UAV can be quickly followed up by tracking the spiral vector. Therefore, UGV_i can obtain the tracking hovering velocity vector $\overline{v}_{tci}^{t_{k-n2}}$ according to the latest multiple UGV center location information,

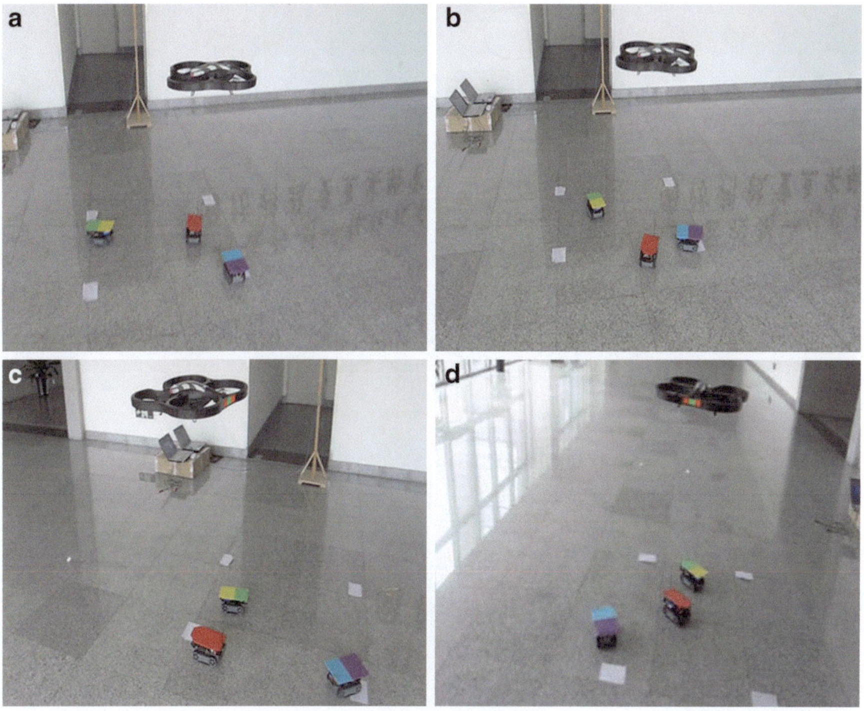

Fig. 6.10 Screenshots of the experiment about heterogeneous coordinated control for multiple UAVs/UGVs

where $0 \leq n_2 \leq n_\tau$, and $n_2 \in \Re$. The vector group of multiple UAV subgroups can be defined as

$$\overline{\boldsymbol{v}}_c^{t_k} = \left(\overline{\boldsymbol{v}}_{c1}^{t_k - n1}, \overline{\boldsymbol{v}}_{c2}^{t_k - n2}, \ldots, \overline{\boldsymbol{v}}_{cNumuav}^{t_k - n_{Numuav}} \right) \tag{6.34}$$

The time delay of multiple UGV subgroup sending the status information to the control center can be defined with τ_{gc}, and the time–event-driven approach is adopted in the control center. The time delay of the control center sending the status information to each UGV can be defined with τ_{gc}, and the time delay of the center of multiple UGV subgroup sending the status information to UAV can be defined with τ_{ca}.

6.2.5.1 Status Buffer of Control Center

$UGV_i(i = 1, 2, \ldots Numugv)$ sends the status information to the control center respectively, and the time-driven approach is adopted in this process. Due to the

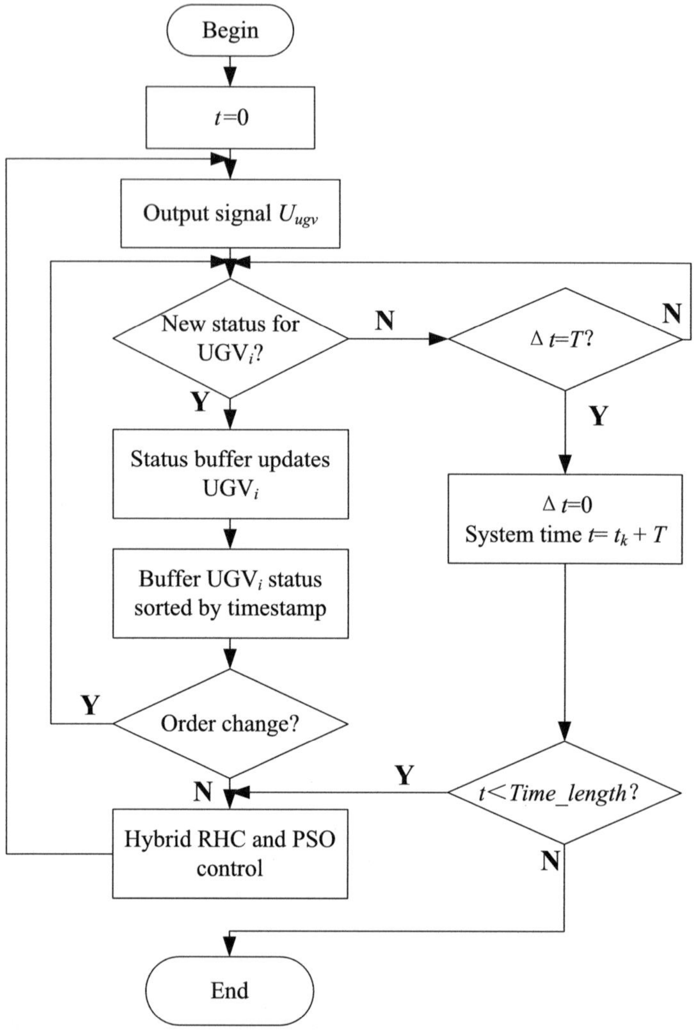

Fig. 6.11 The updating process for status buffer of control center

existence of time delay τ_{gci}, the arrival time is random. So time–event-driven mode is adopted in the control center. When $\tau_{gci} > T$, the control center automatically starts the control algorithm by using the status information of UGV_i.

In the status buffer of the control center, the older state information will be automatically deleted with the advance of the new status information. The updating process for status buffer of control center can be shown with Fig. 6.11. In which, $t = t_k + \Delta t$, $t_k = kT$, $\Delta t \le T$, the simulation time is denoted with $Time\_length$, and the output of control center is U_{ugv}, $i = 1, 2, \ldots, Numugv$.

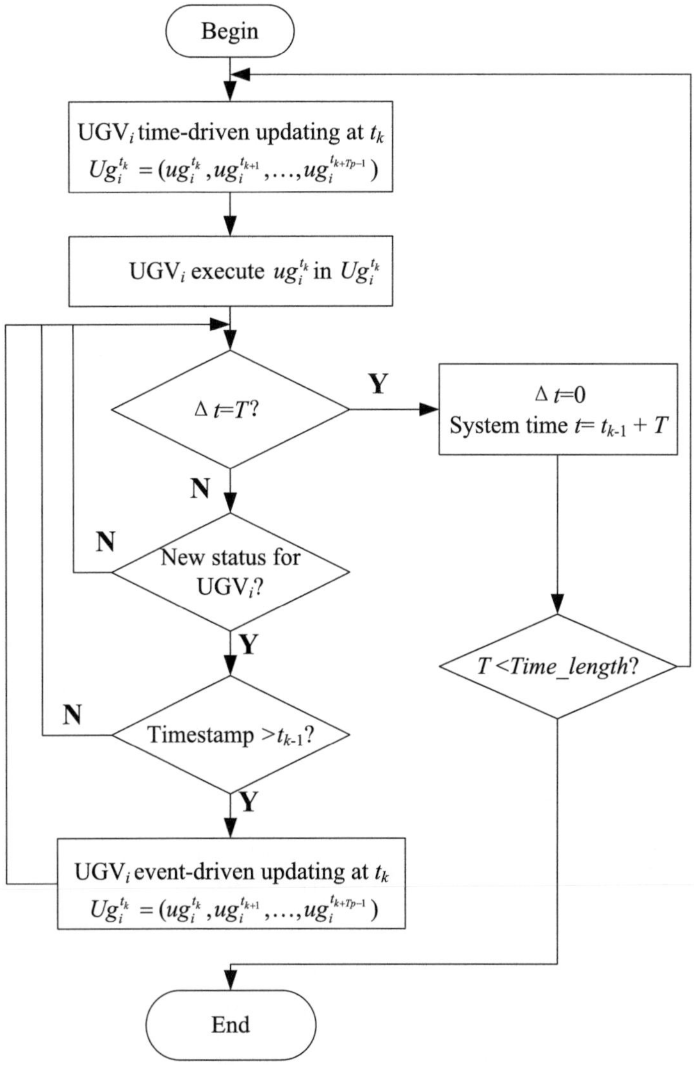

Fig. 6.12 The updating process for UGV$_i$ control input buffer at moment t_k

6.2.5.2 UGV Control Input Buffer

Because there is a time delay τ_{cgi} between control center and UGV_i, it is necessary to set UGV_i control input buffer. In this way, the control sequences can be saved. Based on the maximum network delay $n_\tau T$, the buffer length is set to n_τ. The update of UGV_i control input buffer can be divided into two parts: the time-driven update and the event-driven update. The updating process for UGV_i control input buffer at moment tk can be shown with Fig. 6.12.

6.2.5.3 UAV Center Location Information Buffer

The control center sends the center location information $C_i^{t_k}$ of multiple UGVs to a multiple UAV subgroup by using the time-driven approach. Due to the existence of time delay τ_{cai}, the time of $UAV_i(i = 1, 2, \ldots, Numuav)$ receiving $C_i^{t_k}$ is random. However, the time-driven approach is adopted in UAV_i. When $\tau_{cai} > T$, UAV_i uses the velocity vector instruction $\overline{v}_{ci}$ with the center of the historical status information. The center of the updating buffer can also be divided into two parts: the time-driven update and the event-driven update. In time t_k, the older state information will be automatically deleted with the advance of the new center location information $C_i^{t_k}$. The updating process for UAV_i center location information buffer at moment t_k can be shown with Fig. 6.13.

The transfer timing for multiple UAV subgroup center location information can be shown with Fig. 6.14.

6.3 DE-Based RHC for Multiple UAV Cooperative Search

The search problem has been extensively studied in the literature, starting off with a single-agent problem and further extended to multi-agent search. In military applications, multiple UAV coordinated search is an important means of getting battlefield information in the future war. Compared to the problem of a single searcher, the problem becomes more complex when we consider a team of agents that are cooperatively searching the targets in an area.

For the flight path-planning problem in UAV targets searching, the traditional method is based on search theory, designing search routes covering task areas from the perspective of maximizing the probability of target detection. Such routes are usually fixed pattern, such as scanning-line mode to achieve a complete coverage of the target area. This method is of simple route calculation, fast, and able to guarantee a certain probability of target detection, but the flight route is fixed and the search efficiency is low. Another important method is a dynamic search method based on the search map. The method is based on two-dimensional discrete map to store targets and environmental information. Based on search map information, different strategies for online calculation of the next time search path can be used, such as the random strategy, the local optimal strategy, and the global maximum strategies. These methods can be used in target searching effectively based on real-time detection of information. The difficulty lies in how to quickly calculate the safe search route to the next point. In this study, search map is used for the cooperative area searching of multiple UAVs.

The main concern of this study for multiple UAV search is about how to control multiple UAVs for cooperative search for ground targets. In other words, the problem of cooperation between multiple UAVs is the key for the multiple UAV search problem. Multiple UAV search problem is a complex optimization and

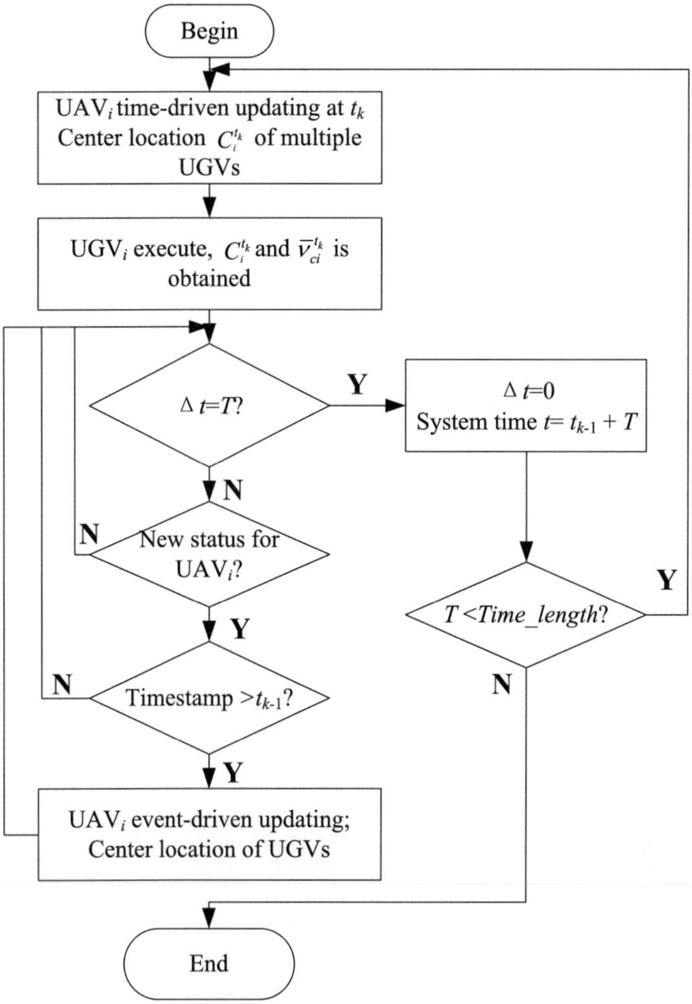

Fig. 6.13 The updating process for UAV$_i$ center location information buffer at moment t_k

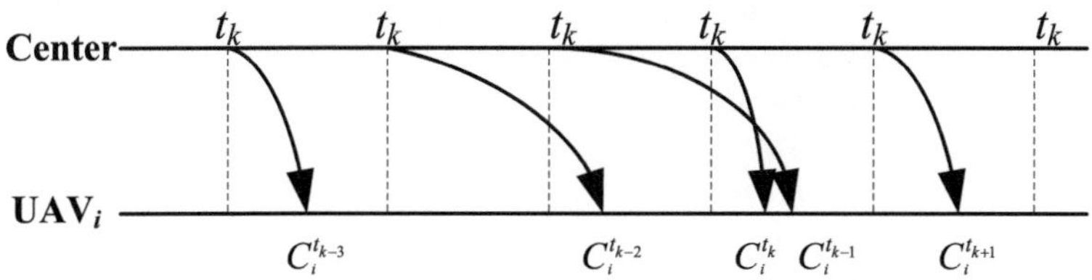

Fig. 6.14 The transfer timing for multiple UAV subgroup center location information

control problem with a large amount of information in process of solution and a high dimension. Recent years, biological swarm intelligence provides a good idea for solving multi-objective UAV distributed coordinate search problem. In view of the flexibility of the intelligent optimization methods based on biological evolution and its advantages in solving high-dimensional problems, in this study, an intelligent optimization method of DE is used for the solution of the multiple UAV cooperative search problem.

Another important issue for multiple UAV cooperative search problem is the requirement of real time and security. Some researchers apply the thought of RHC into the cooperative search problem. Using the online task optimization method based on rolling window, the optimization search strategy can respond quickly for environment changes by optimizing and rolling online. In this study, RHC is used to realize the real time and security during searching process of multiple UAVs.

6.3.1 Model Description for Cooperative Search

6.3.1.1 Some Hypotheses Involving UAV Platform

UAV platform is a direct implementation of the search task and also the controlled object engaged in our study. As the study focuses on the search method in UAV area searching, but not the low-level control of UAV platform, some hypotheses are set in the study:

- The UAV platform is small tactical UAV.
- There is an automatic flight control system for each UAV platform.
- The UAV high-level mission control and the low-level flight control can be considered decoupling.

Besides, in order to reflect the physical characteristics of the UAV, a set of parameters related to the flight performance is constrained:

- *Maximum cruise velocity* $v_{\max}$: It's a basic performance parameter for the UAV platform, which decides the movement pattern of the UAV in the task area.
- *Maximum flight height* $h_{\max}$: To image sensor, the flight height decides the detection range of airborne sensors directly. It affects the effects of UAVs to the target search, detection and identification.
- *Maximum duration time* $t_{\max}$: Decided by the amount of fuel on UAV, it limits the longest time that UAV can perform the task in search area.
- *Minimum turning radius* $R_{\min}$: The minimum turning radius describes the UAV's mobility. Together with the velocity parameter of UAV platform, it decides the flight path in a certain input.

In control of UAV flight track point, a particle model of three degrees of freedom is considered. To facilitate follow-up studies, a discrete form of expression is established to describe the flight control model. When the total number of UAVs is

Nv, for the platform of UAV_i, ($k = 1, 2, \ldots, Nv$), at time k, the dynamic characteristic can be described by the motion model as follows:

$$\begin{cases} x_i\,(k+1) = x_i\,(k) + v_i\,(k)\cos\left(\sigma_i\,(k)\right)\cos\varphi_i\,((k))\cdot t_s \\ y_i\,(k+1) = y_i\,(k) + v_i\,(k)\cos\left(\sigma_i\,(k)\right)\sin\varphi_i\,((k))\cdot t_s \\ z_i\,(k+1) = z_i\,(k) + v_i\,(k)\sin\left(\sigma_i\,(k)\right) \end{cases} \tag{6.35}$$

where t_s is the decision interval, $(x_i(k), y_i(k), z_i(k)) \in R^3$ is the position of UAV_i at time k in the three-dimensional search space, $v_i(k) \in R$ is the velocity of UAV_i at time k, $\varphi_i(k) \in [0, 360)$ is the yaw angle of UAV_i at time k, and $\sigma_i\,(k) \in [0, \sigma_{\max_i})$ is the climb angle of UAV_i at time k.

6.3.1.2 Search Targets Modeling

Search target is the specific object for task of multiple UAVs. Depending on the motion state, search target can be divided into static target and dynamic target. To the static target, there are fixed radar, artillery positions, buildings, roads, bridges, and so on. To the dynamic target, it can be all kinds of vehicles, aircrafts, specific people, etc.

Besides, according to whether it can attack or not, the target can be divided into antagonistic target and no antagonistic target. To the antagonistic target, it includes artillery positions, missiles, and other offensive aircrafts and ships. For this kind of targets, the UAV should avoid entering into their scope of attacks. To the no antagonistic target, there are fixed radar, aircrafts for scout, plants, and so on.

Suppose the targets will not take the initiative to escape the search of UAVs, target elements mainly considered in this study are as follows:

- *Target position state* x_t: It describes the specific location of different targets.
- *Target velocity* v_t: It describes the target speed of movement in space.
- *Target movement pattern*: It describes the variation law of target position and velocity in space, including stationary state, random movement, and deterministic motion (or in a particular trajectory).

Based on the description of target elements above, in the two-dimensional plane, in case the target position is $x_t = (x,y) \in R^2$, then for the static target, there is $x_t(k) \equiv x_t(0)$. Otherwise, for the dynamic target in the two-dimensional space, the direction of motion is denoted as θ_t; in case of discrete time, a simple model of target motion is usually considered:

$$x_t\,(k+1) = x_t\,(k) + \Delta x \tag{6.36}$$

where Δx is the displacement increment of target; in case of deterministic motion, there is $\Delta x = v_t \times t_s$; otherwise, if the target moves in a random way, the displacement increment of Δx is accordingly random. A typical random motion model is

Random Tours. In this case, the target starts form the initial position, $x_t(0)$, along with a random direction, θ_t, which obeys the uniform distribution on the interval of $[0, 2\pi)$, and moves for a random time, *dt*, which obeys the exponential distribution with parameter *d*. Then the position of target $x_t(k)$ is completely random.

6.3.1.3 Environment Information Modeling

Environment information involved in search issues always includes information of targets, other UAV's mission state information, and information of threats. The target information is a key to the multiple UAV cooperative search problem. Here, the study focuses on the description and modeling of target in search environment. Because the environment is dynamic, the uncertainty of targets decides the search problem is essentially a random question. In this situation, the search map model based on probability is a natural choice.

(1) *Basic Search Map (BSM) Model*

The basic idea of search map is to represent the environment as a grid of cells. Suppose the area is divided into $L_x \times L_y$ cells, and each cell in the map associates with a certainty information strut, $P_{ij}(k)$, which describes the general information of environment and target in current cell. The information strut is defined as follows:

$$P_{ij}(k) = \left(p_{ij}(k), \chi_{ij}(k)\right), i \in \{1, \ldots, L_x\}, j \in \left\{1, \ldots, L_y\right\} \tag{6.37}$$

where $p_{ij} \in [0,1]$ is called target occupancy probability (TOP) in grid (i,j) and $\chi_{ij}(k) \in [0,1]$ is environment certainty (EC). We suppose each cell contains at most one target, then $p_{ij}(k) = 0$ represents that the UAV knows nothing about the target in grid (i,j) at time step k, and $p_{ij}(k) = 1$ represents high probability that a target is present in grid (i,j). The $\chi_{ij}(k)$ describes the UAV's determination extent for grid (i,j) at time step k, $\chi_{ij}(k) = 0$ represents UAV knows nothing about information of grid (i,j),and $\chi_{ij}(k) = 1$ represents completely knows the information of this grid.

A schematic diagram of grid partition for search map is shown in Fig. 6.15, where five gray star points represent the target position.

Generally, search map describes the UAV's belief state for existence of target in the mission area and is the direct information that UAV can understand and apply. In cooperative area searching of multiple UAVs, each UAV obtains information of external environment through not only its own sensors but also the communication equipments. The environment information, by the way of the information be obtained, can be divided into three parts: the prior information, initial intelligence from other means of reconnaissance; the probe information which got through the sensors carried by UAV; and communication information from other UAVs. All information can be expressed on the search map. During the search mission, different UAVs share the information on one search map.

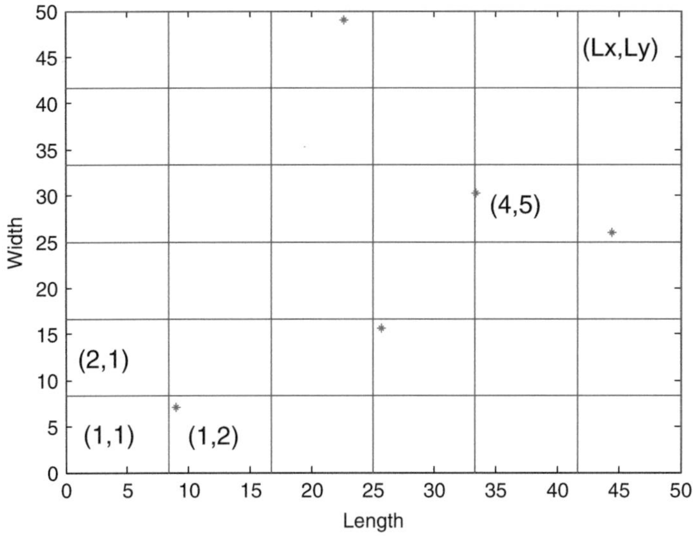

Fig. 6.15 Regional division based on quadrilateral grid

(2) *Extended Search Map (ESM) Model*

Based on information of BSM, single UAV can make its own path decisions. However, the BSM describes only the uncertain information of targets and environment, but not the state information of other UAVs. For cooperative area search of multiple UAVs, cooperation between different agents is a key issue. UAVs must have ability to coordinate their actions and to maximum team search efficiency. To achieve effective multiple UAV cooperation search, the BSM is extended into the extended search map (ESM) based on the Digital Hormone Model (DHM).

The ESM is on basic of BSM and it introduces the hormone information to build a mixed information strut. Then, the information strut defined in (6.37) is extended to (6.38):

$$P_{ij'}(k) = \left(p_{ij}(k), \chi_{ij}(k), H_{ij}(k)\right), i \in \{1, \ldots, L_x\}, j \in \left\{1, \ldots, L_y\right\} \quad (6.38)$$

where $H_{ij}(k)$ is the digital hormone information on grid (i,j) at time k step. The concentration of hormone information is a function of UAV position and time. When UAV moves to grid (i,j) at time k, it generates hormone signal on the related position of search map, and meanwhile, the digital hormone signal is sent to UAVs nearby to impact other UAV's decision on the next time step.

Real hormone has the ability of diffusion and dissemination. The hormone information includes two types of hormones, the activator hormone H_A and inhibitor hormone H_I. The diffusion equation of H_A and H_I are given below:

$$\begin{aligned} \Delta H_A(i,j,k) &= \frac{a_A}{2\pi\sigma^2} e^{-\frac{(x-a)^2+(y-b)^2}{2\sigma^2}} \\ \Delta H_I(i,j,k) &= -\frac{a_I}{2\pi\rho^2} e^{-\frac{(x-a)^2+(y-b)^2}{2\rho^2}} \end{aligned} \quad (6.39)$$

where grid (*a*,*b*) is adjacent to grid (*i*,*j*), a_A, a_I are constants, σ and ρ are the rates of diffusion, respectively, and $\sigma < \rho$ to satisfy the Turing stability condition. With the diffusion of hormone in search map through communication, we get the hormone update function at time *k* by summing up all hormone information from neighboring UAV, where the constant $\tau_H \in [0,1]$ is the rate for dissipation.

$$H_{i\,j}\,(t+1) = \tau_H \cdot H_{i\,j}(t) + \sum_{N_k} (\Delta H_A\,(i,j,k) + \Delta H_I\,(i,j,k)) \qquad (6.40)$$

The initial hormone information on the map for each grid is zero.

6.3.2 DE-Based RHC for Cooperative Area Search

The result of UAV search problem is directly reflected on the UAV's search path. In UAV search problem, the goal of search path planning is to generate the effective trajectory along with an objective function is maximized. In this study, we consider a reward function as the objective function, and the key issue to solve the cooperative area searching of multiple UAVs with DE algorithm is to determine the reward function.

For the cooperative search problem based on RHC, the reward function is related to each UAV's current position $X(k)$ and the following position of track points $[X(k+1|k), X(k+2|k), \ldots, X(k+p|k)]$, in which p is the size of control window. $[X(k+1|k), X(k+2|k), \ldots, X(k+p|k)]$ is the input of the optimization problem.

The reward function for UAV search decision can be described by a composed efficacy *J*. As discussed above, the composed efficacy *J* can be represented as

$$J\left[X(k), X\left(k+1\middle|k\right), X\left(k+2\middle|k\right), \ldots, X\left(k+p\middle|k\right)\right] \qquad (6.41)$$

During the search process, each UAV needs to determine the following *p* track points according to current state and search map information. The goal of search path planning for multiple UAV search is to find the most targets, gain the information on whole search area, and reduce the uncertainty of mission area. Accordingly, the optimization decisions need to reach the following aspects of subgoals:

1. To maximum the probability of finding the target
2. Tend to detect those areas with more uncertainties
3. To realize effective cooperation of multiple UAVs
4. To minimum the cost during the search process

Based on the subgoals introduced above, the composed efficacy at time *k* step, $J(k)$, can be defined as

$$J(k) = \omega_1 \cdot J_T(k) + \omega_2 \cdot J_F(k) + \omega_3 \cdot J_C(k) - \omega_4 \cdot C(k) \qquad (6.42)$$

where $\omega_1, \omega_2, \omega_3$, and ω_4 are corresponding weight; $J_T(k)$, $J_F(k)$, and $J_C(k)$ are, at time k step, three different rewards related to the subgoals introduced above; and $C(k)$ is the search cost at time k step. Based on the ESM information, the definitions of three rewards and the cost are given below:

(1) *Target Finding Reward* $J_T(k)$

The target finding reward describes the possibility of finding targets along the way. During optimization of the UAV path with DE, a number of alternative following track points will be considered at first. However, the algorithm tends to choose the path that reaches the biggest target finding reward as the real path that UAV will fly by.

Suppose for $UAV_i(i=1,2,\ldots,N_v)$, the whole range of sensor detection during time $[k, k+p-1]$ is R_i, the target finding reward $J_T(k)$ at time k can be defined as

$$J_T(k) = \sum_{i=1}^{N_V} \sum_{(m,n)\in R_i} p_{mn}{}^{i}(k) \tag{6.43}$$

where $p_{mn}{}^{i}(k)$ is the probability of target existence in UAV_i's detection scope R_i, related only to the position of UAV_i on search map.

(2) *Reward of Expected to Detect* $J_F(k)$

The search decision tends to make UAV detect the area with small Environment Certainty. The path with smaller EC and bigger probability of target existence gains a bigger reward. The reward of $J_F(k)$ can be calculated by the following equation:

$$J_F(k) = \sum_{i=1}^{N_V} \sum_{(m,n)\in R_i} \left(1 - \chi_{mn}{}^{i}(k)\right) p_{mn}{}^{i}(k) \tag{6.44}$$

where $\chi_{mn}{}^{i}(k)$ is the EC of UAV_i in its detection scope R_i on search map.

(3) *Cooperation Reward* $J_C(k)$

The cooperative of multiple UAVs can avoid excessive repetition detection on a certain area; on the other hand, it can also reduce the risk of collision and ensure the safety of multiple UAV missions. Accordingly, the cooperative reward can be defined as

$$J_{C1}(k) = \sum_{i=1}^{N_V} \sum_{n=0}^{p-1} \left[H\left(x_i\left(k+n\right)\right) - H\left(x_i\left(k+n+1\right)\right)\right] \tag{6.45}$$

where $H(x_i(k+n))$ represents the hormone information on position of UAV_i at track point $x_i(k+n)$.

The other definition is about the overlap degree of tracks between two different UAVs:

$$J_{C2}(k) = \sum_{i=1}^{N_V} \sum_{j=1}^{N_V} \sum_{n=0}^{p-1} f_o \left(\theta_{ij}{}^{p}, d_{ij}{}^{p} \right) \tag{6.46}$$

where $\theta_{ij}{}^{p}$ is the heading angle difference between UAV_i and UAV_j on their pth track points and $d_{ij}{}^{p}$ is accordingly the distance between the two UAV's pth track points. Usually, function f_o can be defined as

$$f_o \left(\theta_{ij}{}^{p}, d_{ij}{}^{p} \right) = \exp^{r_0 \cdot d_{ij}{}^{p} \cdot \cos\left(\theta_{ij}{}^{p}/2 \right)} \tag{6.47}$$

where $r_0 \in R^+$ is adjustable parameter.

The composed cooperation reward can be represented as follows:

$$J_C(k) = J_{C1}(k) + \alpha_C \cdot J_{C2}(k) \tag{6.48}$$

where $\alpha_C \in R^+$ is adjustable parameter.

(4) *Search Cost C*(*k*)

Search cost is the comprehensive cost during process of multiple UAV search mission. It generally performs to be the time-consuming or the fuel consuming in search process. The following equation gives a certain estimate method for search cost:

$$C(k) = \sum_{i=1}^{N_V} \sum_{n=0}^{p-1} \| x_v{}^{i} (k+n) - x_v{}^{i} (k+n+1) \| / v_i (k+n) \tag{6.49}$$

where $x_v{}^{i}(k+n)$ and $x_v{}^{i}(k+n+1)$ are adjacent two track points on track path of UAV_i and $v_i(k+n)$ is the velocity between the two track points.

6.3.3 Experiments

A simulation study is included to illustrate the feasibility of our proposed method for cooperative search for multiple UAVs. The simulation scenario consists of a team of four UAVs searching a 100×100 (50×50 km) cellular environment with five targets and different kinds of threats. The threats are mainly composed of dangerous terrains and enemy threats, which can be shown by the search map in Fig. 6.16. In the search map, M1 represents the mountain, Bw1 denotes the bad weather area, and Fd1 is the forbidden fly area, which are prior information for the UAVs. It is assumed that there is some minor a priori topographical information but no other sources of information on target distribution. The initial distribution of five targets is shown in

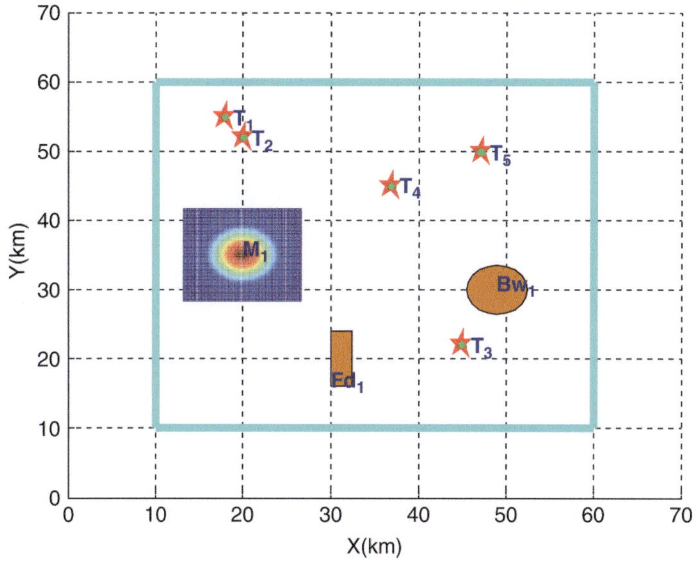

Fig. 6.16 Initial information of search region

Table 6.2 Information of search targets

Target ID	Initial position	Type of target
T1	(8,45)	Static target with unknown initial position
T2	(10,42)	Static target with known initial position
T3	(35,12)	Moving target with prior information of initial position and speed
T4	(27,35)	Moving target with prior information of initial position and velocity
T5	(37,40)	Moving target with prior information of initial position and speed

Fig. 6.16. The target information is shown in Table 6.2. The objective of UAVs is to search the environment so that they can incrementally obtain knowledge of the environment and locate targets with capability of threat avoidance. Four UAVs are initially located at four corners of the search region. For each UAV, the maximum cruise velocity is 0.1 km/s, minimum turning radius is 2 km, and the diameter of detection region for the sensor is 2 km. The search result of our experiment can be shown in Figs. 6.17 and 6.18.

6.4 Conclusions

Multiple UAV/UGV heterogeneous cooperation provides a new breakthrough for the effective application of UAVs and UGVs. On the basis of introduction of UAV/UGV mathematical model, the characteristics of heterogeneous flocking is

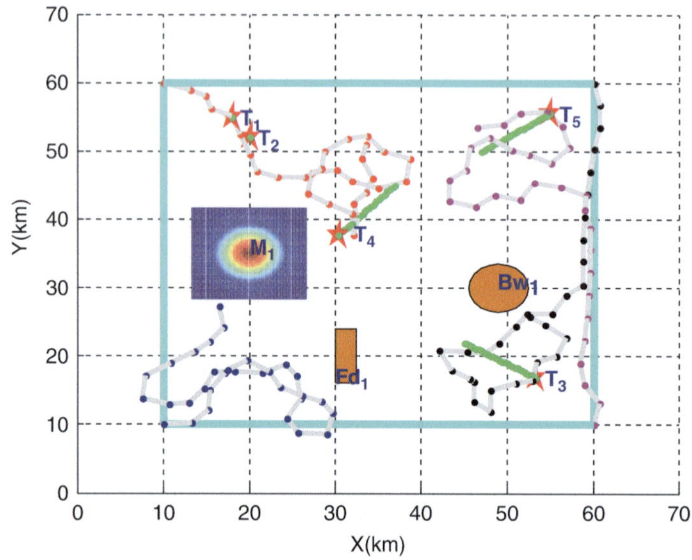

Fig. 6.17 Search result using our proposed method (2 dimensional)

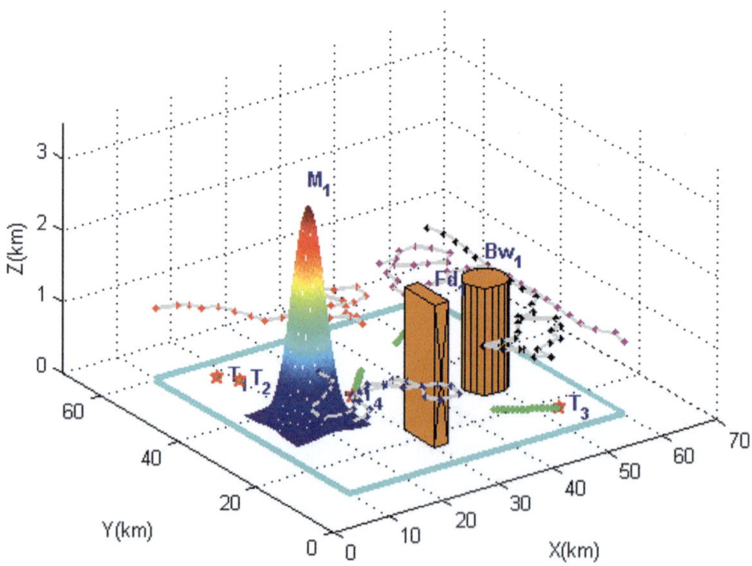

Fig. 6.18 Search result using our proposed method (3 dimensional)

analyzed in detail. Two key issues are considered in multiple UGV subgroups, which are Reynolds rule and VL. RHC with PSO is proposed for multiple UGV flocking, and velocity vector control approach is adopted for multiple UAV flocking. Then, multiple UAV and UGV heterogeneous tracking can be achieved by these two approaches. The feasibility and effectiveness of our proposed method are verified by comparative experiments with artificial potential field method. Besides, we describe a time-delay compensation approach of heterogeneous network control for multiple UAVs and UGVs. The detailed updating process for status buffer of control center, UGV control input buffer, and UAV center location information buffer are also presented.

In Sect. 6.3, a DE-based RHC controller for cooperative area searching of multiple UAVs is presented. The thought of RHC in adopted to satisfy the real-time requirements. Then, the cooperative search problem can be formulated into a function, which is about designing search routes covering task areas from the perspective of maximizing the probability of target detection. Furthermore, an extended search map is used to describe the environment information on the search region. Simulation results demonstrated that the approach we proposed for area searching problem of multiple UAVs is feasible and also effective.

References

Ariyur KB, Fregene KO (2008) Autonomous tracking of a ground vehicle by a UAV. In: Proceedings of American Control Conference, Seattle. IEEE, pp 669–671

Duan H, Liu S (2010) Unmanned air/ground vehicles heterogeneous cooperative techniques: Current status and prospects. Sci China Technolog Sci 53(5):1349–1355

Duan H, Ding Q, Liu S, Zou J (2010a) Time-delay compensation of heterogeneous network control for multiple UAVs and UGVs. J Internet Technol 11(3):379–385

Duan H, Shao S, Su B, Zhang L (2010b) New development thoughts on the bio-inspired intelligence based control for unmanned combat aerial vehicle. Sci China Technolog Sci 53(8):2025–2031

Duan H, Zhang Y, Liu S (2011) Multiple UAVs/UGVs heterogeneous coordinated technique based on Receding Horizon Control (RHC) and velocity vector control. Sci China Technolog Sci 54(4):869–876

Duan H, Luo Q, Ma G, Shi Y (2013) Hybrid particle swarm optimization and genetic algorithm for multi-UAVs formation reconfiguration. IEEE Comput Intell Mag 8(3):16–27

Gowtham G, Kumar KS (2005) Simulation of multi UAV flight formation. In: Proceedings of The 24th Digital Avionics Systems Conference (DASC 2005), USA. IEEE, pp 11.A.13-11–11.A.13-16

Hsieh MA, Cowley A, Keller JF, Chaimowicz L, Grocholsky B, Kumar V, Taylor CJ, Endo Y, Arkin RC, Jung B (2007) Adaptive teams of autonomous aerial and ground robots for situational awareness. J Field Robot 24(11-12):991–1014

Jadbabaie A, Lin J, Morse AS (2003) Coordination of groups of mobile autonomous agents using nearest neighbor rules. IEEE Trans Autom Control 48(6):988–1001

Olfati-Saber R (2006) Flocking for multi-agent dynamic systems: algorithms and theory. IEEE Trans Autom Control 51(3):401–420

Palejiya D, Tanner HG (2006) Hybrid velocity/force control for robot navigation in compliant unknown environments. Robotica 24(6):745–758

Phan C, Liu HH (2008) A cooperative UAV/UGV platform for wildfire detection and fighting. In: Proceedings of Asia Simulation Conference-7th International Conference on System Simulation and Scientific Computing (ICSC 2008), Beijing. IEEE, pp 494–498

Tanner H, Christodoulakis D (2006) Cooperation between aerial and ground vehicle groups for reconnaissance missions. In: Proceedings of 2006 45th IEEE Conference on Decision and Control, San Diego. IEEE, pp 5918–5923

Tanner HG, Jadbabaie A, Pappas GJ (2003) Stable flocking of mobile agents, Part I: Fixed topology. In: Proceedings of 42nd IEEE Conference on Decision and Control, Maui. IEEE, pp 2010–2015

Trentini M, Beckman B (2010) Semi-autonomous UAV/UGV for dismounted urban operations. In: Proceedings of 2010 Defense, Security, and Sensing, Orlando, Florida. SPIE, pp 76921C-76921C-76929

Zhang X, Duan H (2012) Differential evolution-based receding horizon control design for multi-UAVs formation reconfiguration. Trans Inst Meas Control 34(2–3):165–183

Zhang X, Duan H, Yu Y (2010) Receding horizon control for multi-UAVs close formation control based on differential evolution. Sci China Inf Sci 53(2):223–235

Chapter 7
Biological Vision-Based Surveillance and Navigation

Haibin Duan

Abstract From the simplest vision architectures in insects to the extremely complex cortical hierarchy in primates, it is fascinating to observe how biology found efficient solutions to solve vision problems, which may stimulate the emergence of new ideas for the researchers in computer vision. Biological computer vision is an excellent resolution that serves as a low-cost and information-rich source complementing the sensor suite for unmanned aerial vehicle (UAV). For fully autonomous UAV, the capability of autonomous target recognition and visual navigation is of vital importance for a completing a mission in case of GPS signal lost. This chapter mainly focuses on target recognition, image matching, and autonomous visual tracking and landing by taking advantage of the bio-inspired computation, with the aim of dealing with vision-based surveillance and navigation. An artificial bee colony (ABC) optimized edge potential function (EPF) approach is presented to accomplish the target recognition task for low-altitude aircraft. Then a chaotic quantum-behaved particle swarm optimization (PSO) based on lateral inhibition is proposed for image matching. Moreover, implementation of autonomous visual tracking and landing for a type of low-cost UAV (quadrotor) is conducted.

7.1 Introduction

Computer vision is a hot field that includes methods for acquiring, processing, analyzing, and understanding images and high-dimensional data from the real world, which can produce numerical or symbolic information for fully autonomous unmanned aerial vehicle (UAV). Biological vision is a new emerging and challenging area of computer vision. The various physiological components involved

H. Duan and P. Li, *Bio-inspired Computation in Unmanned Aerial Vehicles*,
DOI 10.1007/978-3-642-41196-0_7,

in biological vision are referred to collectively as the visual system and are the focus of much research in psychology, cognitive science, neuroscience, and molecular biology. One of the main challenges of UAV design is the automatic target recognition system for surveillance and navigation. In order to obtain accurate identification results and hence to meet the practical requirements of aircrafts reconnaissance system, the proposed target recognition method must be efficient, stable, and convenient for promotion (Veredas et al. 2009). In many countries, target recognition technology for aircrafts at low altitude is highly confidential, and it is accordingly difficult to see its specific technical details (Duan et al. 2013). Inaccurate detection information will usually directly affect the awareness of the UAV about the battlefield state and target selection. Moreover, uncertain detection information usually influences the recognition system with side effects. In recent years, many methods for air vehicle systems have been proposed to improve the performance of target recognition. Visual detection with ground targets search and assignment were also providing a progress and alternative for UAVs (Bertuccelli and Cummings 2012; Lin et al. 2012).

Edge detection is the fundamental step in edge extraction and object delineation in biological image processing. The goal of edge detection is to mark the points which contain the major information in a digital image (Duan et al. 2011). An effective edge detector can reduce a large amount of data and still keep most of the important feature of the image or object. Many edge detection algorithms have developed based on computation of the intensity gradient vector at which the intensity changes sharply. Among all the methods, shape representation and matching is a very important aspect and has been extensively used for solving object recognition problem (Belongie et al. 2002; Duan et al. 2010a; Liu et al. 2012). Generally, shape matching schemes involve two general steps: feature extraction and similarity measuring (Veltkamp 2001). Edge potential function (EPF) is a newly developed similarity evaluating measure, which was firstly proposed by Minh-Son Dao (Dao et al. 2007). This conception is derived from the potential generated by charged particles and has been proved its feasibility and reliability over Hausdorff distance and Chamfer distance measures.

Target recognition is a key issue to achieve autonomous reconnaissance and attack for aircraft with low altitude (Deng and Duan 2013). In many countries, target recognition technology for UAVs is highly confidential, and it is accordingly difficult to see its specific technical details. In order to obtain accurate identification results and hence to meet the practical requirements of aircrafts reconnaissance system, the proposed target recognition method must be efficient, stable, and convenient for promotion (Veredas et al. 2009). Among all the methods, shape representation and matching is a very important aspect and have been extensively used for solving object recognition problem (Belongie et al. 2002).

7.2 ABC Optimized Edge Potential Function Approach to Target Identification

7.2.1 The Principle of Edge Potential Function

Edge potential function (EPF) was firstly put forward by Minh-Son Dao (Dao et al. 2007). This conception was derived from the potential generated by charged particles and was especially adopted to model the attraction generated by edge structures contained in an image over similar curves.

A set of point charges Q_i in a homogeneous background can generate a potential, the intensity of which depends on the distance from the charges and the electrical permittivity of the medium ε, namely,

$$v\left(\vec{r}\right) = \frac{1}{4\pi\epsilon}\sum_i \frac{Q_i}{\left|\vec{r} - \vec{r_i}\right|} \tag{7.1}$$

where $\vec{r}$ and $\vec{r}_i$ are the observation point and charge locations, respectively. And the exact potential for a position in electric field amounts to the sum of potentials generated by each charged point.

In complete analogy with the above behavior, in our model, (x,y) represents the coordinates of any point of an image, and the ith edge point in the image at coordinates (x_i,y_i) can be assumed to be equivalent to a point charged $Q_{eq}(x_i,y_i)$, contributing to the potential of any image pixel (x,y):

$$EPF\,(x,y) = \frac{Q}{4\pi\epsilon_{eq}} \sum_{(x_i,y_i)\in W} \frac{1}{\sqrt{(x-x_i)^2 + (y-y_i)^2}} \tag{7.2}$$

Minh-Son Dao also outlined several EPF models (Dao et al. 2007), among which we adopt the Windowed EPF (WEPF) model in order to simplify the calculations, as well as improving the robustness of shape matching in cluttered environments. WEPF defines a window W beyond which edge points are ignored, which can be expressed as follows:

$$EPF\,(x,y) = \frac{Q}{4\pi\epsilon_{eq}} \sum_{(x_i,y_i)\in W} \frac{1}{\sqrt{(x-x_i)^2 + (y-y_i)^2}} \tag{7.3}$$

where ε_{eq} is a constant related with the image background situation and Q is equal to the charge of each edge point $Q_{eq}(x_i,y_i)$. Then the edge potential of any pixel of an image can be obtained from an edge map, which is extracted from the image. The edge potential represents a type of attraction field in analogy with the field generated by a charged element.

To complete the model, the searched target template to be matched can be considered as a test object, which is expected to be attracted by a set of equivalent charged points. In this way, the higher the similarity between the searched object and visual objects in the image, the higher the total attraction engendered by the edge field. As a result, EPF can be particularly used as the similarity measure for shape matching problem, for it implicitly includes some important features such as edge position, strength, and continuity, in a unique powerful representation of the edge.

7.2.2 ABC Optimized EPF Approach to Target Identification

The implementation procedure of our proposed ABC optimized EPF approach to target recognition for aircraft with low altitude can be described as follows (Xu and Duan 2010):

Step 1: Image preprocessing:

(1) Obtain the image and convert it into grayscale format.
For other different format images, first convert them into grayscale format in order for further edge detection operation.
(2) Filter the target image to remove the noise.
Conduct filtering operation to the obtained grayscale image in order to mitigate the effect of noise. To this purpose, we applied the median filtering method, which was certified to have an especially good inhibiting effect towards pepper noise and Gaussian noise.
(3) Adopt canny edge extractor to detect the edges of the given image for the sake of obtaining proper binary distribution of the original image.

Step 2: According to the binary edge distribution of the target image and the practical edge potential field function model shown in (7.3), calculate the EPF distribution of the target image.

Step 3: Initialize the parameters of ABC optimization algorithm, such as the population of the bee colony Ns, the number of employed bees Ne, and the number of the unemployed bees Nu, which satisfy the condition shown as follows:

$$N_s = N_e + N_u \tag{7.4}$$

Obviously, a larger N_s will contribute to a larger possibility of finding the best solution of the problem; however, it also means an increased computing complexity of the algorithm. In general, we define $N_e = N_u$, and according to our special problem, we set $N_s = 200$. Denote the largest searching times with *Limit* (50 in our experiments), current iterations with T, and the largest iterations with $T_{\max}$. Initialize the population of geometric transformation parameters, which include the horizontal translation parameter t_x, the vertical translation parameter t_y, the rotation angle parameter angle, and the scaling parameter scale. By considering these operators $c = (\ \mathrm{t_x}, \mathrm{t_y}, \mathrm{angle}, \mathrm{scale}\ )$, the original sketch is iteratively roto-translated and scaled obtaining different instances of it, which are fitted within the potential field to compute a matching index. The goal is to find the optimal combination of parameters, which can provide the best fitness, and to evaluate if the relevant matching index is high enough to determine with a certain degree of confidence the presence of the model in the target image. Initialize the search time of each bee $Bas = 0$, and the starting iteration $T = 1$.

Step 4: According to the geometric parameters of the employed bees, calculate their similarity values respectively based on the defined similarity function as

$$f\,(c_k) = \frac{1}{N^{(c_k)}} \sum_{n^{(c_k)}=1}^{N^{(c_k)}} \left\{EPF\left(x_n^{c_k}, y_n^{c_k}\right)\right\} \tag{7.5}$$

where $c_k = (t_x, t_y, angle, scale)$ is the geometric operator on the kth iteration, $N^{(c_k)}$ represents the number of edge points of the target image contained in the mask image contour under the geometric operator c_k, while $(x_n^{c_k}, y_n^{c_k})$ are their corresponding vertical and horizontal coordinates, and $n^{(c_k)}$ denotes the nth edge point. $f(c_k)$ shows the average potential value calculated along the contour of the mask image, and accordingly, when the fitness function $f(c_k)$ achieves its maximum value, we find the best solution to match the target image.

Step 5: The employed bees search around their current positions to find new solutions, and update their positions if the new fitness value is higher than the original value.

Step 6: The unemployed bees apply the roulette selection method to choose the bee individual that possesses a relatively good fitness value as the leading bee, according to the calculated fitness results of employed bees. Each recruited unemployed bee continues to search new solutions just around the leading bee's solution space, and calculate their fitness values. If the value of the new solution is better than the original value, the unemployed bee converts into an employed bee, which means that update the positions of the employed bees, and continue exploring with *Bas* re-initialized as 0, or else, keep searching around, and its *Bas* value plus one.

Step 7: If the search times Bas is larger than certain threshold *Limit*, the employed bee gives up the solution, and re-search the new food resources, which is realized by re-initializing the geometric parameters and calculating the fitness value.

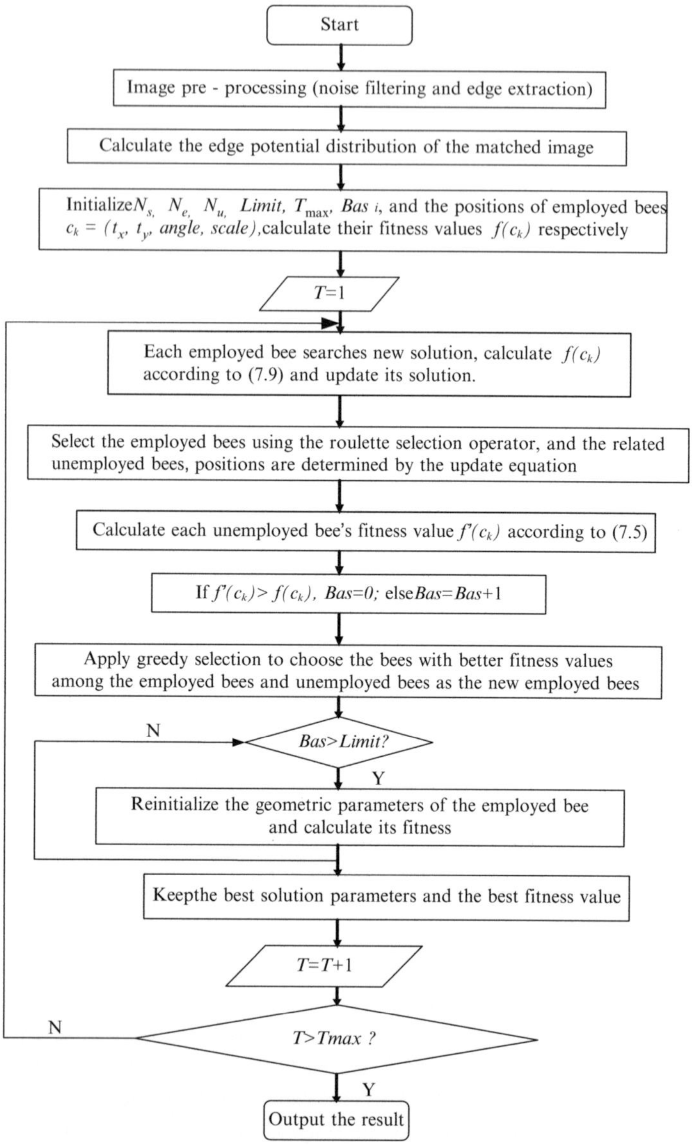

Fig. 7.1 The procedure of our proposed method (Reprinted from Xu and Duan (2010), with kind permission from Elsevier)

Step 8: Store the best solution parameters and the best fitness value.

Step 9: If $T < T_{\max}$, go to Step 5. Otherwise, output the optimal parameters and optimal fitness value.

The detailed procedure can also be shown with Fig. 7.1.

7.2.3 Experiments

In order to investigate the feasibility and effectiveness of the proposed method in this work, a series of experiments are conducted and further comparative experimental results with the GA method are also given.

The initial parameters of ABC algorithm were set as $N_s = 50$, $N_e = 25$, $N_u = 25$, $T_{\max} = 200$, $Limit = 50$.

The 1st experiment (Case 1) is to find an isosceles triangle among a variety of shapes in the original image. After a 330° rotation, 1.2 times scaling, and a [78, 68] translation, the target can be successfully recognized in the image, which means that the best geometric parameters are [78, 68, 330, 1.2]. The experimental results are shown in Fig. 7.2.

The task of the Case 2 is to find a target plane in the airport, and the results are shown in Fig. 7.3.

The results of Case 3 and Case 4 are shown as follows:

Figures 7.2, 7.3, 7.4 and 7.5 are the target recognition results by using our proposed ABC optimized EPF approach. Obviously, the results of the experiments show that our proposed method can recognize the exact positions of targets in the image through the operations of rotation, scaling, and translation.

To compare our identification effect with other approaches, more experiments are conducted by using genetic algorithm (GA). The results by using GA are shown in Fig. 7.6.

From Fig. 7.6, the evolution curve of GA is trapped into stagnancy (local best) from the 20th iteration, while our proposed ABC optimized EPF approach can avoid the local best easily. Furthermore, Table 7.1 shows the performance comparison of our proposed approach and the traditional GA with 14 times experiments.

From the above experimental results, it is clear that our proposed ABC optimized EPF approach is superior to the traditional GA in solving the target recognition problem for aircraft with low altitude.

Figure 7.7 shows the series recognition results of Case 2, Case 3, and Case 4 by using GA, which all contribute to our conclusion that our proposed ABC optimized EPF approach performs better than GA.

7.3 A Chaotic Quantum-Behaved PSO Based on Lateral Inhibition for Image Matching

In this section, we introduce chaos theory and lateral inhibition into the quantum-behaved particle swarm optimization (QPSO). We name our hybrid model as chaotic quantum-behaved particle swarm optimization based on lateral inhibition (LI-CQPSO). It assumes that combining the features of above methods is complementary to PSO. Quantum-behaved PSO, another expression of PSO, is an intelligent algorithm coming from the branch of quantum computing. The chaos

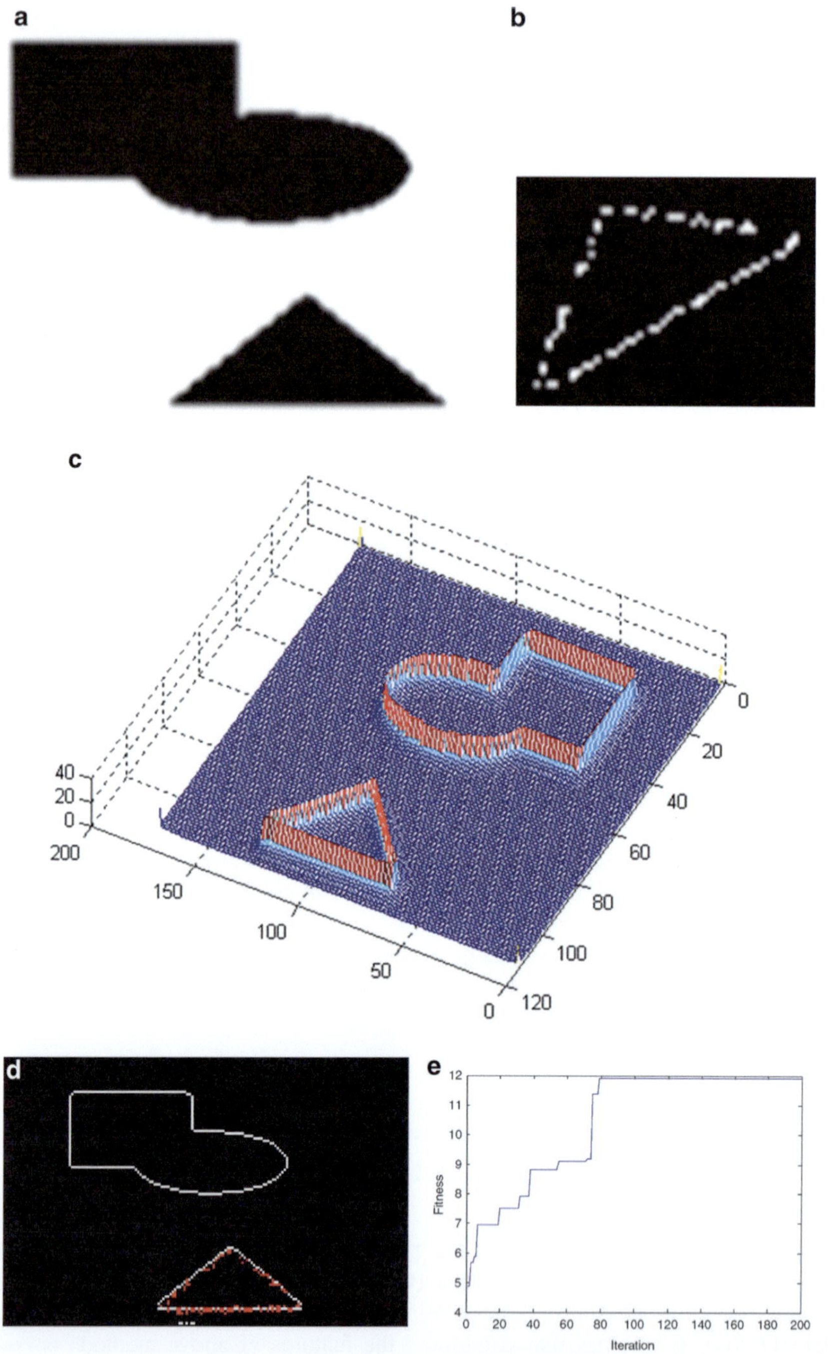

Fig. 7.2 Experimental results of Case 1 by using our proposed method. (**a**) Original image. (**b**) The extracted contour of the target image. (**c**) The edge potential distribution of the original image. (**d**) The target recognition result with ABC optimized EPF. (**e**) The evolution curve of ABC algorithm (Reprinted from Xu and Duan (2010), with kind permission from Elsevier)

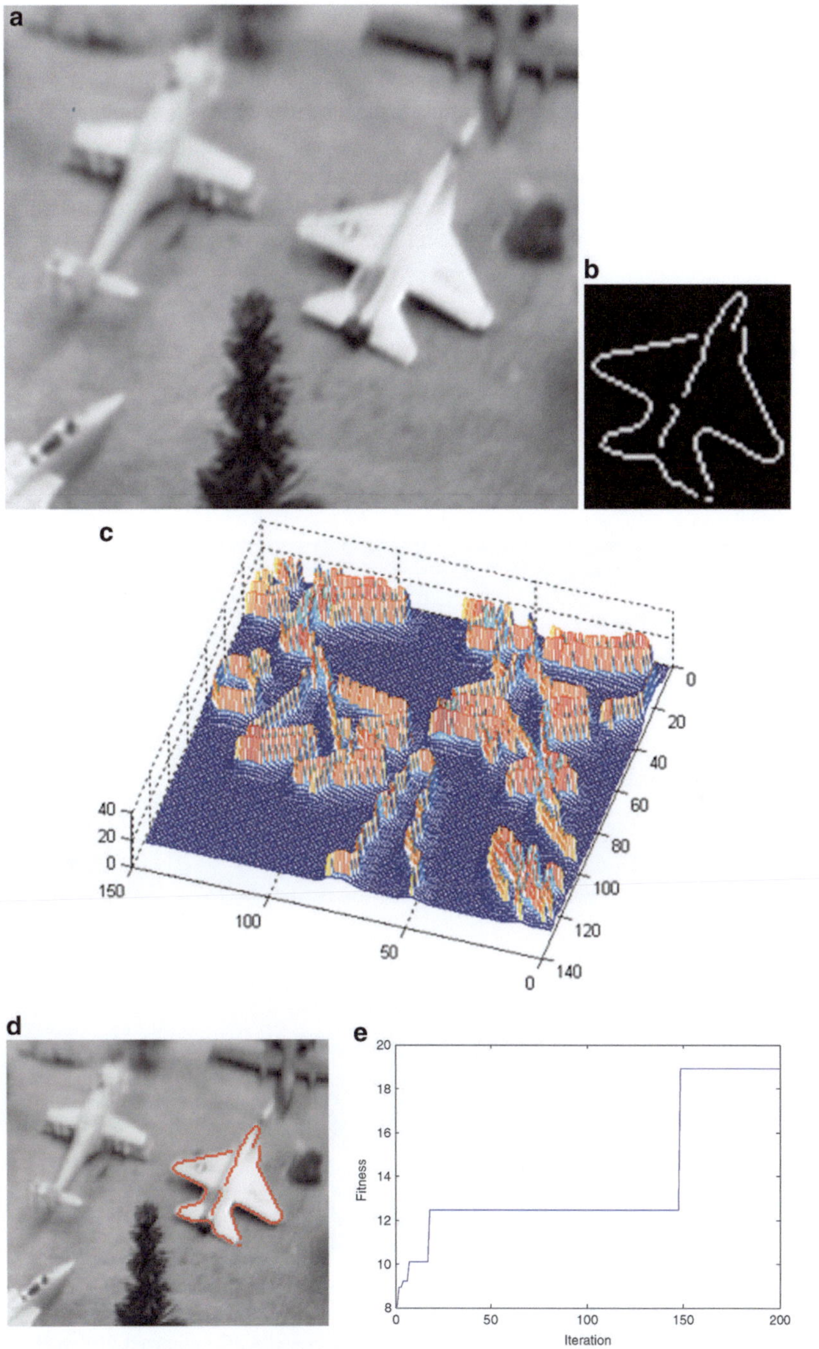

Fig. 7.3 The results of Case 2 by using our proposed method. (**a**) Original image. (**b**) The extracted edge of the identify target. (**c**) The edge potential distribution of the original image. (**d**) The target recognition results with ABC optimized EPF. (**e**) The evolution curve of the ABC algorithm (Reprinted from Xu and Duan (2010), with kind permission from Elsevier)

theory was formulated by Edward Lorentz in 1960, from the study of weather (Chandramouli and Izquierdo 2006). It studies the behavior of dynamic systems that are highly sensitive to initial conditions. The chaotic modeling of particle swarm optimization has been proposed and used successfully in image matching (Duan et al. 2010a). Lateral inhibition (Amari 1977) has been used in the field of image edge extraction, image enhancement, etc. In this section, LI-CQPSO which combines the advantages from the chaos theory, lateral inhibition has been proposed to overcome the problem of the convergent speed and low searching precision of PSO.

7.3.1 The Quantum-Behaved PSO Algorithm

As a variant of PSO, QPSO was proposed in 2004, inspired by quantum mechanics and fundamental theory of particle swarm (Feng et al. 2009). QPSO is characterized

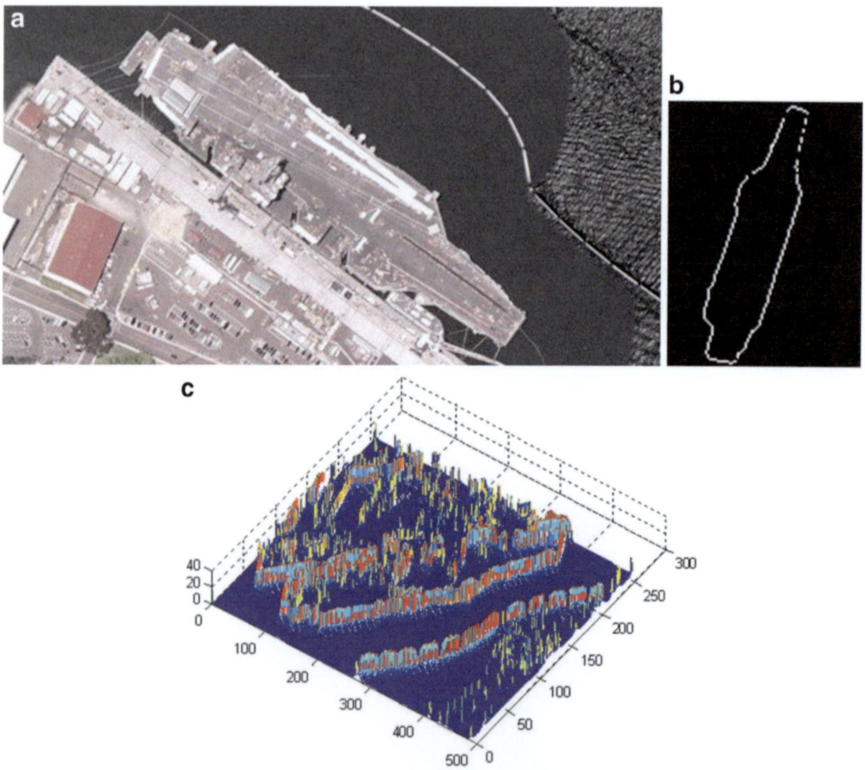

Fig. 7.4 Experimental results of Case 3 by using our proposed ABC optimized EPF method. (**a**) Original image. (**b**) The target sketch. (**c**) The edge potential distribution of the original image. (**d**) The results of target recognition. (**e**) The evolution curve of the ABC algorithm (Reprinted from Xu and Duan (2010), with kind permission from Elsevier)

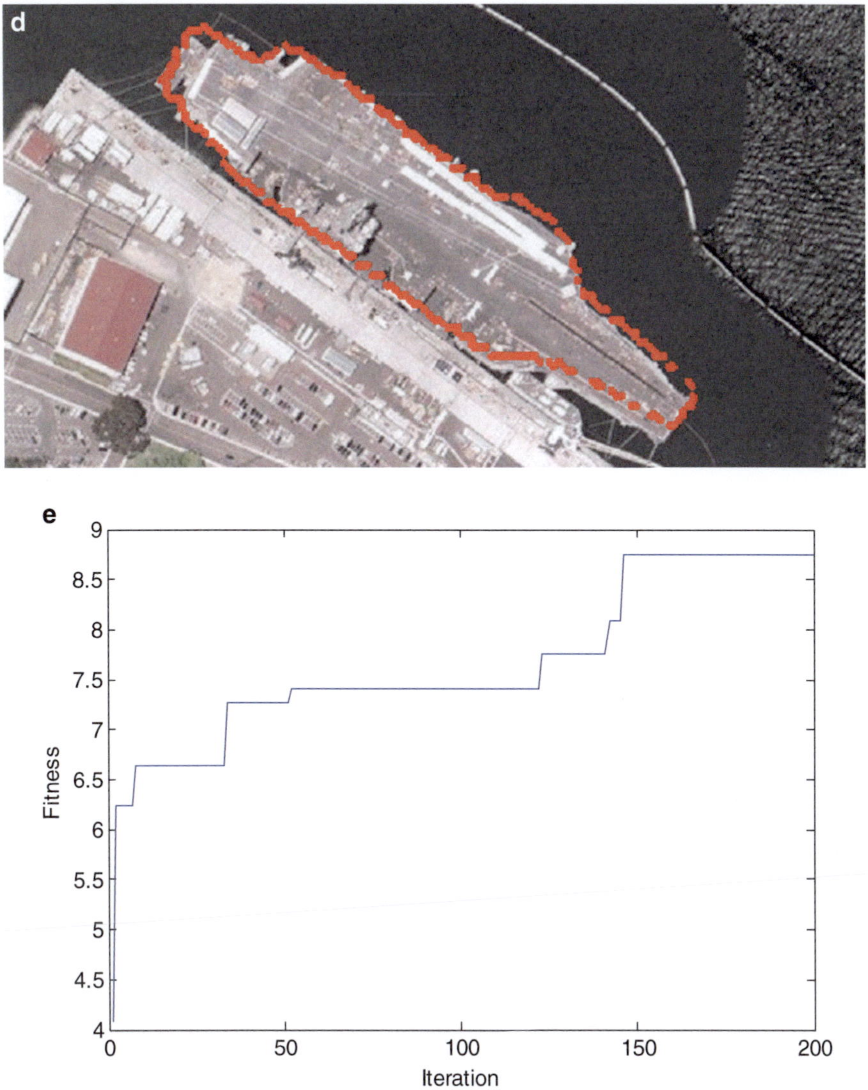

Fig. 7.4 (continued)

by good search ability and fast convergence. According to the uncertainty principle of quantum mechanics, the velocity and position of a particle in quantum world cannot be determined simultaneously (Duan et al. 2010b). Thus, QPSO is different from standard PSO mainly in that the exact values of position and velocity are uncertain. In the QPSO algorithm with M particles in D-dimensional space, the position of the ith particle at the $(t+1)th$ iteration is updated by the following equations (Coelho 2008):

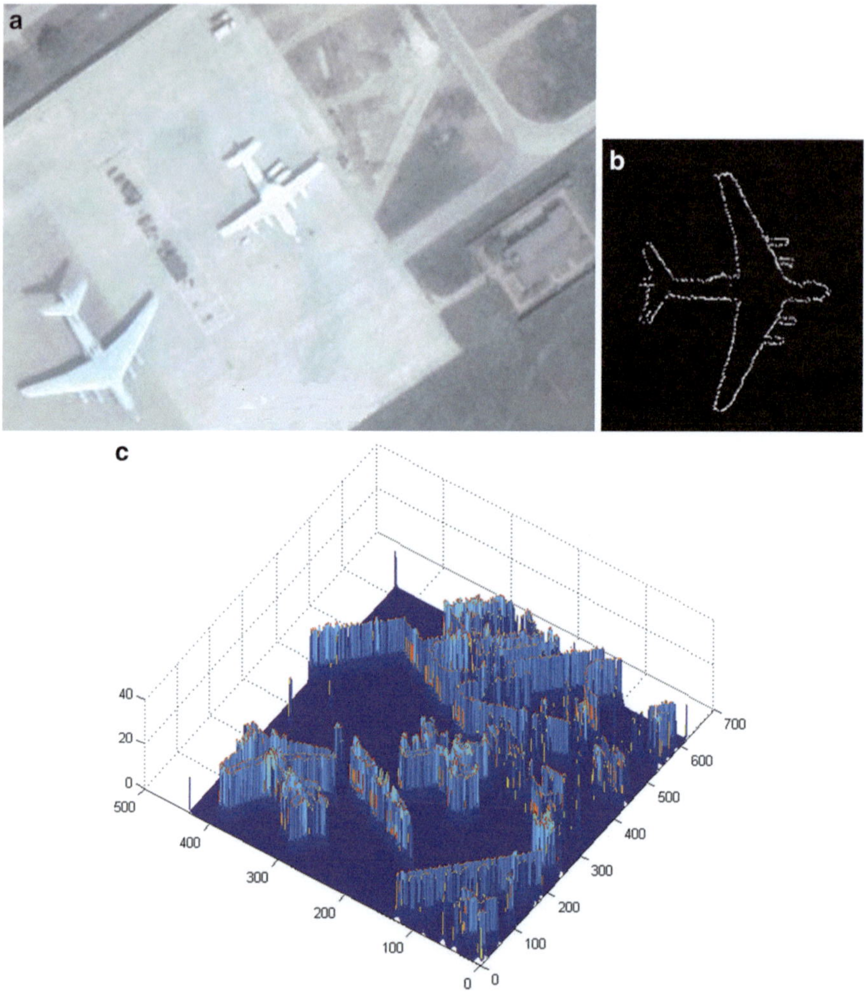

Fig. 7.5 Experimental results of Case 4 by using our proposed method. (**a**) Original image. (**b**) The target sketch. (**c**) The edge potential distribution of the original image. (**d**) The results of target recognition. (**e**) The evolution curve of the ABC optimization algorithm (Reprinted from Xu and Duan (2010), with kind permission from Elsevier)

$$mbest = \frac{1}{M}\sum_{i=1}^{M} p_i \tag{7.6}$$

$$p_i = \phi p_p + (1-\phi)\, p_g \tag{7.7}$$

If rand( ) > 0.5, we can get

$$x_{id}(t+1) = p_{id} - \beta\, |mbest - x(t)|\, In\frac{1}{u} \tag{7.8}$$

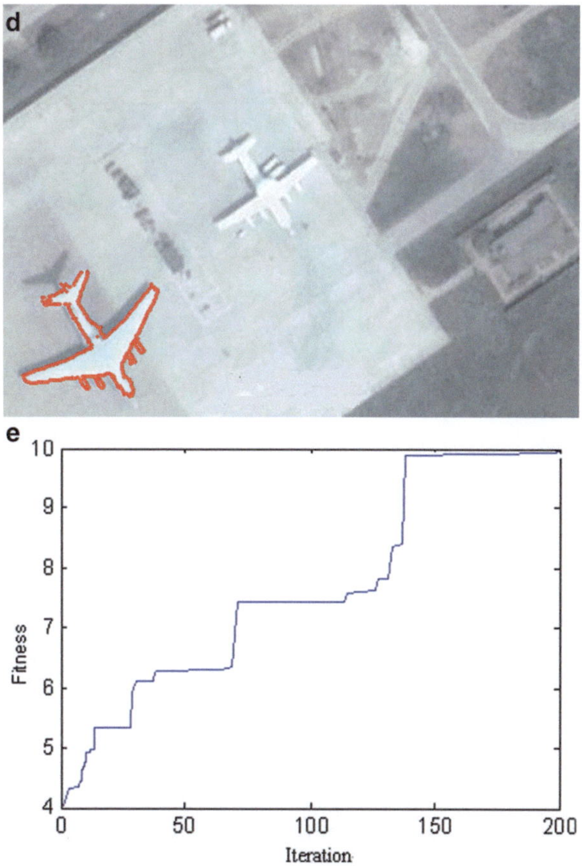

Fig. 7.5 (continued)

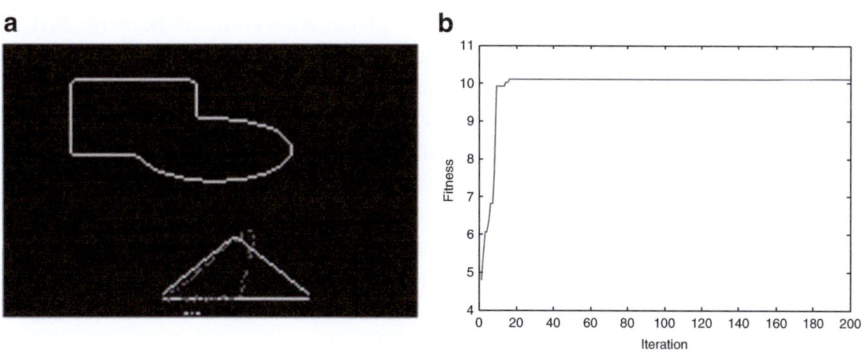

Fig. 7.6 The target recognition results by using GA. (**a**) Experimental result by using GA. (**b**) Evolution curve of GA (Reprinted from Xu and Duan (2010), with kind permission from Elsevier)

Table 7.1 The comparative results using our proposed approach and GA

Our proposed approach		GA	
Best fitness	Best parameters	Best fitness	Best parameters
11.262	[78 68 331 1.2]	11.262	[78 68 331 1.2]
11.262	[78 68 331 1.2]	11.829	[79 69 328 1.2]
11.262	[78 68 331 1.2]	11.262	[78 68 331 1.2]
11.908	[79 69 330 1.2]	10.268	[81 65 189 0.8]
11.272	[79 65 329 1.3]	10.792	[79 68 332 1.2]
10.5326	[79 66 331 1.2]	9.679	[87 67 329 0.8]
11.262	[78 68 331 1.2]	11.262	[78 68 331 1.2]
10.8506	[80 67 333 1.3]	9.282	[77 88 116 0.9]
11.262	[78 68 331 1.2]	10.5153	[77 73 187 0.9]
11.262	[78 68 331 1.2]	11.262	[78 68 331 1.2]
11.262	[78 68 331 1.2]	10.832	[78 67 331 1.2]
10.792	[79 66 332 1.2]	8.679	[87 67 329 0.8]
10.6231	[80 97 115 0.8]	10.597	[79 67 193 0.8]
11.262	[78 68 331 1.2]	10.099	[79 68 329 1.2]

Otherwise, we have

$$x_{id}\,(t+1) = p_{id} + \beta\,|mbest - x(t)|\,In\frac{1}{u} \tag{7.9}$$

where *mbest* denotes mean best position and is the mean value of the pbest positions of all particles. β is a contraction–expansion coefficient in the range of [0,1]. ϕ and u are two uniform random numbers in the interval [0,1]. P_p and P_g have the same meanings as those in the standard PSO.

The QPSO algorithm is superior to the standard PSO mainly in three aspects as follows: Firstly, quantum theory is an uncertain system. More different states of the particles and a wider searching space of the algorithm can be generated in this system. So better offspring are produced. Secondly, the introduction of *mbest* into QPSO is another improvement. In the standard PSO, PSO converges fast, but sometimes the fast convergence happens in the first few iterations and relapses into a local optimal situation easily. While QPSO with *mbest* is introduced, the convergence to average error is lower, because each particle cannot converge fast without considering its colleagues, which makes the frequency of falling into local much lower than PSO. Falling into local means that an algorithm gets converged into local and fails to get a best solution. Lastly, QPSO has fewer parameters than standard PSO, and it is much easier to program and run it. Hence, the performance of the algorithm is significantly improved by QPSO.

Fig. 7.7 Series target recognition results by using GA. (**a**) The target recognition result by using GA for Case 2. (**b**) The target recognition result by using GA for Case 3. (**c**) The target recognition result by using GA for Case 4 (Reprinted from Xu and Duan (2010), with kind permission from Elsevier)

7.3.2 Lateral Inhibition Mechanism

The lateral inhibition mechanism is discovered and verified by Hartline and his research team when they carried out an electrophysiology experiment on the limulus' vision. They found that every microphthalmia of limulus' ommateum is a receptor which it is inhibited by its adjacent receptors and the inhibited effect is mutual and spatially summed. It means that while it is inhibited by its adjacent receptors, it inhibits its adjacent receptors at the same time. And the nearer the adjacent receptors are from each other, the more strongly they inhibit mutually.

In retinal image, the intensively excited receptors in illuminatingly light area inhibit the receptors in illuminatingly dark area more strongly than the latter to the former. Therefore, the contrast and the distortion of sensory information are enhanced. In this way, the important characters of vision scene and the intensity gradient in retinal image, namely, the image's edge, are both strengthened. In this essay, this mechanism is applied to preprocessing the original and the template images to stress the spatial resolution, which can increase the accuracy of the image matching.

After series of electrophysiological experiments, Hartline and his colleagues advanced the following classical lateral inhibition model:

$$r_p = e_p + \sum_{j=1}^{n} k_{p,j} \left(r_j - r_{p,j} \right), p = 1, 2, \ldots, n \tag{7.10}$$

According to different classification criteria, lateral inhibition has many different modified models.

In order to introduce this mechanism to image processing, (7.10) is modified in two-dimensional and gray form, gray value of the pixel (m, n) in image given in (7.11):

$$R\left(m,n\right) = I_0\left(m,n\right) + \sum_{i=-M}^{M} \sum_{j=-N}^{N} \alpha_{i,j} I_0\left(m+i, n+j\right) \tag{7.11}$$

where $\alpha_{i,j}$ is the lateral inhibition coefficient of the pixel (i, j) to the central pixel, $I_0(m,n)$ is the original gray value of pixel (m,n), $R(m,n)$ is the gray value of pixel (m,n) processed by lateral inhibition, and $M \times N$ is the receptive field.

The size of receptive field chosen in this essay is 5×5. Then the competing coefficient of the lateral inhibition network is

$$\begin{aligned} R\left(m,n\right) &= \alpha_0 \times I_0\left(m,n\right) \\ &- \alpha_1 \left[\sum_{i=-1}^{1} \sum_{i=-1}^{1} I_0\left(m+i, n+j\right) - I_0\left(m,n\right) \right] \\ &- \alpha_2 \left[\sum_{i=-2}^{2} \sum_{i=-2}^{2} I_0\left(m+i, n+j\right) - \sum_{i=-1}^{1} \sum_{j=-1}^{1} I_0\left(m+i, n+j\right) \right] \end{aligned} \tag{7.12}$$

And the lateral inhibition modulus satisfies

$$\alpha_0 - 8\alpha_1 - 16\alpha_2 = 0 \tag{7.13}$$

In this section, we choose the following matrix as the modulus:

$$U = \begin{bmatrix} 0.025 & 0.025 & 0.025 & 0.025 & 0.025 \\ 0.025 & 0.075 & 0.075 & 0.075 & 0.025 \\ 0.025 & 0.075 & 1 & 0.075 & 0.025 \\ 0.025 & 0.075 & 0.075 & 0.075 & 0.025 \\ 0.025 & 0.025 & 0.025 & 0.025 & 0.025 \end{bmatrix} \tag{7.14}$$

After combining the modulus template U with (7.14), new gray scales of the image can be obtained. Then finally, the image's edge is extracted by the following equation:

$$I(m,n) = \begin{cases} 0 & R(m,n) \le T \\ 255 & R(m,n) > T \end{cases} \tag{7.15}$$

where T is a user-defined threshold value according to practical situations.

7.3.3 *Chaotic Quantum-Behaved PSO Based on Lateral Inhibition*

The fitness function is often defined to calculate the fitness of every particle according to different situations and real practices. If the image is comparatively huge, the exhaustive operation of this function is extremely time-consuming. To overcome this shortage, we define the following fitness function, suitable for the lateral inhibition processed images:

$$f(m,n) = \frac{1}{KW}\sum_{i=0}^{K-1}\sum_{j=0}^{W-1} I(m+i,n+j) \tag{7.16}$$

where $K \times W$ is the size of the template, (m,n) is the coordinate of the pixel in the original image, and $I(m+i,n+j)$ is the processed gray value of the pixel $(m+i,n+j)$ by (7.15). If the size of the original image is $M \times N$ (corresponding to an image's M row and N line), the range of the coordinate in the original image for matching is $0 \le m \le M-K+1, 0 \le n \le N-W+1$. The maximum value $f(m,n)$ stands for the best solution of the matching.

LI-CQPSO combines the characteristics of random of chaos, efficiency of QPSO, with the accuracy of the lateral inhibition and performs well. Our proposed LI-CQPSO model has the following advantages. Firstly, it has better properties with high accuracy and efficiency than the standard PSO and CQPSO. Secondly, with fewer parameters, this algorithm is much easier to program than the others. These parameters are adjusted to balance between searching accuracy and searching time according to practical problems. At last, the fitness function described in (7.16) decreases the computerized complexity.

The procedure of LI-CQPSO is described as follows:

Step 1: Image preprocessing. Obtain the original image and the template image, and convert them into grayscale format. And filter images to remove the noise. Then set a threshold for the fitness function according to different situations. Preprocess them by the lateral inhibition mechanism. Then save the new matrices of images.
Step 2: Initialization of particles and parameters. Initialize particles' positions and velocities by chaos theory and initialize the parameters of this algorithm, such as maximum iteration and the population of particles.
Step 3: Calculate each particle' fitness value.
Step 4: Compute the mean best position *mbest* of the particles.
Step 5: Select P_p and P_g among particles and update them.
Step 6: Regenerating the next generation. Generate a random number in the range of [0,1], then compute positions of the next generation of particles.
Step 7: Check whether the iteration reaches the maximum iteration. If so, stop the algorithm and output the result, otherwise return to step 3.

The algorithm flow of the above improved PSO is shown below (Fig. 7.8):

7.3.4 Experiments

In order to verify the feasibility and effectiveness of our proposed algorithm in this work, series of comparative experiments with other three PSO-based algorithms are also given. These algorithms are the standard PSO, QPSO, and PSO based on lateral inhibition (LI-PSO), which is introduced lateral inhibition into the standard PSO to extract the edge of the image.

The initial parameters of all these PSO-based methods were set as iter = 80, n = 150; the parameters of the basic PSO were set as $C_1 = 2.05$, $C_2 = 2.05$, $w_{\max} = 1$, $w_{\min} = 0.3$. The threshold for image edge extracting was set as T = 110. The aim of these experiments is to make the template image successfully matched to the original image. The success of matching was determined by the coordinate of the template image fixed in the original image. The experimental results are shown in Fig. 7.9.

In order to compare our proposed method with other approaches, more experiments were conducted by using PSO, QPSO, LI-PSO, and our proposed LI-CQPSO. The comparative results are shown in Fig. 7.10.

From Fig. 7.10, the evolution curves of PSO, QPSO, LI-PSO, and LI-CQPSO are all presented. Among them, the rate of convergence of LI-CQPSO is the fastest

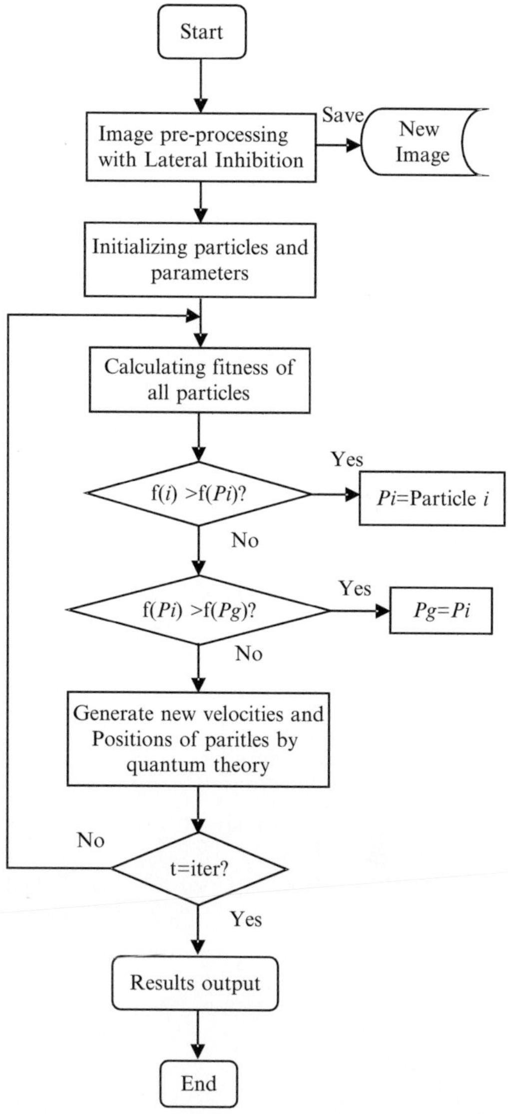

Fig. 7.8 The flowchart of LI-CQPSO (Reprinted from Liu et al. (2012), with kind permission from Elsevier)

apparently, which finds the best solution at the 9th iteration on average, while the other three algorithms need much more iterations. Table 7.2 displays in detail the performance comparison of our proposed approach and the other three algorithms with 50 times experiments. To further prove the performance of our proposed method against standard PSO, QPSO, and LI-PSO, the statistical performances of 50 independent runs are listed in Table 7.2.

From Table 7.2, it turns out that the average cost of our algorithm is 6.5032 s, while the average cost of standard PSO algorithm is 3.9995 and QPSO algorithm is

Fig. 7.9 (**a**) Original image (540 × 359). (**b**) Template image (96 × 91). (**c**) Original image processed by the lateral inhibition. (**d**) Template image processed by the lateral inhibition. (**e**) The final template matching result (Reprinted from Liu et al. (2012), with kind permission from Elsevier)

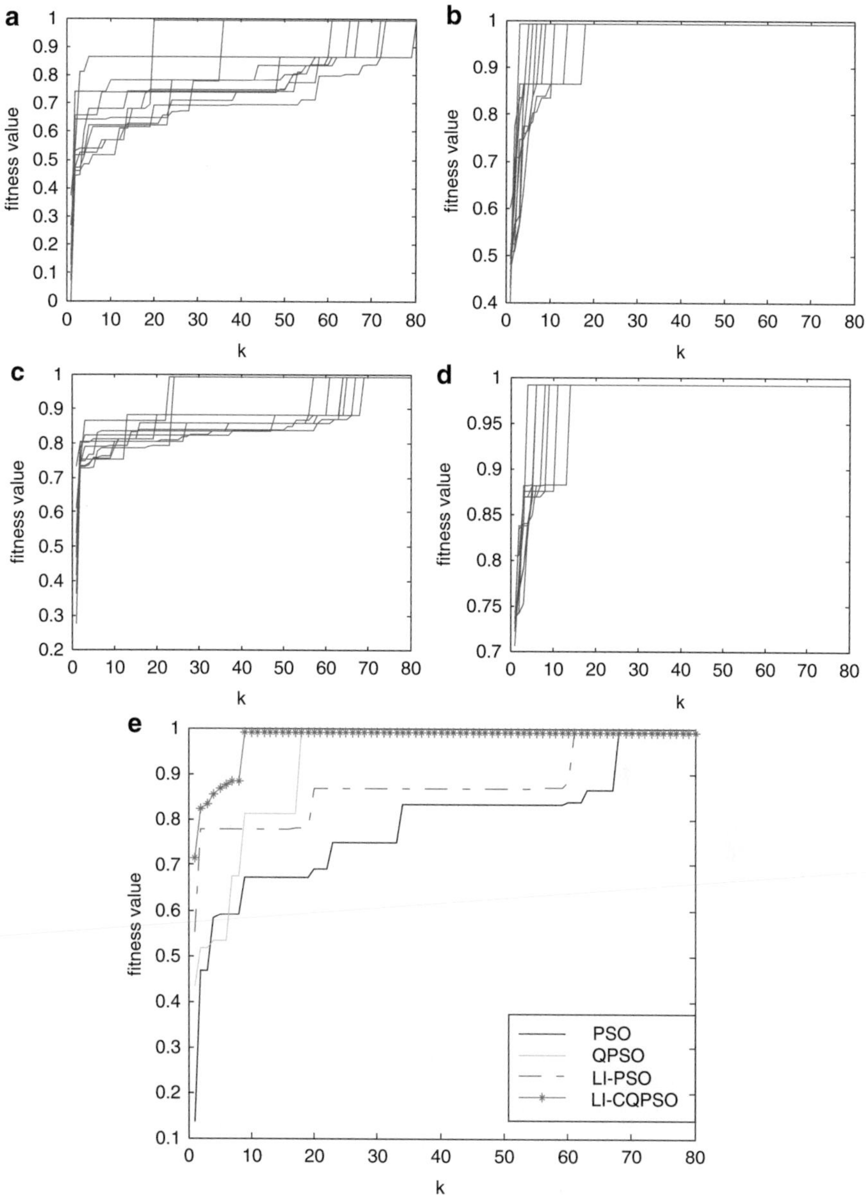

Fig. 7.10 (**a**) Evolution curves of experimental results by using PSO (10 times). (**b**) Evolution curves of experimental results by using QPSO (10 times). (**c**) Evolution curves of experimental results by using LI-PSO (10 times). (**d**) Evolution curves of experimental results by using LI-CQPSO (10 times). (**e**) Iterative curves in comparison with PSO, QPSO, LI-PSO, and LI-CQPSO (Reprinted from Liu et al. (2012), with kind permission from Elsevier)

Table 7.2 Compared results by 4 different algorithms

Test conditions: test 50 times, maximum iteration $iter = 80$			
Population of particles $n = 150$			
Algorithm	Converged iteration	Total time	Correct rate
PSO	68	3.9535 s	78 %
QPSO	10	5.9995 s	92 %
LI-PSO	58	4.0212 s	86 %
LI-CQPSO	8	6.5032 s	94 %

5.9995 s, apparently showing that our method spends more time on image matching. PSO and LI-PSO need less exhaustive matching time but also are converged much slower than QPSO and LI-CQPSO. In fact, the convergent speed of our method is much quicker than the others that means if only considering successful runs, the real time for finding the best matching is much less than 6.5032 s, just equal to $6.5032*8/80 = 0.65032$ s which is less time than the other methods to make the best matching. It is also concluded that chaos theory and lateral inhibition are able to improve the convergent speed of PSO and QPSO. Furthermore, the successful rate of our algorithm shows that our method can find the feasible and optimal matching more stable than the other three algorithms and can effectively solve the image matching problems.

7.4 Implementation of Autonomous Visual Tracking and Landing for Low-Cost Quadrotor

7.4.1 The Quadrotor and Carrier Test Bed

The hybrid test-bed system includes the quadrotor and ground carrier realized by a pushcart. A ground control station is developed to perform image processing and position controlling. It is also used for monitoring and parameter variation in the experiments (Bi and Duan 2013).

AR.Drone is a WiFi-controlled quadrotor with cameras attached to it (one facing forward, the other vertically downward). AR.Drone is developed by Parrot Inc. It is an affordable (usually under $300) commercial quadrotor platform offering an open application programming interface (API) and freely downloadable software development kit (SDK) for developers (Krajník et al. 2011). Many useful pieces of development information can be found on the developers' websites or the official forum.

AR.Drone uses an ARM9 468 MHz embedded microcontroller with 128 M of RAM running the Linux operating system. The onboard downward complementary metal oxide semiconductor (CMOS) color camera provides red, green, and blue (RGB) images in size of 320*240. An inertial system uses a 3-axis accelerometer, 2-axis gyro, and a single-axis yaw precision gyro. An ultrasonic altimeter with a range of 6 m provides vertical stabilization. With a weight of 380 g or 420 g (with "indoor

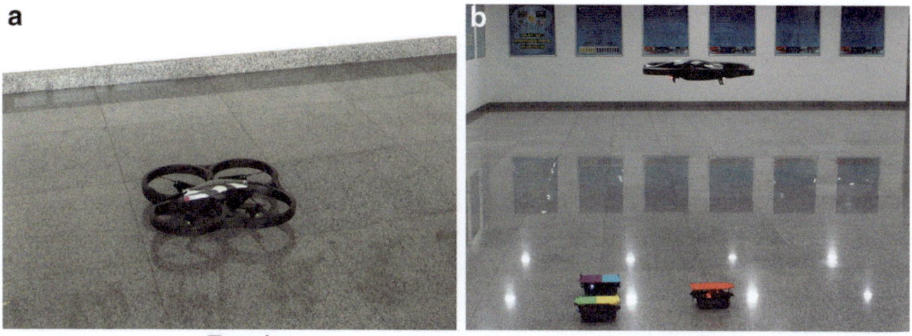

Top view

Side view

Fig. 7.11 Quadrotor AR.Drone picture from different views (Reprinted from Bi and Duan (2013), with kind permission from Elsevier)

Fig. 7.12 Pushcart carrier with the green helipad and two color rectangles (Reprinted from Bi and Duan (2013), with kind permission from Elsevier)

hull"), it can maintain flight for about 12 min with a speed of 5 m/s. Figure 7.11a shows the top view of the quadrotor, and Fig. 7.11b shows the side view of the flying quadrotor.

The carrier in our system is a common pushcart (See Fig. 7.12). The pushcart is powered by man and can move according to a specific path, which is like a moving automobile. The helipad is with a green shape "H," which is a concise copy of standard helipad for helicopter.

The color pattern is printed on a standard A4 paper and fixed on the carrier. We also designed two rectangles in red and blue for the helipad pattern. The helipad is used to determine the position error between the quadrotor and carrier. The two rectangles can help to determine the bearing of the quadrotor by calculating the relative position of the two rectangles.

7.4.2 Computer Vision Algorithm

In this section, a simple and fast RGB filter (available online at http://www.roborealm.com/help/RGB%20Filter.php) is adopted to implement filtering of the noisy signals. The RGB filter uses three values (red, green, and blue) to focus the attention towards the specific colors. RGB filter can diminish all pixels that are not the selected colors. This filter is different from direct RGB channel comparison. The white pixels are also diminished even though they may contain the color selected. Our helipad is pure green, so the RGB filter is accomplished eliminating the red and blue channels by using the following equation:

$$\begin{cases} G = (G - B) + (G - R) \\ B = 0 \\ R = 0 \end{cases} \tag{7.17}$$

where G is the green value, B is the blue value, and R is the red value. Based on (7.17), it is obvious that the value of the white pixels results in zero, and the pure green color ($R = 0$, $G = 255$, $B = 0$) doubles its value. G will be normalized to 0–255 after (7.17). This filter performs better than direct RGB channel comparison in filtering for a particular color as white pixels are removed. It is much robust under different lighting conditions.

The threshold algorithm can produce a binary image after using the RGB filter. A robust implementation is to set the image threshold at a fixed percentage between the minimum and the maximum green values. The percentage is 80 % for default value in our implementation. The threshold can also be manually adjusted according to the different experimental conditions.

The 2-D $(p + q)th$ order moment (Hu et al. 1962) of a density distribution function $f(x,y)$ can be described as

$$m_{pq} = \int_{-\infty}^{+\infty}\int_{-\infty}^{+\infty} x^p y^q f(x, y)dxdy, p, \ q = 0, 1 \tag{7.18}$$

An image can be represented as a discrete function $f(x,y)$. For the $(p + q)th$ order moment of an image, (7.18) can be rewritten as

$$m_{pq} = \sum_i \sum_j i^p j^q f(i, j) \tag{7.19}$$

where i and j correspond to the coordinates in axes x and y, respectively. The center of gravity of an object can be specified as

$$\overline{x} = \frac{m_{10}}{m_{00}} \tag{7.20}$$

$$\overline{y} = \frac{m_{01}}{m_{00}} \tag{7.21}$$

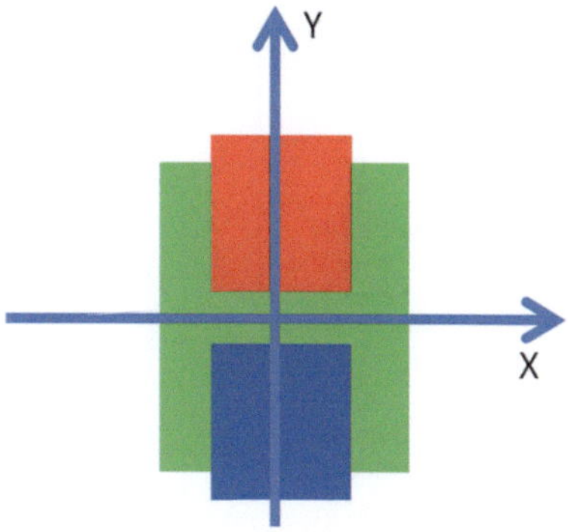

Fig. 7.13 Defined image coordinates (Reprinted from Bi and Duan (2013), with kind permission from Elsevier)

The obtained continuous images are processed on the ground control station for quadrotor. Figure 7.13 shows the defined image coordinate system.

In Fig. 7.13, the origin point is the center of helipad, and the relative position error is represented in pixels. The x -axis means pitch channel, and y -axis means roll channel. The calculated relative position errors in this coordinates are used to generate the control command, and the command is sent to the controller of the quadrotor. The onboard camera and off-board vision algorithms allow the quadrotor to track and land fully automatically without communication between the quadrotor and the carrier.

7.4.3 Control Architecture for Tracking and Landing

The quadrotor position controller is designed as hierarchical control architecture. The low-level attitude control has already been realized in the quadrotor inner-loop controller by the developer. The high-level position control is implemented in our developed ground control station. A finite state machine controls the high-level behavior of the quadrotor. The desired behavior consists of four phases: taking off, hovering, tracking, and landing. The control architecture of quadrotor is shown in Fig. 7.14.

The quadrotor takes off from the helipad on the carrier and holds a fixed height of 0.5 m in our experiments. Commands from the ground control station start the autonomous visual tracking and landing of the quadrotor. While the carrier is stationary, the quadrotor will hover overhead the helipad. The quadrotor must hover right over the center of the helipad in our task. When the carrier is moving, the quadrotor must track overhead the center of the helipad.

The precise autonomous position control is achieved by two independent proportional-integral-derivative (PID) controllers (Ludington et al. 2006). One PID controller is for the pitch channel and the other for the roll channel. The input of the controller is the position errors, which can be obtained by our proposed computer

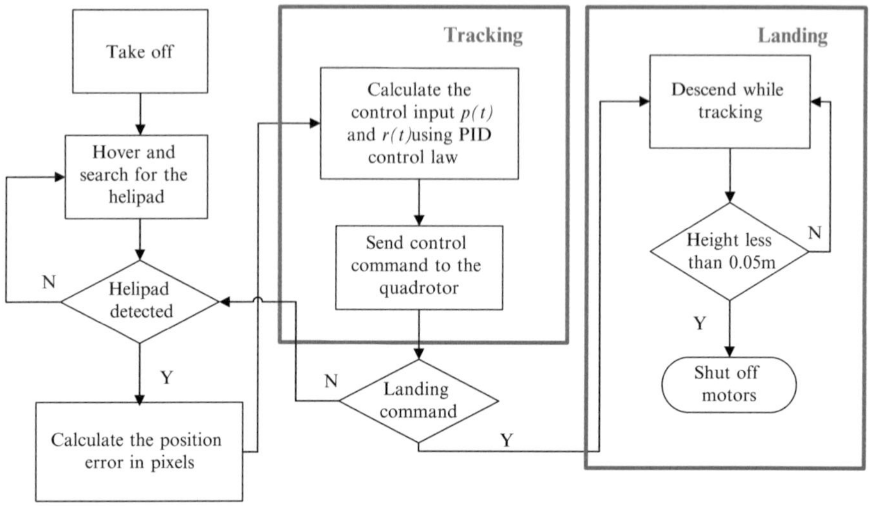

Fig. 7.14 Control architecture of our quadrotor (Reprinted from Bi and Duan (2013), with kind permission from Elsevier)

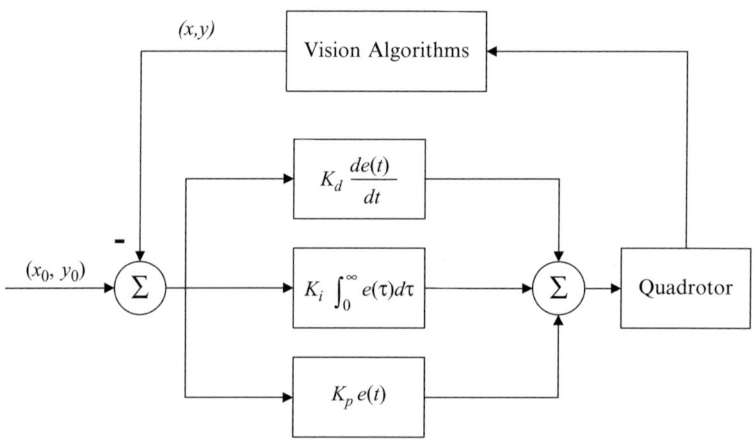

Fig. 7.15 Position PID controller with visual feedback (Reprinted from Bi and Duan (2013), with kind permission from Elsevier)

vision algorithms. The output of the controller is the attitude angle commands. The position error for the corresponding roll and pitch channel can be denoted with $e_r(t)$ and $e_p(t)$. Figure 7.15 shows the PID controllers for our quadrotor's position control with real-time visual feedback.

The roll angle $r(t)$ and pitch angle $p(t)$ can be obtained by

$$r(t) = K_p e_r(t) + K_i \int_0^t e_r(\tau) d\tau + K_d \frac{d}{dt} e_r(t) \tag{7.22}$$

$$p(t) = K_p e_p(t) + K_i \int_0^t e_p(\tau) d\tau + K_d \frac{d}{dt} e_p(t) \tag{7.23}$$

When the quadrotor is in the landing phase, it will maintain a constant descending velocity while keeping track of the helipad. As the position errors in pixels is adopted as the inputs of controllers, height compensation must be considered during landing process. In actual experiments, the landing phase can be separated into three circumstances, which can compensate for the controller outputs. When the quadrotor's height is between 0.3 and 0.5 m, the controller outputs calculated by (7.22) and (7.23) can be used directly. When the quadrotor's height is between 0.05 and 0.3 m, the controller outputs can be reduced according to the two following equations:

$$r(t) = r(t)/1.5; \tag{7.24}$$

$$p(t) = p(t)/1.5; \tag{7.25}$$

If the quadrotor's height is under 0.05 m, the helipad region in the image is too large to use for navigation. The quadrotor will directly shut off motors and land on the helipad quickly. This proposed strategy can avoid the influence of the ground effect, and better performance can be guaranteed effectively.

7.4.4 Experiments

In order to verify the feasibility and effectiveness of our proposed approaches, a custom ground control station for the quadrotor is also developed. Figure 7.16 shows the main interface of our developed ground control station.

The ground control station can receive/send signals and process images transferred from the quadrotor's downward camera at 30 Hz. The station can send control commands at 33 Hz according to the recommendation in the development guide. The ground station is developed in C#, which is referring to the open source.

Different operation modes can be chosen on our ground control station, including "Takeoff," "Land," "Tracking," and "Landing". "Land" means ordinary landing of the quadrotor while "Landing" means the touching down on the specific helipad. "Tracking" starts the process of the quadrotor automatically following the carrier. All the states for monitoring the system are displayed on the ground control station. The PID parameters for the two channels and the threshold for detecting the helipad can also be set on our developed ground control station.

Our computer vision algorithms are implemented on the ground control station. Figure 7.17 shows effect of the proposed vision algorithms. Figure 7.17a is the

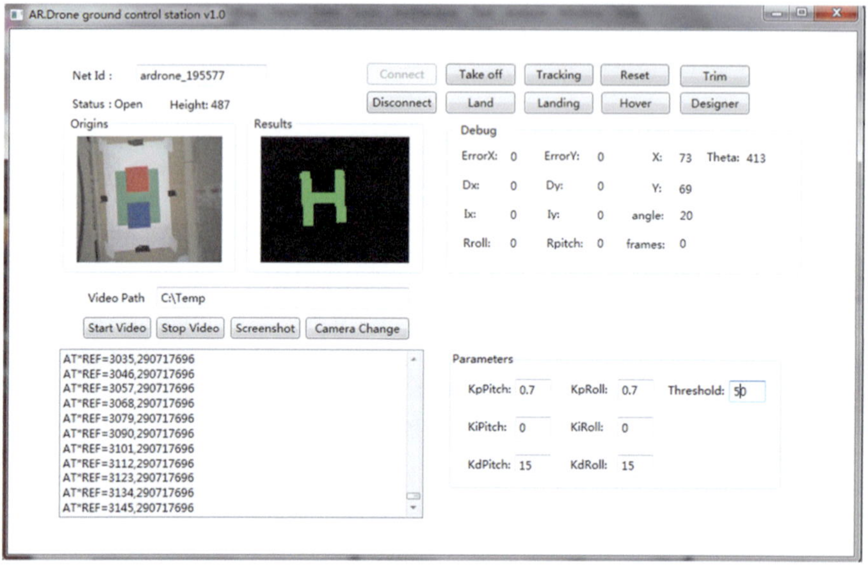

Fig. 7.16 Custom ground control station for our quadrotor (Reprinted from Bi and Duan (2013), with kind permission from Elsevier)

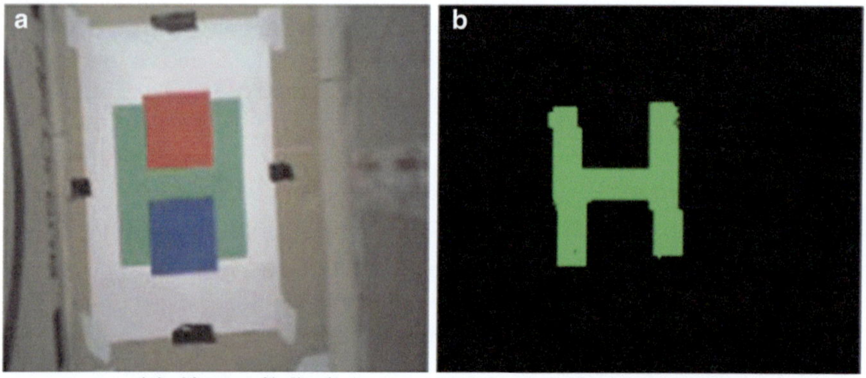

Fig. 7.17 Effect of the vision algorithm (Reprinted from Bi and Duan (2013), with kind permission from Elsevier)

original image of helipad captured from the onboard camera, and Fig. 7.17b shows the resulting image after filtering and thresholding.

The resulting image is displayed in green for user-friendliness. The helipad is extracted effectively from the sample image, which demonstrates that the computer vision algorithms meet requirements of the quadrotor system.

The threshold for helipad detection can be tuned manually if the default value cannot perform effective helipad extraction, which varies according to different

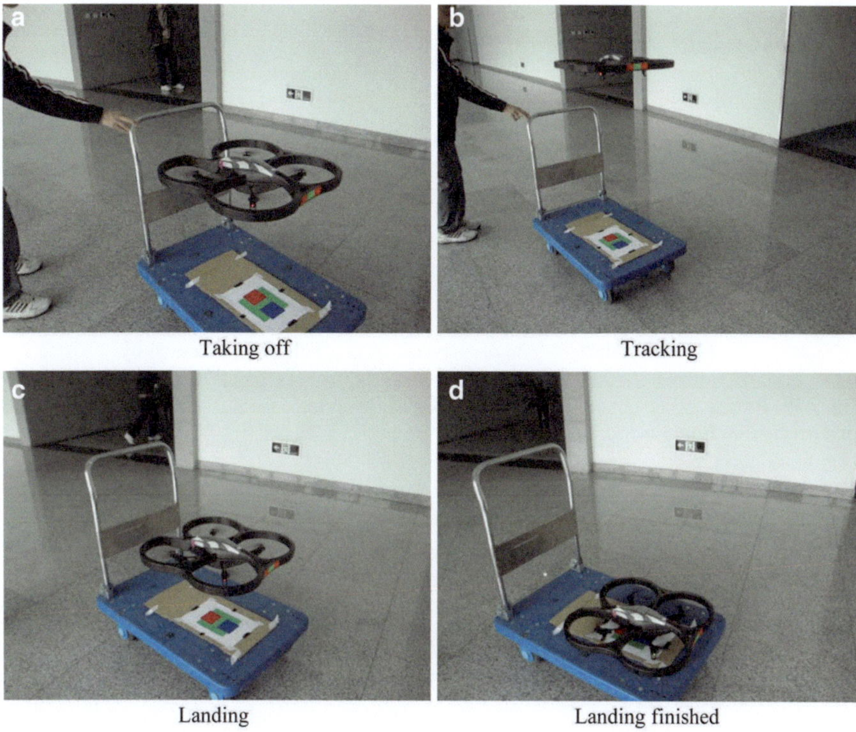

Fig. 7.18 Process of our quadrotor tracking and landing on helipad (Reprinted from Bi and Duan (2013), with kind permission from Elsevier)

lighting conditions. A specific threshold was set, and over 90 % of the helipad region is detected. After setting a threshold, the ground control station starts the tracking and landing procedure. The PID parameters for the pitch and roll channel are the same because of the symmetry of quadrotor's dynamics. In our experiments, the PID parameters can be set as $\{K_p = 0.7, K_i = 0, K_d = 15\}$. The proportional term produces an output value that reduces the current position error. The derivative term helps to slow the velocity of the quadrotor and reduces the magnitude of the overshoot by considering both acceleration and velocity.

Following a large number of flights, the system has proven a good tracking and landing ability. The experimental video's screenshots are shown in Fig. 7.18.

Figure 7.18a shows the taking-off process of our quadrotor from the moving carrier. Figure 7.18b shows the tracking process of our quadrotor after successful detection of the helipad. Figure 7.18c shows the descending of height while keeping tracking the helipad during the landing process. Figure 7.18d shows the precise landing of our quadrotor. The experimental results show the quadrotor using our proposed approaches can complete the tracking and landing on the helipad precisely while the carrier moving at the speed below 0.5 m/s.

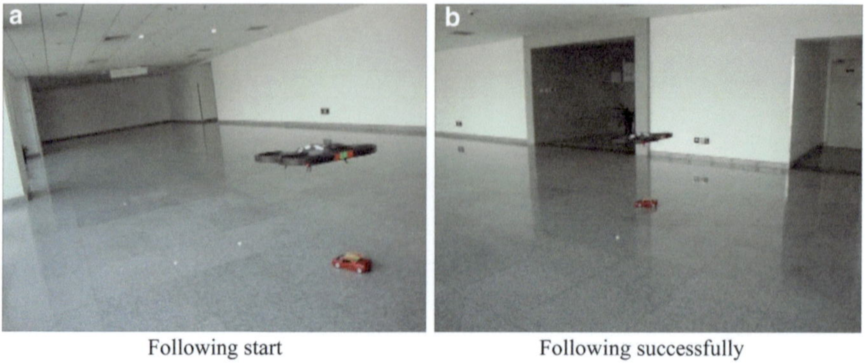

Fig. 7.19 Process of the quadrotor following a toy car (Reprinted from Bi and Duan (2013), with kind permission from Elsevier)

Furthermore, we also apply the quadrotor to follow a remote-controlled toy car by using our computer vision algorithms and position controller. Figure 7.19 shows the successful heterogeneous following process.

7.5 Conclusions

This chapter presented the main applications of biological vision with swarm intelligence for UAVs. The ability of autonomous target recognition and visual navigation is of vital importance for a complete mission in case of GPS signal loss.

In Sect. 7.2, we deal with biological edge detection and target recognition based on ABC for UAVs. The hybrid biological model of saliency-based visual attention and ABC algorithm is established for edge detection of UAVs. The visual attention model can help find the target region fast and accurately, which can reduce the amount of information and make the following processes easier and simpler. The improved ABC algorithm can detect the target edge effectively and avoid the effects of environment and noise. A novel ABC optimized EPF approach to target identify for aircraft with low-altitude flight is proposed in this section. This hybrid method takes advantages of the accuracy and stability for EPF in target shape recognition, and ABC algorithm is adopted to optimize the matching parameters.

For biological image matching, a novel chaotic quantum-behaved particle swarm optimization based on lateral inhibition is presented in Sect. 7.3. Utilizing the periodicity and irregularity of the chaotic variable to initialize the particles and the parameter β of QPSO helps it jump out of the local optimum as well as speeding up the process of finding the optimal parameters. This hybrid method also takes advantages of the accuracy and stability of lateral inhibition for edge extraction in the image preprocessing. Comparative experimental results of the proposed method, standard PSO algorithm, QPSO, and LI-PSO are also given to verify the feasibility and effectiveness of our proposed LI-CQPSO approach, which provide a more effective way for image matching in applications.

What's more, in Sect. 7.4, the exact implementation of a hybrid system consisting of quadrotor and pushcart carrier is presented. The adopted computer vision algorithm is rather simple, fast, and effective under different lighting conditions. The rich vision information can guarantee excellent performance in our designed tracking and controlling system. We have implemented an autonomous visual tracking and landing system with one type of economic UAV(quadrotor). The designed hybrid system demonstrates the usability and fast deployment ability of the miniature quadrotors.

References

Amari S (1977) Dynamics of pattern formation in lateral-inhibition type neural fields. Biol Cybern 27(2):77–87

Belongie S, Malik J, Puzicha J (2002) Shape matching and object recognition using shape contexts. IEEE Trans Pattern Anal Mach Intell 24(4):509–522

Bertuccelli LF, Cummings ML (2012) Operator Choice Modeling for Collaborative UAV Visual Search Tasks. IEEE Trans Syst Man Cybern Part A Syst Humans 42(5):1088–1099

Bi Y, Duan H (2013) Implementation of autonomous visual tracking and landing for a low-cost quadrotor. Optik 124(8):3296–3300

Chandramouli K, Izquierdo E (2006) Image classification using chaotic particle swarm optimization. In: Proceedings of 2006 IEEE International Conference on Image Processing, Atlanta. IEEE, pp 3001–3004

LdS C (2008) A quantum particle swarm optimizer with chaotic mutation operator. Chaos Solitons Fractals 37(5):1409–1418

Dao M-S, De Natale FG, Massa A (2007) Edge potential functions (EPF) and genetic algorithms (GA) for edge-based matching of visual objects. IEEE Trans Multimed 9(1):120–135

Deng Y, Duan H (2013) Hybrid C2 features and spectral residual approach to object recognition. Optik 124(8):3590–3595

Duan H, Deng Y, Wang X, Xu C (2013) Small and dim target detection via lateral inhibition filtering and artificial bee colony based selective visual attention. PLoS ONE 8(8): e72035

Duan H, Xu C, Liu S, Shao S (2010a) Template matching using chaotic imperialist competitive algorithm. Pattern Recognit Lett 31(13):1868–1875

Duan H, Xu C, Xing Z (2010b) A hybrid artificial bee colony optimization and quantum evolutionary algorithm for continuous optimization problems. Int J Neural Syst 20(1):39–50

Duan H, Zhang X, Xu C (2011) Bio-inspired Computing. Science Press, Beijing, Beijing, China

Feng J, Wang B, Sun Z (2009) Niche quantum-behaved particle swarm optimization with chaotic mutation operator. Comput Appl Softw 1(26):50–52

Hu M-K (1962) Visual pattern recognition by moment invariants. IRE Trans Inf Theory 8(2): 179–187

Krajník T, Vonásek V, Fišer D, Faigl J (2011) AR-drone as a platform for robotic research and education. In: Proceedings of Research and Education in Robotics (Eurobot 2011), Prague. Springer Berlin Heidelberg, pp 172–186

Lin F, Dong X, Chen BM, Lum K-Y, Lee TH (2012) A robust real-time embedded vision system on an unmanned rotorcraft for ground target following. IEEE Trans Ind Electron 59(2):1038–1049

Liu F, Duan H, Deng Y (2012) A chaotic quantum-behaved particle swarm optimization based on lateral inhibition for image matching. Optik 123(21):1955–1960

Ludington B, Johnson E, Vachtsevanos G (2006) Augmenting UAV autonomy. IEEE Robot Autom Mag 13(3):63–71
Veltkamp RC (2001) Shape matching: Similarity measures and algorithms. In: Proceedings of 2001 International Conference on Shape Modeling and Applications, Genova. IEEE, pp 188–197
Veredas FJ, Mesa H, Morente L (2009) A hybrid learning approach to tissue recognition in wound images. Int J Intell Comput Cybern 2(2):327–347
Xu C, Duan H (2010) Artificial bee colony (ABC) optimized edge potential function (EPF) approach to target recognition for low-altitude aircraft. Pattern Recognit Lett 31(13):1759–1772

Chapter 8
Conclusions and Outlook

Haibin Duan and Pei Li

Abstract The main focus and emphasis of this monograph has been on the bio-inspired computation in unmanned aerial vehicle (UAV), such as path planning for single UAV and multiple UAVs, formation flight control and formation configuration, heterogeneous cooperative control for multiple UAVs/UGVs, and vision-based surveillance and navigation problems. Despite the fact that we have witnessed significant advances in the UAV in the last few years, new and novel concepts and technologies are required to transcend to higher levels of autonomy. To this end, new development trends, such as small air vehicle, air-breathing hyposonic vehicles, and system integration, are discussed. We attempt to provide insightful sources for the researchers and scholars who have interests in bio-inspired computation for UAVs from three aspects: achieving higher autonomous capability, enhancing the ability to understand and adapt to the environment, and cooperative control of multiple autonomous vehicles.

8.1 Conclusions

Although much of the technology and equipment associated with the unmanned aerial vehicles (UAVs) are relatively new, the concept is old. UAVs have made significant contributions to the warfighting capability of operational forces and also to civil and commercial applications. They greatly improve the timeliness of battlefield information. While reducing the risk of capture or loss of manned assets, they offer advantages for many applications when comparing with their manned counter parts (Duan et al. 2010b; Duan and Liu 2010a; Francis 2012). They preserve human pilots of flying in dangerous conditions that can be encountered not only in military applications but also in other scenarios involving operation in bad weather conditions or near to buildings, trees, civil infrastructures, and other obstacles. Federated systems consisting of multiple unmanned aerial vehicles

The original version of this chapter was revised. A correction to this chapter is available at https://doi.org/10.1007/978-3-642-41196-0_9

H. Duan and P. Li, *Bio-inspired Computation in Unmanned Aerial Vehicles*,
DOI 10.1007/978-3-642-41196-0_8, © Springer-Verlag Berlin Heidelberg 2014

performing complex missions present new challenges to the control community. UAVs must possess attributes of autonomy in order to function effectively in a "system-of-systems" configuration. Coordinated and collaborative control of UAV swarms demands new and novel technologies that integrate modeling, control, communications, and computing concerns into a single architecture. Typical application domains include reconnaissance and surveillance missions in an urban environment, target tracking and evasive maneuvers, search and rescue operations, and homeland security. Major technological challenges remain to be addressed for such UAV swarms or similar federated system-of-systems configurations to perform efficiently and reliably. Excessive operator load, autonomy issues, and reliability concerns have limited thus far their widespread utility. The systems and controls community is called upon to play a major role in the introduction of breakthrough technologies in this exciting area.

The autonomy has been and may continue to be the bottleneck of obtaining ability to make decisions without human intervention (Clough 2005; Francis 2012). To some extent, the ultimate goal in the development of autonomy technology is to teach machines to be "smart" and act more like humans. During the last few years, we have witnessed significant advances in the unmanned aircraft and unmanned systems state of the art. Yet, new and novel concepts and technologies are required for a more widespread use of such critical assets, not only for military but also for commercial and other applications such as homeland security, rescue operations, forest fire detection, and delivery of goods, to name just a few applications. It is generally believed that the popularity of artificial intelligence in the 1980s and 1990s, such as expert systems, neural networks, machine learning, natural language processing, and vision, gives new hope to obtain fully autonomous capability. However, it remained to be seen whether future development of autonomy will benefit from artificial intelligence and how it will drive the autonomy forward.

By introducing the collaboration of bio-inspired computation and control problems in UAVs, such as path planning for single UAV and multiple UAVs, formation flight control and formation configuration, heterogeneous cooperative control for multiple UAVs\UGVs, and vision-based surveillance and navigation problems, this monograph tries to highlight the way toward full autonomy. The main focus and emphasis of this monograph has been on the bio-inspired computation in UAVs. The main objectives pursued have been on addressing the question of how to achieve higher autonomous capability by taking advantage of bio-inspired computation. The main goals of this monograph are to develop innovative and novel concepts, techniques, and solutions to meet the more and more difficult requirement in executing challenging applications.

8.2 New Trends in UAV Development

8.2.1 Small Air Vehicles

An increase in the number and use of small air vehicle system (SUAS) platforms has been evident for a number of years (Ammoo and Dahalan 2006). The SUAS community even hosts its own focused conferences focused on its own unique issues and opportunities. Small vehicles offer a number of significant advantages over their larger brethren, not the least of which is their (typically) lower acquisition cost. For ISR applications, this applies to the platform and payload sensor(s) alike. Smaller platforms with a similarly reduced infrastructure mean a smaller logistics footprint, a factor which lowers operations and maintenance (O&M) costs, as well. Recent advances in vehicle control and novel propulsion schemes have resulted in system performance that rivals that of the larger platforms, e.g., daylong endurance. Moreover, the platforms are inherently as survivable as the larger kin due to their fundamentally small size and resultant relative low observability. Newer "quiet propulsion" options, including fuel cell and fuel cell-electric hybrid solutions make them even harder to detect. Current systems such as the ScanEagle, Raven, and Fury have developed user constituencies that value their operational attributes and capabilities.

More advanced designs should make them almost ideal for "stand-in" operations, where their lower sensor resolution is compensated for by proximity to target. The added advantages of their more flexible operational footprint, including reduced crew size, will make them even more attractive in a downsizing defense environment. Potential civil and commercial applications of these small platforms (e.g., traffic surveillance and management, law enforcement, power line monitoring) abound, as well.

The smallest in this class are termed micro- and nano-air vehicles, with some designs small enough to fit in the palm of a hand. Although they are the least mature from an operational perspective, the development community has been quite active over the last decade. Benefitting from recent research in low Reynolds number aerodynamics, these ultratiny systems are pioneering new methods of physical integration and component synergy. AeroVironment's Hummingbird was named one of the 50 top inventions of 2010 by Time Magazine.

Perhaps the most fascinating of these is the emerging nano-unmanned aircraft system (UAS) class. It consists of tiny autonomous or remotely controlled air vehicles able to operate indoors and outdoors and to transition between these environments. The US Defense Advanced Research Projects Agency (DARPA) recently concluded a multiyear technology development program on nano-air vehicles that culminated in a successful flight demonstration. The size/weight constraints on nano-class vehicles result in designs that use relatively thin airfoils. The outcome is that many nano-UAVs operate in the less understood ultralow Reynolds number aerodynamics flow regime, which makes high-fidelity modeling and simulation a significant challenge. Limited sensing and communication capabilities mean that onboard vehicle state estimation, a vital ingredient for a successful control strategy, continues to be problematic. In recent years, the use of motion capture systems such as Vicon Bonita has enabled off-board state sensing/estimation for laboratory flight experiments (an excellent example is MIT's RAVEN indoor flight environment). While the use of these systems enables researchers to focus on vehicle dynamics, trajectory generation, and control, off- board sensing is less useful in real-world UAV missions. Another nano-UAV challenge associated with constraints on airfoil size/shape and the size, weight, and power (SWAP) limitations is that control authority is quite limited. This makes stable and robust flight problematic. Furthermore, the development of effective flight control algorithms that are computationally inexpensive, while fully accommodating limitations on control surface type, size, and actuation modality, remains a tough problem. On the whole, the nano-UAV challenge is to develop robust and low computational footprint tools that enable the effective operation of this class of vehicles in spite of their physical and computational limitations.

8.2.2 Air-Breathing Hypersonic Vehicles

The necessity for a reliable and cost-effective access to space for both civilian and military applications has spurred a renewed interest in hypersonic vehicle (Duan and Li 2012; Sun et al. 2013), as witnessed by the success of NASA's (National Aeronautics and Space Administration) scramjet-powered X-43A, a substantial experimental vehicle that flew in 2004 and 2005. On May 3, 2013, it was extensively reported that the Boeing X-51 Waverider had launched successfully from 50,000 ft and had accelerated using a rocket to Mach 4.8 at which point it separated from the rocket and ignited its scramjet. It then accelerated further to Mach 5.1 and climbed to 60,000 ft before shutting down its engine and intentionally crashing into the Pacific. It's reported the engine ran for in excess of 240 s and the aircraft covered over 260 miles. The integration of the airframe and the scramjet engine results in strong coupling between propulsive and aerodynamic forces. The slender geometries and light structures cause significant flexible effects. Large-scale variations of altitude and velocity lead to uncertainties in the aerodynamic parameters. Consequently, controller design is a key issue in making air-breathing hypersonic flight feasible and efficient. The dynamics of hypersonic vehicles is highly nonlinear, coupled, and partly unpredictable due to the facts such as strong nonlinearity and high flight altitude, which makes the system modeling and flight control extremely challenging.

Owing to the dynamics' enormous complexity, the longitudinal dynamics equations developed by Bolender and Doman are employed to the modeling, control design, and simulations for hypersonic vehicle by most researchers. Compared with traditional flight vehicles, the longitudinal model of hypersonic vehicle is known to be unstable, non-minimum phase with respect to the regulated output and affected by significant model uncertainty. Therefore hypersonic vehicles are extremely sensitive to changes in atmospheric conditions as well as physical and aerodynamic parameters. It is essential to guarantee stability for the system and provide a satisfied control performance to improve the safety and reliability.

Conventionally, the research on flight control design for this class of vehicles focused primarily on mitigating the effect of the structural flexibility on the vehicle stability and control performance by means of linear robust design methods applied on linearized models. A linearized model is established around a trim point for a nonlinear, dynamically coupled simulation model of the hypersonic vehicle, thus linear controller can be designed in a neighborhood of the operating point. In this way, the linear control theories, such as decentralized control, linear quadratic regulator (LQR) approach, and gain scheduling method, have been researched to accomplish the difficulties in the control design. In recent years, contributions have begun to address the design of nonlinear controllers on nonlinear vehicle models as well. Owing to the nonlinearity and the coupling, many researchers take the dynamics information into consideration for controller design to provide the good control performance.

Although the approaches aforementioned provide robust performance under significant changes in flight conditions and fuel level in experimental simulation, it is far from satisfactory. There are three main concerns we should take into consideration in reaching our goal to provide satisfactory design of guidance and control systems for hypersonic vehicle: stability, performance, and robustness. However, few presented schemes meet the requirements or some of them just meet it with a simplified model. So there are some fundamental and essential issues in control designs that are worth probing further despite of the great progresses we have made in the last several years. The typical challenges of controller design for hypersonic vehicles can be generalized as eight aspects: input/output coupling, unstable/non-minimum phase, parameter variation, flexible modes, control constraints, tightly integrated airframe engine configuration, less known aerothermodynamic effects of hypersonic speeds, and lack of adequate flight and ground test data. The first issue that needs to be addressed is robustness, which is a key issue for a reliable controller for hypersonic vehicle, including the stability robustness and the performance robustness. To meet range-safety requirements in the event of failures, reconfigurable control allocation is necessary to take the problem of fault-tolerant control designs for flight critical components into consideration. And the complex interactions between elements of a hypersonic vehicle will require a tightly integrated design process to achieve the optimal performance necessary to meet space access mission objectives.

8.2.3 Design from the Perspective of System Integration

The UCAS, in particular, has been in various development stages for over a decade in its erratic odyssey to operational acceptance. Initiated and technologically evolved during a series of related DARPA programs from 1994 to 2006, the UCAS concept is based on a system-of-systems architecture that evolved into flexible and affordable approach to satisfying demanding multimission needs (Francis 2012). The concept has also raised significant concerns regarding its operational employment, especially its use in lethal scenarios. Although neither intended nor implemented, the idea that this armed robot would make lethal decisions has been a perception that the program has had to overcome almost since its inception.

In 2003, a joint program which combined multiple and somewhat disparate service interests was created. This joint unmanned combat air systems (JUCAS) program also programmatically joined two major airframe contractors and their respective air vehicle designs in what was to be a significant system integration challenge. Both the Boeing X-45C and Northrop Grumman X-47B designs represented state-of-the-art platforms in many respects, but the system-of-systems aspect of the development was in its infancy and especially daunting.

JUCAS was conceived to conduct extremely hazardous missions in otherwise denied airspace. The air vehicle designs were challenged to exhibit exceptional range, endurance, and persistence for fighter class configurations. Both aircraft designs were tailless and focused on achieving exceptionally low observability. Global operation was a given requirement, so the system was architected to provide flexible operation at very long distances between the platforms and their remote human crews. As such the systems design featured interoperable platforms, as well as collaborative CONOPS and tactics, including "group autonomy."

Another important consideration in UCAS design involved integration with other systems and capabilities. The extreme global reach (range/radius) and persistence of these platforms dictated a need for in-flight infrastructure support that was extensive. Airborne support assets included refueling tankers, potential communications relay platforms, and collaborative sensing aircraft. Of even more significance was the need for supporting spaceborne assets. Communications and navigation support (e.g., GPS) needs were obvious, providing connectivity to control stations, a host of terrestrially based platforms, as well as supporting airborne elements. In addition, complementary sensing systems, especially those providing broad area coverage capabilities, came to be viewed as highly beneficial in many mission scenarios. The global reach of space systems could help assure the same for UCAS or, for that matter, any other UCAS with similar aspirations. The relative sanctuary of space would provide a higher level of survivability for the system as a whole. The suite of space-based capabilities came to be viewed as an important infrastructure element of the JUCAS. As such, UCAS and similar concepts can serve as catalysts for true air and space integration.

The Common Operating System (COS) is the centerpiece of the system architecture introduced by Francis (2012), which aims at creating an interoperable capability and optimize the impact of available and emerging technologies. Analogous to the concept of the contemporary computer system, COS treats other system elements as functions and the hardware as "peripherals." The COS was conceived to manage all critical nonplatform-based system functions, including control and management of system resources, interplatform information exchange, communications bandwidth allocation and overall quality of service, an integrated perspective of battle space awareness, autonomous multivehicle operations, and other interplatform functionality. It is important that this approach provides the architectural underpinning that allowed one to think of the air vehicles and other major system elements as "nodes in the network."

It is important to note that the concept's implementation did segregate the major software and hardware developments in a manner that permitted maturation of each of the key system elements at its own developmental pace. Platform-unique or endemic functions (e.g., FCS inner loops, engine throttle controls) were architecturally excluded from the COS, also segregating developmental efforts in a logical manner, with consideration for other system attributes such as survivability and security. The concept allows for a more independent evolution of hardware and software, of vehicles and other platforms, and it promotes more rapid maturation of the system as a whole, as a result. The COS architecture was conceived to promote rapid, arbitrary reconfiguration of the system elements, as well as permitting creation of platform packages uniquely tailored to the mission(s) at hand. As such, the concept has applicability to a wide variety of potential multiplatform aerospace applications, including airspace management, disaster preparedness, and homeland security network operations.

8.3 Further Extensions of Bio-inspired Intelligence in UAVs

8.3.1 Achieve Higher Autonomous Capability

8.3.1.1 Classical Decision Cycle Known as the "OODA Loop"

An integrated suite of information technologies is required to achieve autonomous capability as we know it today. To understand how they are integrated, it is instructive to employ a construct known as the "OODA loop" (see Fig. 8.1): a

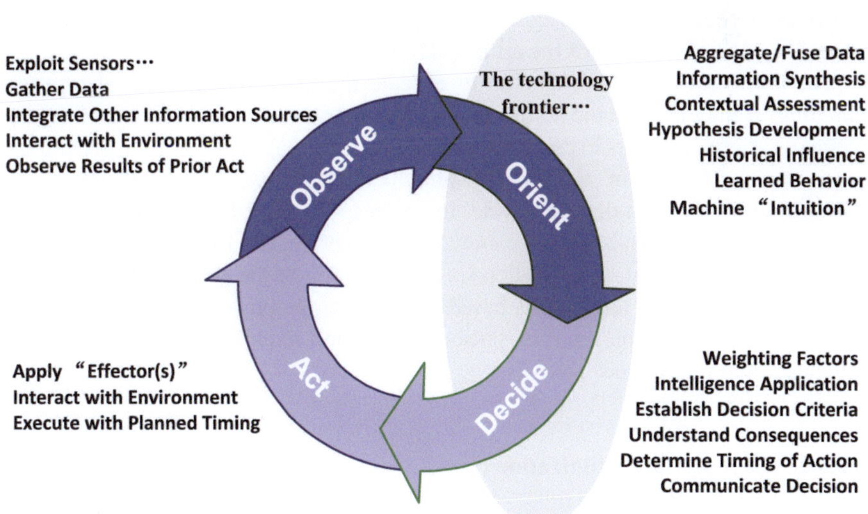

Fig. 8.1 "OODA" process

simple model for the classical decision cycle that was introduced by USAF Col. John Boyd in the late 1960s and intended to support discussions of concepts related to classical air combat. The four elements of this description of the decision cycle, namely, Observe, Orient, Decide and Act, are the same actions as those that any human would perform in achieving even the most common objectives in daily life, and in that order. Typically, this cycle is repeated over and over, sometimes even subconsciously, in the quest to achieve the desired outcome of a task. In these cases, sensory input data is continually refined as the OODA process is repeated. For an autonomous system, machine-based elements must perform the same functions to achieve a desired result. In that case, the "Observe" function is carried out by one or more sensors in much the same manner as human sensors, while the "Act" function is executed by one or more effectors that interface with the rest of the system. The "Orient" and "Decide" functions are the purview of the system's computational capability, and their successful execution is dependent on the relevance, adequacy, and accuracy of the software algorithms which underpin their functionality. These functions can be very complex depending on the tasks being undertaken and involve a plethora of subfunctions which are necessary to their successful execution.

The "Orient" function is arguably the more complex of the two, since it can involve the aggregation and fusing of asynchronous data, the synthesis of disparate information sources, the contextual assessment of the situation and environment, and the development of relevant hypotheses necessary for the decision-making step. It may also have to account for historical influences and learned behavior in the process. In more advanced, future applications, it may even involve the need to apply some form of intuition, as we understand it. The term most commonly associated with this step in modern robotics is "perception." Although often associated with visual imagery, the term can be applied more broadly to other forms and classes of information. Perception algorithms that exploit fused multiphenomenological data are at the edge of the state of the art in today's systems.

The ability of a machine system to execute the "Decide" step can also prove daunting. It may involve the ability to establish relevant decision criteria, correctly weigh a variety of factors in creating the best decision-making algorithm, accurately determine the timing of actions, and anticipate the consequences of actions in anticipation of the next decision cycle. In one respect, these latter two steps reflect the frontier of machine-based autonomy as we know it. The logic employed in the decision step is critical, since it dictates the character of the other elements of the cycle. For example, if the logic is based on physics, the physical variables used in the equations dictate the types of sensor information required and how it is to be integrated and interpreted.

8.3.1.2 Beyond the Optimization

Bio-inspired computation, short for biologically inspired computation, is the use of computers to model the living phenomena and simultaneously the study of life to improve the usage of computers, which has attracted a lot of researchers' attention (Bonabeau et al. 1999). A variety of bio-inspired models have been proposed to successfully solve many real-world problems (Kennedy and Eberhart 1995; Dorigo et al. 1996; Storn and Price 1997; Karaboga and Basturk 2007). Some of the phenomena are also known as swarm intelligence, inspired by the social behavior of gregarious insects and other animals. The emergent behavior of multiple unsophisticated agents interacting among themselves and with their environment leads to a functional strategy that is useful to achieve complicated goals in an efficient manner. Just as we have explained, ants, which know little about the environment, are capable of finding the shortest path from their nest to food sources (Duan et al. 2011a). Bees perform waggle dances to convey useful information on nectar sources to their hive mates. A number of desirable properties also exist in swarm intelligence models, which include feedback, self-organization and adaptation to changing environments, and multiple decentralized interactions among agents to work collaboratively as a group in completing complex tasks.

From the computational point of view, bio-inspired computation models are largely stochastic search algorithms. Although the rigorous theoretical analysis for most of the bio-inspired computation methods has not been conducted and the current study in this field is still in the experimental and preliminary application stage, the bio-inspired computation methods have already found their applications in many typical fields. They are useful for undertaking distributed and multimodal optimization problems. The search process is robust and efficient in maintaining diversity. A mechanism to impose a form of forgetting is also adopted in some swarm intelligence algorithms such that the solution space can be explored in a comprehensive manner. Thus, the algorithms are able to avoid convergence to a locally optimal solution, and, at the same time, to arrive at a global optimized solution with a high probability.

We can learn more than the optimization algorithms from the so-called swarm intelligence. The interaction among the agents and feedback mechanism are the basic elements that result in the emergence of dynamic patterns at the colony level. These patterns can be material or social and lead the colony to structure its environment and solve problems. The most interesting properties of these self-organized patterns are robustness (the ability for a system to perform without failure under a wide range of conditions) and flexibility (the ability for a system to readily adapt to new, different, or changing requirements). Robustness results from the multiplicity of interactions between individuals that belong to the colony. This ensures that, if one of the interactions fails or if one of the insects misses its task, their failure is quickly compensated by the other insects. This also promotes stability of produced patterns, whereas individual behaviors are mostly probabilistic. Flexibility of self-organized systems is well illustrated by the ability of social insects to adapt their collective behaviors to changing environments and to various colony sizes. These adaptations can occur without any change of the behavioral rules at the individual level. For instance, in the case of the selection of the shortest path in ants, a geometrical constraint applied on one of the two alternative paths increases the time needed by the ants to come back to their nest through this path and thus biases the choice toward the other path without any modification of the insects' behaviors.

Robustness in adaptability to environmental change for UAVs is necessary; the future need is to be able to adapt and learn from the operational environment because every possible contingency cannot be programmed a priori. This adaptation must happen fast enough to provide benefits within the adversary's decision loop, and the autonomy should be constructed so that these lessons can be shared with other autonomous systems that have not yet encountered that situation. Yet even in a hostile, dynamic, unstructured, and uncertain environment, this learning must not adversely affect safety, reliability, or the ability to collaborate with the operator or other autonomous systems. Although such capabilities are not currently available, the emergence mechanism of robustness and flexibility in biological colony may provide us many entirely new ideas.

8.3.1.3 Bio-inspired Hardware

Bio-inspired hardware (BHW) is also named "evolvable hardware (EHW)," which refers to hardware that can change its architecture and behavior dynamically and autonomously by interacting with its environment, and bio-inspired hardware has been proposed as a new method for designing systems for complex real-world applications (Duan et al. 2012). At present, almost all bio-inspired hardware uses an evolutionary algorithm (EA) as the main adaptive mechanism. In early 1960s, Neumann, the father of the computer, firstly proposed the great concept of developing a general-purpose machine, which has the capacity of self-reproduction and self-repairing. However, it was Garis who made the first move to investigate the design of evolving circuits. In his paper, Garis suggested the establishment of a new field of research called evolvable hardware (bio-inspired hardware).

In view of this emerging field, it is expected to have a great impact on space exploration and defense applications; the NASA and Department of Defense (DoD) of the United States have shown great interest in this field. The first NASA/DoD workshop on evolvable hardware was held in California on July 19–21, 1999, and the same workshops are held each year now. The aim of NASA/DoD is to develop a series of bio-inspired hardware for space shuttles, spacecrafts, space probes, satellites, strategic aircrafts, and nuclear submarines.

Each candidate circuit for UAVs can be either simulated or physically implemented in a reconfigurable device. Typical reconfigurable devices are field-programmable gate arrays (FPGA) (for digital designs) or field-programmable analogue arrays (for analogue designs). In its most fundamental form, a bio-inspired algorithm manipulates a population of individuals where each individual describes how to construct a candidate circuit. Each circuit is assigned a fitness, which indicates how well a candidate circuit satisfies the design specification. The bio-inspired algorithm uses stochastic operators to evolve new circuit configurations from existing ones. Over time the evolutionary algorithm will evolve a circuit configuration that exhibits desirable behavior. This definition can also be illustrated with the following simple equation.

$$\text{BHW} = \text{Bio-inspired Algorithms} + \text{FPGA}$$

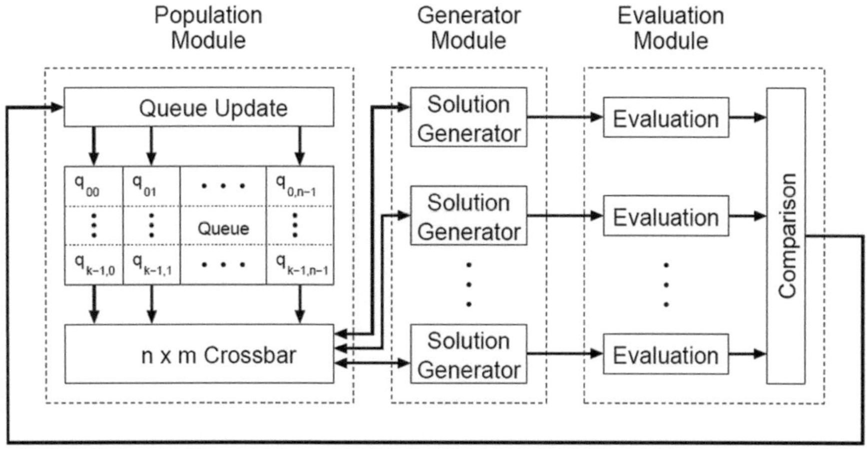

Fig. 8.2 The bio-inspired hardware-based high-level autonomous control architecture

The bio-inspired hardware-based high-level autonomous control architecture for UAV is illustrated in Fig. 8.2.

Generally, the bio-inspired hardware for UAVs consists of three main modules: population, generator, and evaluation. The population module manages all communication between the queue and the generator module. At the end of the current iteration, the population module receives the best solution from the evaluation module, which is then inserted into the queue. The evaluation results of the solutions (from the solution module) are collected in a comparison block, which chooses the best solution of the current iteration. The advantages of this bio-inspired hardware include fast processing speed, self-repairing, self-organization and self-adaptation.

Although adaptive BHW has made great progresses in the last several years, there are some fundamental and interesting issues that are worth probing further in this field (Yao and Higuchi 1999; Haddow and Tyrrell 2011). One of the goals of the early pioneers of the field was to evolve complex circuits. That is, to push the complexity limits of traditional design and, as such, find ways to exploit the vast computational resources available on today's computation mediums. However, the scalability challenge for BHW continues to be out of reach. Scalability is a challenge that many researchers acknowledge, but it is also a term widely and wrongly used in the rush to be the one to have solved it. Another important challenge is the measurements for the BHW. In traditional electronics, for example, terms such as functional correctness as well as area and power costs are general metrics applied. Other metrics applied may relate to the issue being resolved, e.g., reliability. However, the application of area metrics, especially gate counts, is often applied but may be questioned. Further challenge with metrics, highlighted in the above description, is the ability to compare solutions between traditional and evolved designs. Another area of interest for BHW is device production challenges. Defects arising during production lead to low yield, i.e., fewer "correct devices" and thus more faulty devices, devices that pass the testing process but where undetected

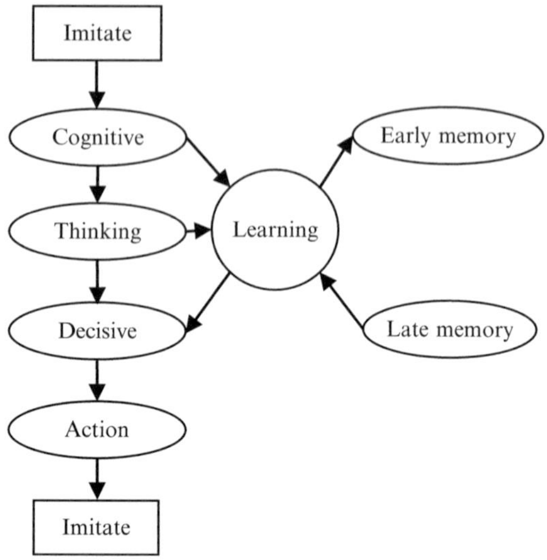

Fig. 8.3 Information processing procedure of the artificial brain

defects lead to faults during the device's lifetime. There are many ways that evolution may be applied in an attempt to correct or tolerate such defects. However, the challenge remains: a need for real defect data or realistic models. Fault models are difficult to define due to the strong dependence between the design itself and the actual production house due to the complex physical effects that lead to defects.

8.3.1.4 Artificial Brain-Based High-Level Autonomous Control for UAVs

The concept of artificial brain in the field of autonomous control mainly refers to the hardware development of artificial brain similar to a human brain, which has the ability of cognition (Duan et al. 2010b). The idea of evolution is adopted in artificial brain, which has a pre-memory capacity by continuing learning. The artificial brain is "cognitive," "thinking," and "decisive" and can actively response to external environments accordingly. The information processing process of an artificial brain can be illustrated by Fig. 8.3.

Combined with artificial intelligence and control theory, an artificial brain can reproduce the decision-making process of the real brain by using computer. Artificial brain-based controller can enable UAV to have higher intelligence. There are two implementation ways for the artificial brain: lifelike modeling and social modeling. Artificial brain-based controller has two major functions: control and learning. The former means the artificial brain-based controller can control a variety of UAV movements; the latter refers to learn the relevant specific knowledge from the outside environments. Of course, the artificial brain-based controller must acquire some knowledge during the control of UAVs.

An artificial brain-based UAV adopts the theoretical aspects of artificial life and artificial tools in the controller implementation. The artificial tools mainly include

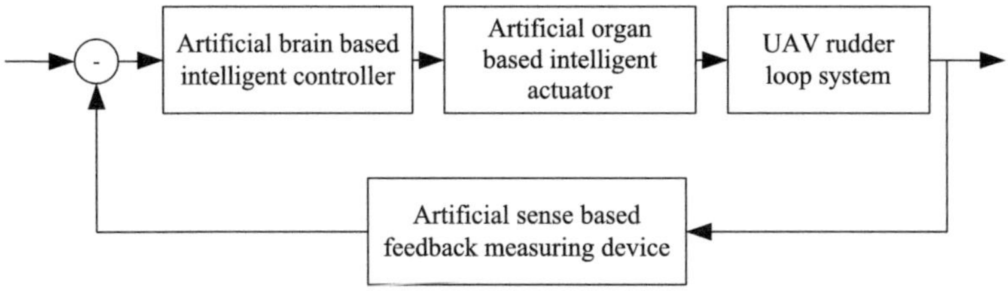

Fig. 8.4 Artificial brain-based UAV

"artificial brain," "artificial senses," "artificial organs," "artificial workers," and "artificial animals." The basic controller scheme of the artificial brain-based UAV can be illustrated by Fig. 8.4.

(1) Artificial brain-based intelligent controller: Computer software, hardware, or light mechanical/electrical materials are adopted to develop a variety of simulated "natural brains" of the brain model, which can act as the key controller of future UAV. As the artificial brain with high-level intelligent thinking, the UAV equipped with artificial brain-based intelligent controller has strong autonomy
(2) Artificial sense-based feedback measuring device: Artificial senses are a novel type of sensors simulating human or animal sensing organs. Artificial sense-based feedback measuring device is formed by a variety of intelligent sensors, which can make UAV with the visual, auditory, and multisensory information integration and multimode data-mining capabilities.
(3) Artificial organ-based intelligent actuator: Artificial organ is a piece of equipment which can simulate the effects of human or animal organs, such as artificial arms, artificial legs, and artificial hearts. The UAV equipped with the artificial organs can simulate control mechanism and regulation functions of the human body, hands, and feet. In this way, the two-way regulation and coordinate control can be implemented.

8.3.2 Enhance the Ability to Understand and Adapt to the Environment

8.3.2.1 Intelligent Multisensor Data Fusion

To operate in complex and uncertain environments, the autonomous system must be able to sense and understand the environment. This capability implies that the autonomous system must be able to create a model of its surrounding world by conducting multisensor data fusion (MDF) and converting these data into meaningful information that supports a variety of decision-making processes. The perception system must be able to perceive and infer the state of the environment from limited information and be able to assess the intent of other agents in the environment. This understanding is needed to provide future autonomous systems with the flexibility and adaptability for planning and executing missions in a complex, dynamic world. Although such capabilities are not currently available, recent advancements in computational intelligence (especially neuro-fuzzy systems), neuroscience, and cognition science may lead to the implementation of some of the most critical functionalities of heterogeneous, sensor net-based MDF systems. The following developments will help advance these types of processing capabilities:

1. *Reconfigurability of sensor weighting*: When a heterogeneous sensor net is used for an MDF system, each sensor has a different weight for different applications. As an example, regardless of whether a dissimilar MDF methodology is used to identify an object, an image sensor has much higher weight than radar. On the other hand, when an MDF methodology is used to measure a distance from the sensor to an object, a rangefinder or radar has a much higher weight than an image sensor.
2. *Adaptability of malfunctioning sensors and/or misleading data*: Even if an MDF methodology is used to identify an object, an image sensor cannot perform if it is faced to the sun. Data from the image sensors either will be saturated or need to be calibrated. Additionally, the image sensor data needs to be continuously calibrated if the weather is cloudy and changing because the measured data will be different based on shadows and shading. Therefore, the environment of a heterogeneous sensor net is a key parameter to be considered for design and implementation of an MDF system.
3. *Intelligent and adaptive heterogeneous data association*: Heterogeneous, sensor net-based MDF systems must process different data simultaneously, such as one-dimensional radar signals and two-dimensional imaging sensor data. As the combination of heterogeneous sensors change, the data combination is changed. Therefore, adaptive data association must be performed before conducting MDF and data input to the decision-making module.
4. *Scalability and resource optimization of self-reconfigurable fusion clusters*: The limiting factor of an MDF system is the scalability of self-reconfiguring the fusion cluster to adapt to a changing battlefield and/or the malfunction of one or more sensors. As the number of sensors used for a sensor net increases, the combinatorial number of reconfigurations exponentially increases. To manage such complexity, the MDF system will require a highly intelligent, fully autonomous, and extremely versatile reconfigurable algorithm, including sensor resource management and optimization. Great progress has been made in sensor management algorithms and cross-cued sensor systems, but true optimization is an elusive goal that is currently unavailable. Such capability can be obtained only from intelligent computing technology, which is currently in its infancy.

8.3.2.2 Battlespace Awareness with Greater Autonomy

Battlespace awareness is a capability area in which unmanned systems in all domains have the ability to significantly contribute well into the future to conduct intelligence, surveillance, and reconnaissance (ISR) and environment collection-related tasks. To achieve this, unmanned systems development and fielding must include the tasking, production, exploitation, and dissemination (TPED) processes required to translate vast quantities of sensor data into a shared understanding of the environment. Because unmanned systems will progress further with respect to full autonomy, onboard sensors that provide the systems with their own organic perception will contribute to battlespace awareness regardless of their intended primary mission. This capability area is one that lends itself to tasks and missions being conducted collaboratively across domains, as well as teaming within a single domain.

Due to the uncertainties of input data and knowledge for complicated combat situation assessment, it is necessary to make reasoning from the incomplete, uncertain, and imprecise data. Therefore, how to represent and reason becomes one of the key issues in situation assessment under complicated combating environments.

Bayesian network is a knowledge representation tool that encodes probabilistic relationships among variables of interest. This representation has two components: (a) a graphical structure, or more precisely a directed acyclic graph, and (b) a set of parameters, which together specify a joint probability distribution over the random variables. Over the last decade, the Bayesian network has become an increasingly important area for research and application in the entire field of artificial intelligence. Bayesian network has now become a popular representation for encoding uncertain expert knowledge in expert systems. More recently, researchers have developed methods for learning Bayesian networks from data. The techniques that have been developed are new and still evolving, but they have been shown to be remarkably effective for some data-analysis problems.

Bayesian network is an NP-hard problem, and swarm intelligence is a type of efficient methods for solving NP-hard problems. Therefore, Bayesian network can be integrated with swarm intelligence for solving the UAV situation assessment problem under complicated combating environments.

The abovementioned procedure makes effective use of the heuristic information in the problem domain and is very suitable to solve large-scale Bayesian network learning problems. This procedure can provide an effective approach for UAV situation assessment under complicated combating environments.

8.3.3 Cooperative Control of Multiple Autonomous Vehicles

8.3.3.1 Cooperative Control for Multiple UAVs

Modern military systems are becoming increasingly sophisticated, with a mixture of manned and unmanned vehicles being used in complex battlefield environments (Duan et al. 2009; Zhang et al. 2010; Duan et al. 2013b). Traditional solutions involve a centralized resource allocation, followed by decentralized execution. More modern battlespace management systems are considering the use of cooperative operation of large collections of distributed vehicles, with location computation, global communication connections, and decentralized control actions. Compared to autonomous vehicles that perform solo missions, greater efficiency and operational capability can be realized from teams of autonomous vehicles operating in a coordinated fashion. It is easy to understand that a group of UAVs is more capable than a single UAV, since the workload can be divided among the group. Research involving multiple UAV coordination is being undertaken from coordinated path planning for multiple UAVs, to the controller design of formation flight and formation reconfiguration, or to opening new application domains. This field is still in its infancy and many exciting new approaches are being explored for different applications.

The cooperative control problem is characterized by three attributes: complexity, information structure, and uncertainty. A hierarchical decomposition can be used to address the attribution of cooperative control problem. One control level and decision levels can be introduced in this hierarchical decomposition. At decision level 1 is the vehicle agent that does path planning, trajectory generation, and maintains models of terrain, threats, and targets. At decision level 2 is the sub-team agent which coordinates the activities of any tasks that require more than one vehicle to accomplish. If the sub-team has more than one task, then the agent apportions the vehicles to the tasks. The team agent, at decision level 3, is responsible for meeting the mission objective, determining sub-objectives for the sub-teams, and apportioning resources and tasks. Take an example to explain this hierarchical control system, which is about coordinating multiple vehicles to jointly reach a target area while minimizing combined exposure to radar. The hierarchical decomposition approach is pursued with simultaneous attack coordination at the sub-team agent level 2. The optimal trajectory generation for a single vehicle is addressed at the UAV planning agent level 1 to minimize exposure, while a timing constraint is imposed at level 2. Each vehicle independently plans their path. This hierarchical control system is by construction distributed and redundant, since there is no leader or line of succession, and hence is fault tolerant. The agent hierarchy is the same on each vehicle. All vehicles arrive at the same decisions; therefore, conflict situations are avoided such as having two vehicles attack the same target when only one is needed. The hierarchical decomposition for multiple UAV cooperative control is shown in Fig. 8.5.

To enable the applications for multivehicle systems, various cooperative control capabilities need to be developed, including formation control, rendezvous, attitude alignment, flocking, foraging, task and role assignment, payload transport, air traffic control, and cooperative search (Fig. 8.6). Execution of these capabilities requires that individual vehicles share a consistent view of the objectives. So we briefly describe a few applications of consensus algorithms for multivehicle coordination problem. The first one is rendezvous problem, which requires that a group of vehicles in a network rendezvous at a time or a location is determined through team negotiation. Consensus algorithms can be used to perform the negotiation in a way that is robust to environmental disturbances such as nonuniform wind for a team of UAVs. So consensus algorithms are needed to guarantee that all vehicles reach consensus on a rendezvous objective such as a rendezvous

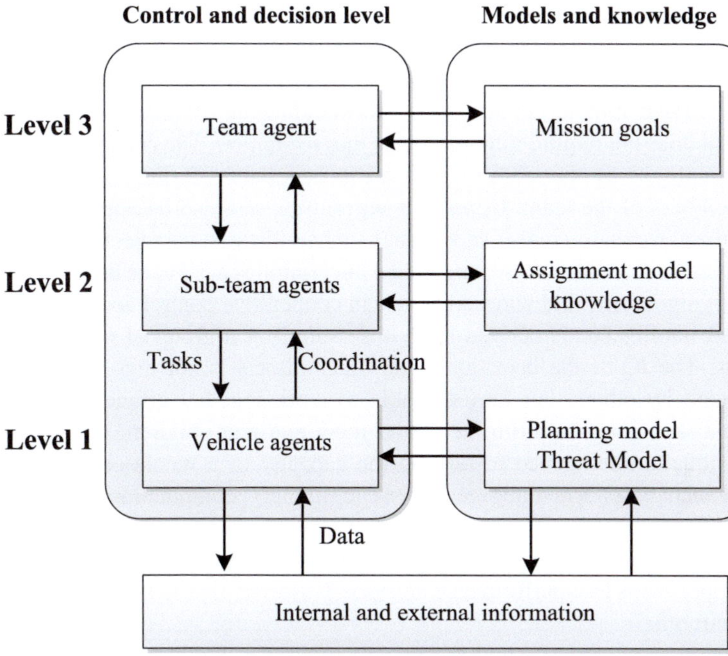

Fig. 8.5 Hierarchical decomposition for multiple UAV cooperative control

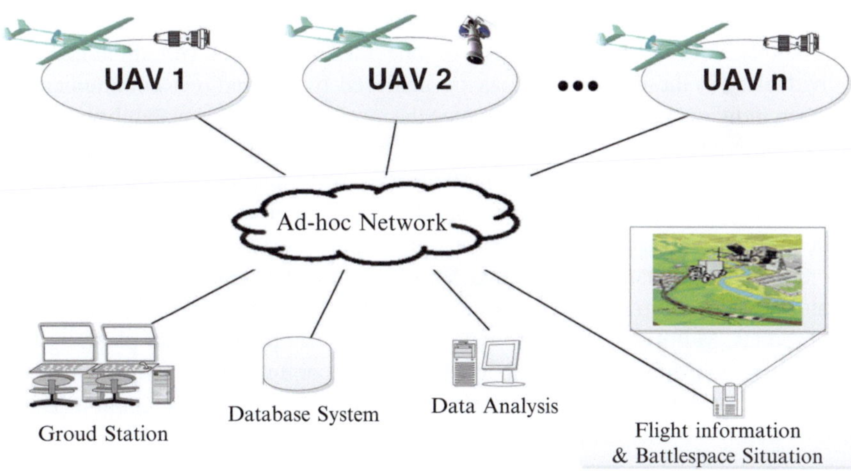

Fig. 8.6 Schematic diagram of cooperative control for multiple UAVs

time or rendezvous location. The second one is formation stabilization problem, which requires that vehicles collectively maintain a prescribed geometric shape. This problem is relatively straightforward in the centralized case, where all team members know the desired shape and location of the formation. On the other hand, in the decentralized formation stabilization problem, each vehicle knows the desired formation shape, but the location of the formation needs to be negotiated among team members. The information state for this problem includes the center of the formation. Each vehicle initializes its information state by proposing a formation center that does not require it to maneuver into formation. The consensus algorithm is then employed by the team of vehicles to negotiate a formation center known to all members of the team. Consensus algorithms can also be applied to execute decentralized formation maneuvers. Issues such as disturbance rejection, time delay, communication or sensor noise, and model uncertainties need to be addressed before consensus algorithms find widespread use in cooperative control applications.

Swarm intelligence is derived from observation of a group of insects by some scientists. The IQ of the bees, ants, and other insects is not high, and no one is commanded by others, but they are able to work together efficiently. They can build up a strong and beautiful nest, find food, and rear children. It is far beyond the individual's intelligence to rely on the capacity of a whole community. One major branch of swarm intelligence is evolutionary computation such as GAs, ACO, PSO, and ABC. If we apply the swarm intelligence to the future UAV systems, the combating efficiency can be improved significantly, even the UAVs are equipped with low-performance sensors. However, this is unreachable for the general aircrafts.

The main characteristics and advantages of swarm intelligence are as follows: (1) The cooperative individuals in each groups are distributed, and this makes the individuals adapt much easier to the current working state; (2) the robustness of the system can be guaranteed by the noncentral control scheme, and the solution may not be affected by the failure of one individual or a certain number of individuals; (3) the cooperation can be achieved by the indirect communication between individuals, which can guarantee the system with better scalability; and (4) as the communication overhead is still very small with the increase of individual number and the capacity of each individual is very simple, the exact execution time for each individual is relatively short. The abovementioned advantages have made swarm intelligence one of the most important research directions in the field of intelligent information processing. Swarm intelligence can be applied to combinatorial optimization problems, knowledge discovery, communication networks, and robotics. Multiple UAV coordination involving cooperative path planning and replanning, cooperative formation control and reconfiguration, cooperative task assignment, or other opening new application domains may be a potential research area by taking advantage of swarm intelligence.

8.3.3.2 Heterogeneous Control for Multiple UAVs/UGVs

UAVs can be used to cover large areas searching for targets. However, sensors on UAVs are typically limited in their accuracy of localization of targets on the ground. On the other hand, unmanned ground vehicles (UGVs) can be deployed to accurately locate ground targets, but they have the disadvantage of not being able

Fig. 8.7 Artistic concept of real-life autonomous convoy operation with the heterogeneous UAVs and UGVs

to move rapidly or see through such obstacles as buildings or fences (Duan and Liu 2010b; Duan et al. 2010a; Duan et al. 2011b). The use of multiple collaborative vehicles is ideally suited for many complicated tasks in dynamic and uncertain environment. And heterogeneous cooperative techniques can widen the application fields of unmanned aerial or ground vehicles and enhance the effectiveness of implementing detection, search, and rescue tasks. The use of multiple UAVs in concert with UGVs affords a number of synergies. First, UAVs with cameras and other sensors can obtain views of the environment that are complementary to views that can be obtained by cameras on UGVs. Second, UAVs can avoid obstacles while keeping UGVs in their field of view, providing a global perspective and monitoring the positions of UGVs while keeping track of the goal target. This is especially advantageous in two and a half dimensions where UAVs can obtain global maps, and the coordination of UAVs and UGVs can enable efficient solutions to the mapping problem. Third, if UAVs can see the UGVs and the UGVs can see UAVs, the resulting three-dimensional sensor network can be used to solve the simultaneous localization and mapping problem, while being robust against failures in sensors like GPS or errors in dead reckoning. In addition to this, the use of UAVs and UGVs working in cooperation has received a lot of attention for defense applications because of the obvious tactical advantages in such military operations as scouting and reconnaissance. The artistic concept of autonomous convoy operation with heterogeneous UAVs and UGVs is illustrated in Fig. 8.7 (McCook and Esposito 2007).

Although multiple UAV and UGV heterogeneous cooperation has made great progresses in the past years, there are still some fundamental and interesting issues in this specific field that are worth probing further. (1) The current progresses in this field are rather superficial; the dynamic and time-varying topology of the multiple UAV and UGV flocking system is a promising direction. Furthermore, how to construct a more complex environment with multifunctional vehicle systems is also an interesting research direction. (2) For the flocking movement and formation control of multiple UAVs and UGVs, there are various methods, and the stability and robustness of these approaches need further analysis. A more in-depth theoretical foundation of heterogeneous cooperative control should be established before rushing into large-scale multiple UAV and UGV heterogeneous system implementation. (3) In heterogeneous UAV and UGV systems, it is rather difficult to design the communication structure which meets the requirements of real-time communication among UAVs and UGVs. There are strong connections with cooperative control, but usually it is less concerned with the data rate through a communication channel than with the patterns of information flow among the networked agents. Implementation issues in both hardware and software are at the center of successful deployment of networked control systems. Data integrity and security are also very important and may lead to special considerations in control system design even at an early stage. (4) Future framework to analyze and design algorithms for the pursuit evasion games with sensor networks should include a more extensive comparison in order to evaluate possible trade-offs between the two. Also, the algorithms can be extended more rigorously when there are more pursuers than evaders, and coordinated maneuvering of pursuers allows the capture of "fast and smart" evaders similarly as observed in mobs of lions hunting an agile pray.

In addition to understanding the environment, unmanned systems must also possess the ability to collaborate through the sharing of information and deconfliction of tasking. Collaborative autonomy is an extension of autonomy that enables a team of unmanned systems to coordinate their activities to achieve common goals without human oversight (Duan et al. 2013a). This trend in autonomy will continue to reduce the human role in the system. Autonomously coordinated unmanned systems may be capable of faster, more synchronized fire and maneuver than would be possible with remotely controlled assets. This trend will lead to a shift toward strategic decision-making for a team of vehicles and away from direct control of any single vehicle. The ability to collaborate is one of the keys to reducing force structure requirements. The collaborative autonomy that is developed must be scalable to both larger numbers of heterogeneous systems and increased mission and environment complexity. Collaborative autonomy must be able to adapt to the air, ground, and maritime traffic environment and to changes in team members, operators, and the operational environment.

References

Ammoo M, Dahalan M (2006) Micro air vehicle: technology review and design study. In: Proceedings of the 1st Regional Conference on Vehicle Engineering & Technology, Kuala Lumpur, Malaysia. Available online: http://eprints.utm.my/318/1/MohdShariffAmmoo2006_MicroAirVehicleTechnologyReview.pdf

Bonabeau E, Dorigo M, Theraulaz G (1999) Swarm intelligence: from natural to artificial systems. Oxford University Press, New York

Clough B (2005) Unmanned aerial vehicles: autonomous control challenges, a researcher's perspective. J Aerosp Comput Inf Commun 2(8):327–347

Dorigo M, Maniezzo V, Colorni A (1996) Ant system: optimization by a colony of cooperating agents. IEEE Trans Syst Man Cybern B Cybern 26(1):29–41

Duan H, Li P (2012) Progress in control approaches for hypersonic vehicle. Sci China Technol Sci 55(10):2965–2970

Duan H, Liu S (2010a) Non-linear dual-mode receding horizon control for multiple unmanned air vehicles formation flight based on chaotic particle swarm optimisation. IET Control Theory Appl 4(11):2565–2578

Duan H, Liu S (2010b) Unmanned air/ground vehicles heterogeneous cooperative techniques: current status and prospects. Sci China Technol Sci 53(5):1349–1355

Duan H, Zhang X, Wu J, Ma G (2009) Max-min adaptive ant colony optimization approach to multi-UAVs coordinated trajectory replanning in dynamic and uncertain environments. J Bionic Eng 6(2):161–173

Duan H, Ding Q, Liu S, Zou J (2010a) Time-delay compensation of heterogeneous network control for multiple UAVs and UGVs. J Internet Technol 11(3):379–385

Duan H, Shao S, Su B, Zhang L (2010b) New development thoughts on the bio-inspired intelligence based control for unmanned combat aerial vehicle. Sci China Technol Sci 53(8):2025–2031

Duan H, Zhang X, Xu C (2011a) Bio-inspired Computing. Science Press, Beijing, Beijing, China

Duan H, Zhang Y, Liu S (2011b) Multiple UAVs/UGVs heterogeneous coordinated technique based on Receding Horizon Control (RHC) and velocity vector control. Sci China Technol Sci 54(4):869–876

Duan H, Yu Y, Zou J, Feng X (2012) Ant colony optimization-based bio-inspired hardware: survey and prospect. Trans Inst Meas Control 34(2–3):318–333

Duan H, Luo Q, Ma G, Shi Y (2013a) Hybrid particle swarm optimization and genetic algorithm for multi-UAVs formation reconfiguration. IEEE Comput Intell Mag 8(3):16–27

Duan H, Luo Q, Yu Y (2013b) Trophallaxis network control approach to formation flight of multiple unmanned aerial vehicles. Sci China Technol Sci 56(5):1066–1074

Francis MS (2012) Unmanned air systems: challenge and opportunity. J Aircr 49(6):1652–1665

Haddow PC, Tyrrell AM (2011) Challenges of evolvable hardware: past, present and the path to a promising future. Genet Program Evolvable Mach 12(3):183–215

Karaboga D, Basturk B (2007) A powerful and efficient algorithm for numerical function optimization: artificial bee colony (ABC) algorithm. J Glob Optim 39(3):459–471

Kennedy J, Eberhart R (1995) Particle swarm optimization. In: Proceedings of IEEE International Conference on Neural Networks, Piscataway. IEEE, pp 1942–1948

McCook CJ, Esposito JM (2007) Flocking for heterogeneous robot swarms: a military convoy scenario. In: Proceedings of Thirty-Ninth Southeastern Symposium on System Theory (SSST'07), Macon. IEEE, pp 26–31

Storn R, Price K (1997) Differential evolution – a simple and efficient heuristic for global optimization over continuous spaces. J Glob Optim 11(4):341–359

Sun H, Yang Z, Zeng J (2013) New tracking-control strategy for airbreathing hypersonic vehicles. J Guidance Control Dyn 36(3):846–859

Yao X, Higuchi T (1999) Promises and challenges of evolvable hardware. IEEE Trans Syst Man Cybern Part C Appl Rev 29(1):87–97

Zhang X, Duan H, Yu Y (2010) Receding horizon control for multi-UAVs close formation control based on differential evolution. Sci China Inf Sci 53(2):223–235

Author Correction to: Bio-inspired Computation in Unmanned Aerial Vehicles

Author Correction to:
H. Duan and P. Li, *Bio-inspired Computation in Unmanned Aerial Vehicles*, https://doi.org/10.1007/978-3-642-41196-0

The below listed corrections have been done in the following pages of the current version:

In page 3 – Line 08, the sentence "used in the 1991 Gulf War" has been replaced by "used in the 1991 Gulf War (Wikipedia website, 2013)."

In page 8 – Line 18, the sentence "the autonomic system" has been replaced by "the automatic system"

In Page 14 – Line 29, the sentence "It relies on four basic ingredients:" has been replaced by "It relies on four basic ingredients (reprinted from Garnier et al. (2007), with permissions from Springer Nature):"

In Page 27 – Line 15, the sentence "to system engineers" has been replaced by "to system engineers (Xargay et al., 2012)"

In Page 27 – Line 31, the sentence "the onboard sensors" has been replaced by "the onboard sensors (Xargay et al., 2012)"

The updated version of these chapters can be found at
https://doi.org/10.1007/978-3-642-41196-0_1
https://doi.org/10.1007/978-3-642-41196-0_2
https://doi.org/10.1007/978-3-642-41196-0_3
https://doi.org/10.1007/978-3-642-41196-0_4
https://doi.org/10.1007/978-3-642-41196-0_5
https://doi.org/10.1007/978-3-642-41196-0_8
https://doi.org/10.1007/978-3-642-41196-0

H. Duan and P. Li, *Bio-inspired Computation in Unmanned Aerial Vehicles*,
https://doi.org/10.1007/978-3-642-41196-0_9,

In Page 31 Fig. 1.7, the caption "Particle Swarm Intelligence" has been replaced by "Particle Swarm Optimization"

In Page 33 – Behind Line 41, an entry was missing in the reference list. The missing reference has been added now as given below:

Wikipedia website (2013). Unmanned aerial vehicle. https://en.wikipedia.org/wiki/Unmanned_aerial_vehicle. Accessed 22 Jun 2002

In Page 39, the last line "path e_1 and path e_1" has been replaced by "path e_1 and path e_2"

In Page 40, the last line "from v_s to v_e" has been replaced by "from v_s to v_d"

In Page 41 – Line 19, the sentence "(initially the memory contains only the start city)" has been replaced by "(initially the memory contains only the start city) (Dorigo and Stützle, 2003)"

In Page 48 – Line 7, the sentence "(which can effectively be seen as a step size." has been replaced by "which can effectively be seen as a step size. (Poli R et al., 2007)"

In Page 51, Line 4, the sentence "according to the problems under consideration." has been replaced by "according to the problems under consideration. (Grosan and Abraham, 2011)"

In Page 62, line 5, the equation has been updated as $x^* : f(x^*) = \min_x f(x)$

In Page 62, the equation 2.20 has been replaced as

$$u_{ji} = \begin{cases} v_{ji} \ , \textit{if randb} \leq CR \ \textit{or} \ j = \textit{randr} \ , & j = 1, \ldots, D \\ x_{ji} \ , \textit{if randb} > CR \ \textit{or} \ j \neq \textit{randr} \ , & j = 1, \ldots, D \end{cases}$$

In Page 68, Behind Line 16, an entry was missing in the reference list. The missing reference has been added now as given below:

Dorigo M, Stützle T (2003) The ant colony optimization metaheuristic: algorithms, applications, and advances. In: Glover F, Kochenberger GA (eds) Handbook of Metaheuristics. Springer, Boston, MA, pp 250–285

In Page 68, Behind Line 34, an entry was missing in the reference list. The missing reference has been added now as given below:

Grosan C, Abraham A (2011) Swarm intelligence. In: Intelligent systems: a modern approach. Springer, Berlin, Heidelberg, pp 409–422

In Page 69, Behind Line 11, an entry was missing in the reference list. The missing reference has been added now as given below:

Poli R, Kennedy J, Blackwell T (2007) Particle swarm optimization: an overview. Swarm Intell 1(1): 33–57.

In Page 80, Line 12, the sentence "Q and R R are diagonal positive matrix" has been replaced by "Q and R are diagonal positive matrix"

In Page 102, fig. 4.1 has been updated with the dotted line guided the UAV position which has been placed on the triangle.

In Page 114 – The sentence "Nc_max" in step 1 and step 11 has been replaced with "Ncmax" for consistency.

In Page 116 – The sentence "Nc_max" in step 1 and step 11 has been replaced with "Ncmax" for consistency.

In Page 155, for fig. 5.8 – The equation "4.x" has been updated as "5.x"

In Page 250, Line 9 – the sentence "Air-Breathing Hyposonic Vehicles" has been updated as "Air-Breathing Hypersonic Vehicles"

In Page 263, for fig. 8.6 – the word "AD hoc" has been updated as "Ad-hoc"

Author Biography

Prof. Haibin Duan, is currently a full professor of School of Automation Science and Electrical Engineering, Beihang University (Beijing University of Aeronautics and Astronautics, BUAA), Beijing, China. He received the Ph.D. degree from Nanjing University of Aeronautics and Astronautics (NUAA) in 2005. He was an academic visitor of National University of Singapore (NUS) in 2007, a senior visiting scholar of The University of Suwon (USW) of South Korea in 2011. He is currently an *IEEE Senior Member*, committee member of *Technical Committee on Guidance, Navigation and Control (TCGNC), Chinese Society of Aeronautics and Astronautics(CSAA)*, committee member of *Chinese Association for Artificial Intelligence(CAAI)*, and committee member of *Chinese Association of Automation(CAA)*. He is the Editor-in-Chief of "*International Journal of Intelligent Computing and Cybernetics (IJICC)*". He has published over 60 peer-reviewed papers in international journals and 3 monographs. His current research interests include bio-inspired computation, multi-UAVs cooperative control, and biological computer vision. Prof. Duan's academic website is: http://hbduan.buaa.edu.cn His contacting email is: hbduan@buaa.edu.cn

H. Duan and P. Li, *Bio-inspired Computation in Unmanned Aerial Vehicles*,
DOI 10.1007/978-3-642-41196-0, © Springer-Verlag Berlin Heidelberg 2014

MIX
Papier aus verantwortungsvollen Quellen
Paper from responsible sources
FSC® C105338

If you have any concerns about our products,
you can contact us on
ProductSafety@springernature.com

In case Publisher is established outside the EU,
the EU authorized representative is:
Springer Nature Customer Service Center GmbH
Europaplatz 3, 69115 Heidelberg, Germany

Printed by Libri Plureos GmbH
in Hamburg, Germany